Bioproducts for Health II

Bioproducts for Health II

Editors

Manuela Pintado
Ezequiel Coscueta
María Emilia Brassesco

MDPI • Basel • Beijing • Wuhan • Barcelona • Belgrade • Manchester • Tokyo • Cluj • Tianjin

Editors

Manuela Pintado
Centro de Biotecnologia e Química Fina
Universidade Católica Portuguesa
Porto
Portugal

Ezequiel Coscueta
Centro de Biotecnologia e Química Fina
Universidade Católica Portuguesa
Porto
Portugal

María Emilia Brassesco
Centro de Biotecnologia e Química Fina
Universidade Católica Portuguesa
Porto
Portugal

Editorial Office
MDPI
St. Alban-Anlage 66
4052 Basel, Switzerland

This is a reprint of articles from the Special Issue published online in the open access journal *Molecules* (ISSN 1420-3049) (available at: www.mdpi.com/journal/molecules/special_issues/Bioproducts_II).

For citation purposes, cite each article independently as indicated on the article page online and as indicated below:

LastName, A.A.; LastName, B.B.; LastName, C.C. Article Title. *Journal Name* **Year**, *Volume Number*, Page Range.

ISBN 978-3-0365-6993-2 (Hbk)
ISBN 978-3-0365-6992-5 (PDF)

Contents

About the Editors

Manuela Pintado

Associate Professor Manuela Pintado works at the Faculty of Biotechnology of the Portuguese Catholic University (ESB-UCP). She is also the Associate Director of ESB-UCP and the Director of the Centre for Biotechnology and Fine Chemistry (CBQF), a research unit and Associate Laboratory of ESB-UCP. Manuela Pintado is also the President of the European Biotechnology Research Institute (EBRI). In 1991, she earned her B.Sc. in Pharmacy (Clinical Analysis and Pharmacy) from the University of Porto. In 1999, she completed her Ph.D. in Biotechnology (in Food Science and Engineering) from the Portuguese Catholic University. Manuela Pintado leads the Bioactive and Bioproducts Research Laboratory. Her research focuses on developing and validating bioactive ingredients, by-products and new resources valorization and applying these developments to other biotechnological fields. Her goal is to develop high-value bioproducts, including novel functional foods. She is a prestigious author of high impact in her scientific area, with over 550 published articles in refereed international journals, 88 handbook chapters, and 62 patents. (https://orcid.org/0000-0002-0760-3184).

Ezequiel Coscueta

Ezequiel Coscueta is a Junior Researcher at CBQF and an Invited Assistant Professor at ESB-UCP. He earned his Ph.D. in Biological Sciences from the Universidad Nacional de Rosario in 2018 and his B.Sc. in Biotechnology from the Universidad Nacional del Litoral in 2013. Ezequiel Coscueta focuses on developing sustainable technologies for the Bioactive and Bioproducts Research Laboratory at CBQF, specifically interested in the circular economy approach, green chemistry, peptidomics, and nanotechnology, all applied to the food and nutraceutical industries, with a focus on gastrointestinal health. He has authored over 25 international peer-reviewed articles, contributed to five handbook chapters, edited four high-impact international journal issues, and holds one patent. Ezequiel Coscueta has also led groups and work teams related to innovation and sustainable development, such as Smart Waste Young Professionals. (https://orcid.org/0000-0003-4496-2001).

María Emilia Brassesco

María Emilia Brassesco is a Junior Researcher at CBQF with a Ph.D. in Chemical Sciences (awarded in 2019) and a B.Sc. in Biotechnology (completed in 2014). She is the Project Manager for European Projects at the Bioactive and Bioproducts Research Laboratory at CBQF. She has authored 17 articles in peer-reviewed international journals and contributed four chapters to handbooks. María Emilia Brassesco was the Editor of three special issues in high-impact international journals. Her expertise lies in biotechnology processes, particularly protein purification and bioactive compounds, with experience in the food and nutraceutical industries. Her main interests are in green chemistry and the circular economy approach. (https://orcid.org/0000-0001-9027-4643).

Preface to "Bioproducts for Health II"

In order to build a promising future, health and sustainability must be interelated. Marine, forestry, agriculture, and food systems are important sources of bioproducts used in health applications. To explore the potential of such sources for the development of natural products capable of biological activities, it is necessary to develop new technologically sustainable strategies. Despite the range of natural compounds already available, there is a need to identify bioactive molecules (e.g., polysaccharides, proteins and peptides, polyunsaturated fatty acids, and polyphenols) from different natural sources with positive health properties, including antihypertensive, antidiabetic, anti-obesity, antimicrobial, anti-atherosclerotic, antioxidant, antithrombotic, immune-modulatory, relaxing, and satiety-inducing effects. The Second Edition of this Special Issue aimed to identify and gather works on the latest varied sources of bioproducts, the biological and functional activities of these bioactive compounds, their mechanisms of action, and the methods used for extraction and purification, without losing our focus on alignment with the concept of green technology.

Manuela Pintado, Ezequiel Coscueta, and María Emilia Brassesco
Editors

molecules

MDPI

Review

Targeting HIF-1α by Natural and Synthetic Compounds: A Promising Approach for Anti-Cancer Therapeutics Development

Rituparna Ghosh †, Priya Samanta †, Rupali Sarkar, Souradeep Biswas, Prosenjit Saha, Subhadip Hajra * and Arijit Bhowmik *

Department of Cancer Chemoprevention, Chittaranjan National Cancer Institute (CNCI), Kolkata 700026, India
* Correspondence: subhadiphajra@cnci.ac.in (S.H.); arijitbhoumik@gmail.com (A.B.)
† These authors contributed equally to this work.

Abstract: Advancement in novel target detection using improved molecular cancer biology has opened up new avenues for promising anti-cancer drug development. In the past two decades, the mechanism of tumor hypoxia has become more understandable with the discovery of hypoxia-inducible factor-1α (HIF-1α). It is a major transcriptional regulator that coordinates the activity of various transcription factors and their downstream molecules involved in tumorigenesis. HIF-1α not only plays a crucial role in the adaptation of tumor cells to hypoxia but also regulates different biological processes, including cell proliferation, survival, cellular metabolism, angiogenesis, metastasis, cancer stem cell maintenance, and propagation. Therefore, HIF-1α overexpression is strongly associated with poor prognosis in patients with different solid cancers. Hence, pharmacological targeting of HIF-1α has been considered to be a novel cancer therapeutic strategy in recent years. In this review, we provide brief descriptions of natural and synthetic compounds as HIF-1α inhibitors that have the potential to accelerate anticancer drug discovery. This review also introduces the mode of action of these compounds for a better understanding of the chemical leads, which could be useful as cancer therapeutics in the future.

Keywords: HIF-1α; hypoxia; metastasis; angiogenesis; cancer stem cells; natural compounds; synthetic drugs

Citation: Ghosh, R.; Samanta, P.; Sarkar, R.; Biswas, S.; Saha, P.; Hajra, S.; Bhowmik, A. Targeting HIF-1α by Natural and Synthetic Compounds: A Promising Approach for Anti-Cancer Therapeutics Development. *Molecules* **2022**, *27*, 5192. https://doi.org/10.3390/molecules27165192

Academic Editor: Manuela Pintado

Received: 22 February 2022
Accepted: 24 March 2022
Published: 15 August 2022

1. Introduction

Inadequate oxygen supply to tumor tissues creates hypoxia, one of the most significant clinical conditions responsible for cellular proliferation, angiogenesis, metastasis, and propagation of cancer stem cells (CSCs). In the case of hypoxia, oxygen concentration in the normal human renal cortex and brain tissues changes from 9.5% to ~1.3% and ~4.6% to ~1.7%, respectively [1,2]. However, an adequate amount of oxygen and nutrition supply is highly needed for the growth of any multicellular organ, as well as solid tumors. However, oxygen supply becomes restricted when distance between a tumor cell and blood vessel exceeds 70 μm [3]. This condition is known as diffusion-limited hypoxia. Additionally, chronic hypoxia leads to a restricted supply of oxygen, which depends on severe structural and functional abnormalities of the tumor microvessels (perfusion-limited O_2 delivery), a deterioration of the diffusion geometry (diffusion-limited O_2 delivery), tumor-associated, and therapy-induced anemia [4]. In hypoxic conditions, reactive oxygen species (ROS) production increases in tumor cells [5]. Generation of ROS triggers damage of cellular biomolecules such as lipids, proteins, DNA, and RNA that leads to cell death [6]. Despite this, cancer cells find smart ways to escape cell death and ultimately survive in the unfavorable conditions [7].

The most common survival features of cancer tissues include their augmentation toward metabolic reframing, angiogenesis, and metastasis. These responses are mediated by hypoxia-inducible factor (HIF) family of proteins, the expression of which becomes altered upon hypoxia. The HIF pathway remains conserved from *Caenorhabitis elegans* to human and becomes activated to maintain oxygen homeostasis in hypoxic conditions [8]. Hypoxia-inducible factor was first discovered by Semenza and coworkers, and they described the presence of hypoxia response element (HRE), which is conserved by every target gene of HIFs [9]. HIF is a heterodimeric protein with a basic helix loop helix structure that contains one α and one β subunit. In addition to the ubiquitously expressed HIF-1, the HIF family consists of two other members called HIF-2 (also called EPAS1, MOP2, or HLF) and HIF-3. HIF-1 and HIF-2 share many functional and structural similarities such as DNA recognition, binding and heterodimerization with HIF-1β (ARNT). HIF-2α is a constitutively expressed cytoplasmic protein and mainly found in vascular endothelial cells. Various studies reported that it is also up-regulated in cancer cells. Another member of this family, HIF-3α, functions mainly as a negative regulator of HIF-1α. An alternatively spliced variant of HIF-3α called IPAS directly binds to HIF-1α and prevents its binding with HIF-1β which results in the formation of an abortive complex. In this way, HIF-3α prevents HIF-1α from forming an active transcription factor and in turn hinders binding of HIF-1α to the HRE sequence of hypoxia-induced genes [10]. Multiple studies reported the prominent role of HIF-1α in malignant cells. Hence, in this review, we present insights into the complexity of cellular metabolism, metastasis, tumor angiogenesis, and survival of cancer stem cells induced by HIF-1α.

The role of HIF-1α in the apoptosis of cancer cells is controversial. In some types of cancer, HIF-1α triggers apoptosis, whereas in some cases it induces cell proliferation. Usually, normal as well as cancerous cells die owing to the depletion of oxygen. In certain cases, HIF-1α triggers apoptosis in cancer cells via regulation of pro-apoptotic, anti-apoptotic and apoptotic proteins. However, sometimes, HIF-1α also arrests cell growth in the G1/S phase to escape cell death [11]. Consequently, activation of cell cycle arrests as well as induction of apoptosis encourages cancer cells to be resistant in radiation and chemotherapy.

Along with this, in the case of angiogenesis, hypoxia triggers the formation of new blood vessels as well as sprouting from existing blood vessels to increase the supply of oxygen. Expression of angiogenic markers such as VEGF, TGF-β, PDGF-β, plasminogen activator-1 (PAI-1), erythropoietin (EPO) and GLUT-1 are regulated by HIF-1α [12]. HIF-1α is known to be involved in all steps of blood vessel formation. However, tumor vasculature has different morphological features from normal blood vessels [12]. Hypoxia induces the imbalance between pro and anti-angiogenic factors production, which leads to enhanced, rapid, and chaotic blood vessel formation [13,14]. However, other than involvement in angiogenic progression HIF-1α also plays an important role in other hypoxia-induced hallmarks of cancer, such as induction of metastasis [15,16]. HIF-1α regulates a broad range of genes involved in epithelial to mesenchymal transition, disrupting basement membrane of surrounding tumor tissue and invasion [16,17]. In this way, hypoxia helps the cells of the primary tumor to escape the hypoxic region and migrate to a distal site to form a secondary tumor by regulating HIF-1α. In addition, adaptations of these new characteristics make a cancer cell more aggressive in hypoxic conditions.

It is reported that hypoxia leads to clonal selection of tumor cells [18,19]. In hypoxic conditions, HIF-1α promotes enrichment of cancer stem cells (CSCs) to adopt a different way to survive [19]. CSCs are a subpopulation of cells within tumor. These cells have self-renewing as well as self-differentiation properties [20]. In addition, CSCs have altered gene expression to gain drug resistant properties [20]. Various cell surface markers such as CD44, CD24, CD29, CD90, CD133, epithelial-specific antigen (ESA), and aldehyde dehydrogenase1 (ALDH1) have been used to isolate and recognize CSCs from different tumors [21]. Although the mechanism of hypoxic controls of cancer is not fully revealed, it is reported that CSCs usually reside in the hypoxic region of the tumor [22]. Danet et al., cultured human hematopoietic stem cells (HSCs) under hypoxic conditions and showed that they can promote their ability to repopulate when they are transplanted to nonobese diabetic (NOD)/severe combined immunodeficiency (SCID) mice [23]. Ezashi et al., reported that embryonic stem cells are maintained by low oxygen tension and significantly block spontaneous cell differentiation [24]. However, recent studies suggested that hypoxia directly regulates cancer stem cell-related pathways [25,26]. Low oxygen tension may convert non-stem cancer cells into cancer stem cell-like status with increased self-renewing capacity as well as induction of essential stem cell factors, such as OCT4, Nanog, and c-MYC [25]. HIF-1α is critical for cancer stem cell maintenance, as knockdown of HIF-1α in cancer stem cells leads to reduce self-renewal capacity, increases apoptosis, and attenuates tumorigenesis [19,26].

Therefore, targeting HIF-1α can be an effective way to control various crucial hallmarks of cancer. Several drugs have already been designed to combat cancer by targeting HIF-1α. Some of these drugs are in clinical trial whereas, some show great potential in vitro and in vivo. Along with the synthetic chemotherapeutic drugs, some natural compounds have high probability to block HIF-1α pathways [27]. Henceforth, a brief description of different potential anti-cancer natural compounds and synthetic chemotherapeutic agents targeting HIF-1α are discussed in this review.

2. Structure and Regulation of HIF-1α

HIF, a heterodimeric protein, is comprised of two subunits, i.e., HIF-1α and HIF-1β, with a molecular weight of 120 kDa and 91–94 kDa, respectively [28]. HIf-1α contains a basic helix loop helix (BHLH) and PAS domain, which is named after three proteins first recognized in drosophila, i.e., Per, ARNT, and Sim. Both HIF-1α and HIF-1β possess PAS domain which is situated before BHLH domain near the N-terminal end. Moreover, HIF-1α contains three other important domains such as N-terminal transcriptional activation domain (NAD), C terminal transcriptional activation domain (CAD) and an oxygen dependent domain (ODD). HIF-1α binds to DNA through the basic domain of BHLH, whereas HLH domain is essential for dimerization of HIF-1α and HIF-1β [29]. In cells, HIF-1α is tightly regulated depending on the availability of oxygen, whereas HIF-1β (also known as ARNT) is constitutively expressed and remains abundant irrespective of alterations in oxygen tension. Furthermore, the stability of the HIF-1α protein mainly depends on the post translational alterations [30]. In normoxic conditions, HIF-1α possesses a pair of sequences in the C terminal portion of the protein resulting a shorter half-life and instability [31,32]. In the presence of oxygen, proline hydroxylase (PHD) promotes hydroxylation of two conserved proline residues (Pro402 and Pro564 in human) present in LXXLAP motif of ODD [30]. As a result of this hydroxylation, the von Hippel-Lindau protein (pVHL) recognizes and binds to HIF-1α, which in turn facilitates its degradation by poly ubiquitination [33]. In contrast, in hypoxic conditions, PHD becomes inactive and HIF-1α protein escapes the degradation. Hence, HIF-1α translocates into the nucleus followed by dimerization with ARNT and becomes activated to function as an effective transcription factor [34,35]. Along with these, p300 and CREB-binding protein (CBP) act as important transcriptional regulators of HIF-1α and in hypoxic conditions, it binds to the carboxy-terminal transactivation domain (CTAD) of HIF-1α. In the presence of oxygen, FIH1 (factor inhibiting HIF1) hydroxylates the asparagine residue at 803 positions in the CTAD region. This prevents the binding of p300/CBP in this region which in turn helps in the binding of pVHL–elongin–cullin-2 complex to ODD that augments proteasomal degradation of HIF-1α by 26S proteasome (Figure 1) [36].

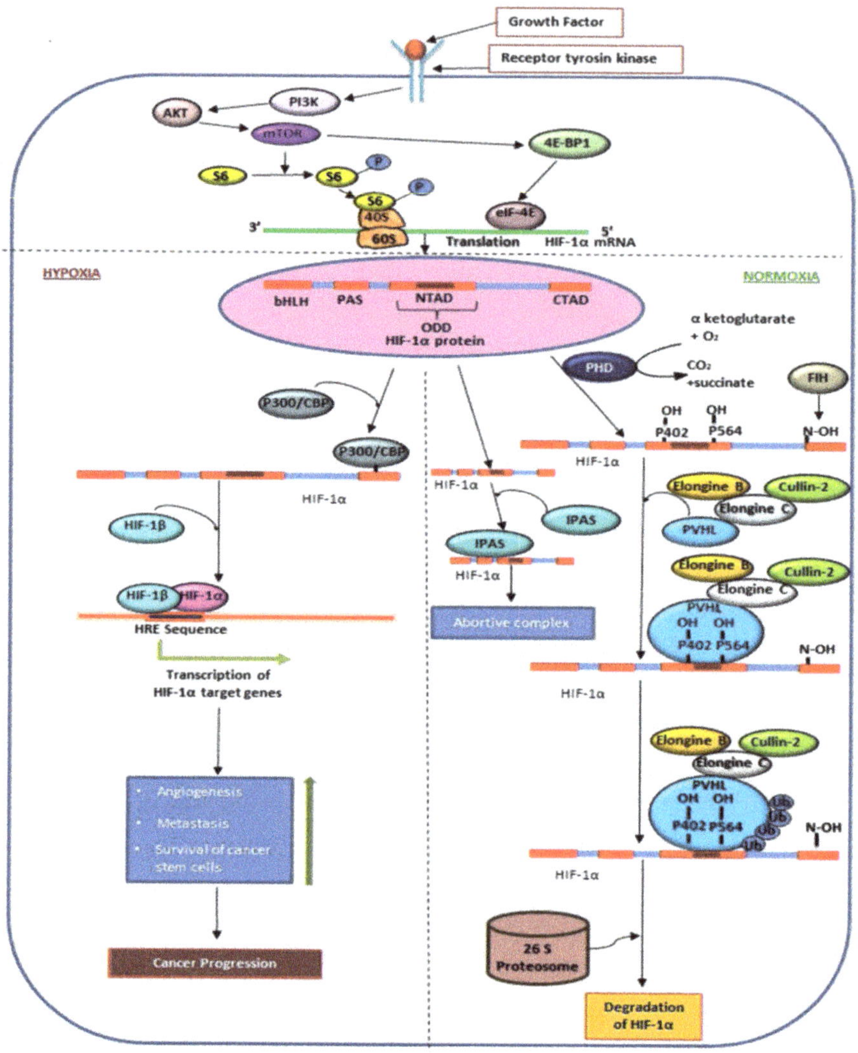

Figure 1. Molecular mechanism of regulation of HIF-1α in hypoxic and normoxic conditions. The figure represents the induction of HIF-1α translation via PI3K/AKT/mTOR pathway. On binding to growth factors at receptor tyrosin kinase PI3K becomes activated which further induces AKT and mTOR pathway activation followed by phosphorylation of S6. HIF-1α synthesis is induced by eIF-4E which binds to HIF-1 α upon activation by 4E-BP1 which is a downstream signaling molecule of mTOR. In the presence of oxygen, Pro-402, Pro-564 in ODD and Asn-803 in CTAD are hydroxylated by PHD and FIH. As represented in the figure, hydroxylation at Asn-803 prevents binding of P300/CBP to HIF-1 α in normoxic conditions, whereas, hydroxylation at Proline residues allow VHL- elongine-C-elongine-B-Cullin-2 complex to bind at ODD of HIF-1α followed ubiquitination of HIF-1α via 26 proteasome. Expression of HIF-1α is also regulated by IPAS, a variant of HIF-3 which binds with HIF-1α to form an abortive complex. In hypoxic conditions, P300/CBP binds at CTAD which prevents degradation of HIF-1α. HIF-1α enters nucleus and forms active transcription factor by binding with HIF-1β in order to transcribe genes for angiogenesis, metastasis, and survival of cancer stem cells in tumor tissue.

Along with the regulation via ubiquitin-proteasome pathway, several growth factors are responsible for translational regulation of HIF-1α, such as insulin, IGF-1, IGF-2, EGF, v-SRC, endothelin-1, ADM, erythropoietin, and cytokines. These growth factors can induce synthesis of HIF-1α protein via PI3K/AKT/mTOR pathway by binding at the Tyrosine kinase receptor. PI3K activates its downstream regulator mTOR via AKT. mTOR phosphorylates eukaryotic translation initiation factor 4E (eIF-4E) binding protein (4E-BP1) and disrupt integrity of eIF-4E which is required for inhibiting cap-dependent mRNA translation of HIF-1α. In addition, mTOR promotes phosphorylation of the ribosomal protein S6, and induces HIF-1α translation (Figure 1) [9,37]. Bypassing proteasomal degradation facilitates HIF-1α translocation into the nucleus and formation of heterodimer with HIF-1β. This heterodimer specifically binds to a 5′-RCGTG-3′ hypoxia-responsive element (HRE) sequence in the promoter or enhancer of various hypoxia-inducible genes, which includes erythropoietin, vascular endothelial growth factor, glucose transporters, and glycolytic enzymes, as well as genes involved in iron metabolism, and cancer cell and cancer stem cell survival [38].

3. Role of HIF-1α in Cancer Progression

In response to both hypoxic stress and oncogenic signals, HIF-1α becomes activated and controls different mechanisms involved in cancer cell survival and proliferation resulting in the formation of vascular tumors with metastatic potential [39]. Other than these, a variety of mechanisms regulate HIF-1α mediated cancer stem cell propagation and maintenance [40]. For better understanding role of HIF-1α in cancer cell metabolism, angiogenesis, metastasis, and survival of cancer stem cells are discussed here.

3.1. Role of HIF-1α in Cellular Metabolism

Lowering the oxygen level in cancer cells not only causes HIF-1α gene over expression but also influences the fluctuation of cellular metabolic homeostasis [41,42]. Expression of HIF-1α affects the rate of different metabolic pathways such as glycolysis, glycogenolysis, neoglucogenesis, β-oxidation, citric acid cycle, etc. [43–46]. On the other hand, through these metabolic alterations, HIF-1α promotes insulin resistance and obesity in most of the cancer patients [47,48]. In hypoxia, HIF-1α acts as a transcription factor and regulates oncogenic metabolism by two ways, i.e., by promoting anaerobic glycolysis and by suppressing TCA cycle or mitochondrial oxygen consumption [49]. HIF-1α induces some glycolytic enzymes and transporter-like aldolase A, mitochondrial pyruvate kinase 1 (PDK1) and glucose transporters (GLUTs) [50–53]. In the cancer cells, HIF-1α not only stimulates the induction of GLUT1, GLUT3 and GLUT4 transporter to uptake blood glucose, but also enhances glycolytic breakdown of intracellular glucose by transactivating phosphofructokinase 1 (PFK1) and aldolase (Figure 2) [52,54–56].

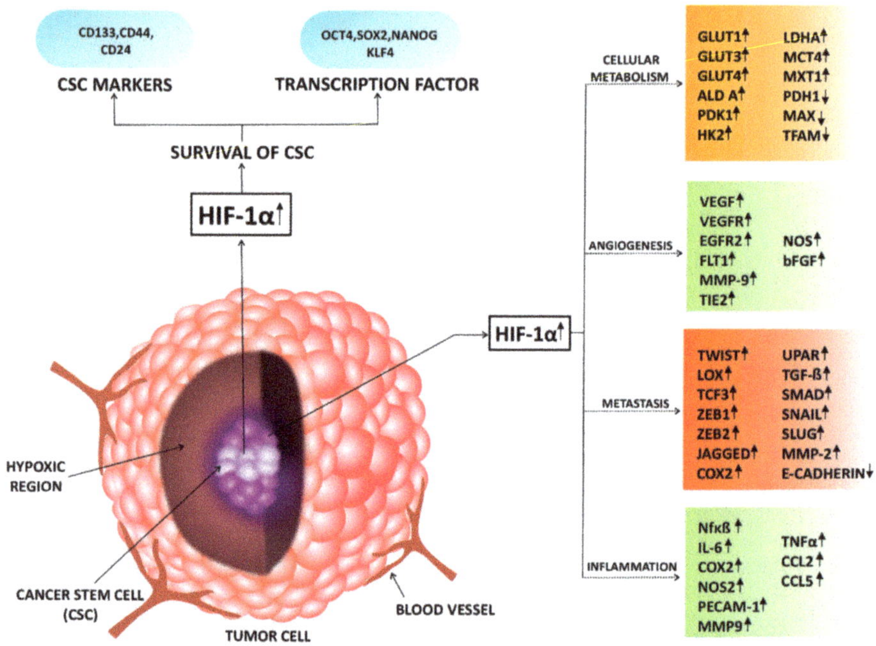

Figure 2. Representation of target genes of HIF-1α involved in tumor progression: Figure represents tumor tissue with hypoxic region at its core. Cancer stem cell (CSC) population resides at the core region. In hypoxic tumor tissue, HIF-1α helps in regulation of genes involved in cellular metabolism [41–66], angiogenesis [67–86], metastasis [87–102], CSC propagation & maintenance [103–111] and cancer inflammation [112–122] which are listed in the figure. Upward arrows (↑) indicate increased expression of the protein and Downward arrows (↓) indicate decreased expression of the protein.

Another glycolytic pathway member, Hexokinase 2(HK2), is a rate limiting enzyme involved in phosphorylation of glucose to glucose-6-phosphate and also reported to be up-regulated by HIF-1α in hypoxic cancer cells [57,58]. In addition to these, HIF-1α can trigger the expression of lactate dehydrogenase A (LDHA) which leads to conversion of glycolytic end products pyruvate and NADH to NAD+ and lactate [59]. This NAD+ is needed for another cycle of glycolysis which is essential for oncogenic metabolism. It is also evident that in cancer cells HIF-1α induces the expression of monocarboxylate transporter 4 MCT4, a plasma membrane binding transporter, which is known for its lactate extruding activity followed by pyruvate to lactate conversion [60,61]. The other mechanism of metabolic alteration by HIF-1α is the down-regulation of pyruvate dehydrogenase (PDH) expression by pyruvate dehydrogenase kinase 1 (PDK1) phosphorylation which successively blocks TCA cycle [62]. Generally, PDH mediates the breakdown of pyruvate to acetyl-CoA and CO_2 in mitochondria. Hence, glycolysis dependency of cancer cells increases due to inhibition of PDH mediated breakdown of pyruvate to acetyl-CoA and CO_2 in mitochondria [62]. Alteration of lipid biosynthesis is another approach of HIF-1α mediated metabolic regulation of cancer cells. In mitochondria, pyruvate produces acetyl-CoA that converts to citrate and translocates to cytoplasm and produces acetyl-CoA and oxaloacetate by the enzyme ATP citrate lyase. Although this acetyl-CoA is used for lipid biosynthesis in endoplasmic reticulum in normoxic conditions, its production is hindered by HIF-1α [62]. Therefore, for cancer cells, it is essential to expedite metabolic functions using alternative sources of fatty acid precursors. Hence, uptake of extracellular fatty acid is prompted by HIF-1α dependent activation of peroxisome proliferator activated receptor gamma (PPARγ) gene [46]. Activation of this gene promotes glycerolipid biosynthesis in cancer cells by inducing genes such as fatty acid binding proteins (FABP) 3, 4, and 7 [63]. Apart from these, phenomena reversion of TCA cycle occurs using HIF-1α induced isocitrate dehydrogenase (IDH) in cancer cells to maintain de novo synthesis of fatty acid [64]. Along with these, HIF-1α is also able to up-regulate glutaminase 1 (GLS1), an essential enzyme for glutamine metabolism that controls production of α-ketoglutarate, which is a main precursor molecule of de novo fatty acid synthesis [65]. Moreover, reduction in mitochondrial biogenesis to regulate oxygen consumption in cancer cell is a significant part of HIF-1α mediated metabolic alteration. This process of mitochondrial biogenesis requires transcription factor A (TFAM), which has a direct trans-activator such as MYC and MAX. However, MXI1, a member of MYC family and a negative regulator of MAX and MYC, is up-regulated by HIF-1α (Figure 2). This leads to suppression of mitochondrial biogenesis and lower oxygen consumption in HIF-1α over expressed cancer cell [66]. Hence, from this part of the review, it is evident that HIF-1α has significant role in metabolic regulation as well as survival and growth of cancer cells.

3.2. *Role of HIF-1α in Regulating Angiogenesis*

HIF-1α expression is the key mediator of angiogenesis in physiological conditions as well as patho-physiological conditions. As with the development of any multicellular organism, tumor development also depends on the adequate amount of oxygen and nutrients supply through the blood vessels. In tumor proliferation beyond 1–2 mm, oxygen and nutrients cannot diffuse properly to the core, which creates hypoxia. This hypoxic condition triggers activation of HIF-1α resulting in the overexpression of HIF regulated genes such as VEGF, metalloproteinases, chemokines to stimulate angiogenesis, and endothelial cells recruitment (Figure 2) [67–69]. HIF-1α expression is also regulated by mutations in tumor suppressor genes, such as VHL, p53, and PTEN [70–72] Moreover, overexpression of oncogenes that includes v-SRC, EGFR, HER2, and subsequent signaling through the phosphatidylinositol-3-kinase (PI3K) and mitogen-activated protein (MAP) kinase pathways activate HIF-1α expression [52–54]. In the above cases, HIF-1α ultimately promotes angiogenesis by regulating broad range of genes including VEGF [73,74]. Deficiency in HIF-1β/ARNT in hepatoma cells results in a less vascular, slow growing tumor as well as reduced VEGF expression compared to the tumors produced from wild-type cells [54]. Loss of VHL causes in constitutive HIF-1α activation and increases VEGF expression results in more hemorrhagic tumors with higher microvessel density in teratocarcinomas and fibrosarcomas compared to the tumors derived from wild-type cells [75,76]. Similarly, deletion of HIF-1α also leads to decreased VEGF expression and defective vascularization of tumors in nude mice [77]. HIF-1α also regulates the broad range of genes such as angiopoietins and VEGF receptors involved in the regulation of angiogenesis. Tang et al., reported that HIF-1α can down-regulate VEGFR-2 by regulating a VEGFR-1/VEGF/VEGFR-2 autocrine loop, which is essential for its post transcriptional induction in hypoxic conditions [78]. Loss of HIF-1α also disrupts this feedback mechanism, which is responsible for proper vasculogenesis, endothelial cell proliferation, tube formation, and growth of a solid tumor in vivo [78]. It was reported that endothelial cells cannot migrate properly to the hypoxic region due to the loss of HIF-1α [78]. Another study showed that HIF-1α regulates angiogenesis by controlling the VEGF/FLT1 signaling pathway in neuroblastoma cells [79]. FLT1, a "fms-like tyrosine kinase" receptor has a crucial role in regulating angiogenesis, migration of endothelial cells and cell survival in several types of cancer including prostate, colon, pancreas, glioblastoma, lymphoma, leukemia, and mesothelioma. FLT1 owns a binding site for HIF-1α. Moreover, VEGF/FLT1 activates HIF-1α via activation of ERK1/2 resulting in an alteration in VEGF level in tumor cells. This autocrine feedback loop is also associated with up-regulation of a potent anti-apoptotic protein, BCL-2 and an angiogenic factor bFGF indicating a significant role of HIF-1α in angiogenesis [79]. In addition, HIF-1α induces VEGF by recruitment of bone marrow-derived CD45+ cells which secretes metalloproteinase-9 (MMP-9) in the tumor site. A recent study reported that HIF-1α recruits bone marrow-derived CD45+ myeloid cells containing Tie2+, VEGFR1+, CD11b+, and F4/80+ subpopulations by inducing stromal-derived factor 1α (SDF1α) [80].

Moreover, several studies reported that the regulation of Nitric oxide synthases (NOS) is mediated by HIF-1α [81,82]. NOS isoforms produce nitric oxide (NO) which helps in survival of endothelial cell via inhibition of apoptosis induced by up-regulating caspase signaling [81,82]. NOS contains HRE site at its promoter. Quintero et al., reported that NOS is positively correlated with HIF-1α in squamous cell carcinoma [81]. In addition, inhibition of production of NO by the NOS inhibitor, L-NMMA prevents stabilization of HIF-1α in oral squamous carcinoma cell line [81]. HIF-1α also regulates angiogenesis via PI3K/AKT signaling pathway in human breast tumor and glioblastoma cells [83,84]. In addition, the PI3K/AKT signaling pathway encourages tube formation of HUVEC cells induced by bFGF activated by HIF-1α [85]. PDGF-A, PDGF-B, and EGF bind to the platelet-derived growth factor receptor (PDGFR) and epidermal growth factor receptor (EGFR), respectively. These bindings promote HIF-1α synthesis via the PI3K/AKT signaling cascade resulting in angiogenesis in cancer [86]. It is reported that activation of the EGFR/PI3K/AKT/mTOR pathway can increase VEGF expression by up-regulating HIF-1α [85]. Another study reported that PI3K/mTOR pathway increases HIF-1α protein levels without altering HIF-1α mRNA levels [85].

Moreover, Ravi et al., demonstrated that loss of p53 promotes angiogenesis in tumor xenografts in nude mice by HIF-1α regulation [71]. P53 helps in proteasomal degradation of HIF-1α mediated by Mdm-2 resulting in down-regulation of VEGF expression in colon carcinoma [71].

3.3. *Role of HIF-1α in Metastasis*

Malignant cells achieve a unique ability to disseminate from primary lesion to distal organ to form a secondary tumor. This phenomenon is called metastasis. This important hallmark of cancer is responsible for 90% of cancer-related lethality in patients [87]. Epithelial to mesenchymal transitions (EMT) is one of the most important characteristics gained by malignant cells to fulfill the goal metastasis. Loss of expression of epithelial marker proteins and gain of expression of mesenchymal marker proteins allow epithelial cells to detach from their neighboring cells by disrupting cell-cell attachment which helps cancer cell migration to distant sites. It is reported that hypoxia can stimulate metastasis by inducing EMT [88,89]. Yang et al., reported that Twist, an important mesenchymal marker, is directly regulated by HIF-1α [88]. Twist is a BHLH transcription factor and plays a pivotal role in regulation of transcription factors such as Snail in metastasis [88]. HIF-1α activates histone deacetylase 3 (HDAC3) which then binds to the promoters of CDH1 and junction plakoglobin (JUP) followed by transcription of Snail. HIF-1α promotes metastasis in cancer tissues via SMAD and non-SMAD signaling pathways by up-regulating TGF-β expression. Phosphorylated TGF-βRI binds with TGF-βRII, which in turn activates SMAD signaling pathway. This pathway regulation results in HIF-1α activation that causes the transcription of several EMT inducing genes. Non-SMAD signaling pathway is also triggered by HIF-1α via PI3K/AKT/mTOR signaling network. Along with this, WNT/β-catenin also plays a significant role in hypoxia-induced EMT by HIF-1α in malignant cells. In addition, HIF-1α activates the hedgehog signaling pathway, which helps in EMT of cancer cells [89]. HIF-1α is inversely correlated with the expression of E-cadherin in ovarian cancer cell lines, SKOV3 and OVCAR3 as well as in vivo showing an attachment between hypoxia and metastasis. In addition, regulation of E-cadherin expression is mediated by an up-regulation of Snail via HIF-1α in malignant cells [90]. Along with this, Lysyl oxidase (LOX) is also up-regulated in hypoxia. Erler et al., validated that HIF-1α regulates LOX, an extracellular matrix protein which helps in formation of premetastatic niche mediated by bone marrow cell recruitment in hypoxic conditions via a functional hypoxia-responsive element [91]. Their study reported the important role of HIF-1α in hypoxia-induced metastasis. Expression of LOX also has a significant correlation with hypoxia in breast cancer patients with ER-negative tumors and head and neck cancer patients [91]. HIF-1α also promotes EMT by regulating other genes involved in this process such as TCF3, ZEB1, and ZEB2 (Figure 2) [92]. In hypoxic conditions, ZEB1-MYB-E-cadherin signaling plays a key role in the activation of EMT. Moreover, HIF-1α also induces Jagged2, cyclooxygenase-2 (COX-2), and urokinase receptor (uPAR) which participate in metastasis via regulating EMT [15–17]. Not only via inducing EMT, HIF-1α also induces metastasis by regulating proteins involved in invasion. Cancer cells invade surrounding tissues by disrupting the basement membrane. Various reports suggest that Matrix metalloproteinases (MMPs) are involved in degradation of components of extracellular matrix. HIF-1α induces MMP-2 and MMP-9 expression which promotes invasion by degrading type IV collagen, an important component of the basement membrane (Figure 2) [92,93]. HIF-1α also induces expression of TGF-β in hypoxic conditions, which in turn activates its downstream regulators such as Smads, Snail, Slug, and Twist (Figure 2). On the other hand, HIF-1α also inhibits E-cadherin expression by restricting activity of its upstream regulator TGF-β [94,95]. To be metastatic, cancer cells have to extravasate into the tissue after its migration to distal organ where they have to proliferate again to form a secondary tumor. In hypoxic conditions, HIF-1α induces expression of receptor tyrosine kinase MET. MET is the specific receptor for Hepatocyte growth factor (HGF) which is a pleiotropic cytokine also known as scatter factor-1. This specific binding promotes invasion and extravasion in cancer tissues. However, this metastatic phenomenon is also accelerated through HIF-1α synthesis resulted by a positive feedback loop maintained by MET overexpression [96]. Additionally, HIF-1α has an important role in metastatic progression through up-regulation of NF-κB expression. Thus, accumulation of NF-κB causes overexpression of genes such as MMP-2/MMP-9 and activates urokinase-type plasminogen activator and chemokines such as stromal-derived factor-1alpha (SDF-1α). The receptors of SDF-1α, CXCR4 is involved in homing and migration of cancer cells [97,98]. Overexpression of inhibitor of *κB* (I*κB)* significantly inhibits CXCR4 expression resulting in inhibition of SDF-1α mediated cellular migration [99]. Moreover, NF-κB is also reported to repress E-cadherin and activate ZEB1 protein which are the downstream regulators of HIF-1α pathway indicating that NF-κB and HIF-1α both depends on each other to induce EMT in tumor [100–102]. Therefore, HIF-1α not only has a prominent role in EMT, it can also regulate metastasis to accelerate the virulence of different cancers.

3.4. Role of HIF-1α in Cancer Stem Cell Proliferation and Maintenance

Cancer stem cells (CSCs) play an important role in cancer recurrence, metastasis, and therapy resistance. Low concentration of oxygen in cells or tissues, referred to as hypoxia, are one of the most invasive microenvironmental stresses. Hypoxia is the most common feature of solid tumors. Hypoxia is associated with many aspects of biological processes during tumor development and progression, such as cell survival, invasion, angiogenesis, and cellular metabolic alterations [19]. HIF-1α acts as a master transcription factor, can be stably expressed under hypoxia, and acts as a significant molecule to regulate the development of CSCs but the mechanism remains indistinct. Studies revealed that HIF-1α is related to the production of CSC markers [19]. The data indicates that HIF-1α can induce the production of multiple stem cell markers, such as OCT4, SOX2, Nanog, and Krüppel-like factor 4 (KLF4) (Figure 2) [19]. Additionally, the silencing of HIF-1α can hinder the progression of cancer by inhibiting the expression of stem cell markers. In the case of glioma, breast cancer, and prostate cancer, HIF-1α activates pro-survival pathways such as Notch, wingless, INT-1 (WNT), and the Hedgehog pathway, which are important for CSC maintenance, which leads to radioresistance and repopulate CSCs during or after treatment [103]. HIF-1α binds to the CD133 promoter and promotes the production of $CD133^+$ glioma, colon, and pancreatic CSCs via OCT4 and SOX2. In turn, CD133 promotes HIF-1α expression and its translocation to the nucleus under hypoxic conditions. A recent study suggests that HIF-1α expression can be down-regulated by microRNA such as miR-935 in a feedback loop which in turn may inhibit glioma development [103]. There is a different opinion which is that expression of hypoxia-induced HIF-1α leads to a decrease in CD133 expression in gastrointestinal cancer cells that overexpress CD133.

Under normoxic conditions, expression of HIF-1α is suppressed by the inhibition of mTOR signaling in CD133-overexpressing gastrointestinal cancer cells [104]. Under hypoxia while promoting differentiated cell phenotypes, HIF-2α silencing inhibits CSC phenotypes and is complementary to existing DNA alkylating treatments to inhibit glioma CSC activity. HIF-1α binds directly to the CD47 promoter to facilitate gene transcription, which helps to escape phagocytosis of macrophages and maintains the stem phenotype of breast CSCs. Endogenous HIF-1α promotes CD24 expression, as well as tumor formation and metastasis. In breast CSCs, the stability of Nanog mRNA through the transactivation of RNA demethylase ALKBH5 is increased by HIF-1α, and is involved in encoding N6-methyladenosine demethylase. 4-trimethylaminobutyraldehyde dehydrogenase (ALDH1A1), a subtype of aldehyde dehydrogenase, is associated with the self-renewal, metastasis, and resistance of cancer cells, is regulated by HIF-1α in breast cancer. Sequentially, ALDH1A1 promotes expression of HIF-1α via retinoic acid signaling. Zhang et al., reported the interrelation between CD47 and the cancer stem cell phenotype, but the molecular mechanisms of CD47 regulation have not been determined [105]. On the other hand, HIF-1α directly activates transcription of CD47 gene in hypoxic breast cancer cells. CD47 expression is enriched for cancer stem cells, and the depletion of cancer stem cell is led by deficiency of CD47. High CD47 expression is related to increased HIF-1α target gene expression. Thus, CD47 expression is fatal for breast cancer phenotype that is mediated by HIF-1α. The number of breast CSCs decreases upon inhibition of CD47 expression which is resulting in increased phagocytosis of breast cancer cell [105]. In prostate cancer samples, the co-localization of HIF-1α, OCT4 and Nanog suggest that the production of CSCs may be regulated by HIF-1α by regulating stem factors. In cervical cancer cells, OCT4B, an isoform of OCT4 promotes neovascularization by up-regulating HIF-1α production. In addition, HIF-1α also inhibits the expression of epithelial marker proteins, which can be confirmed by the use of HIF-1α inhibitors. Due to its association with neovascularization, HIF-1α can be used as a malignant marker of chondrosarcoma. Under hypoxic conditions, HIF-1α, a direct or indirect upstream regulator of the CSC marker proteins (Figure 2), may become a novel target to inhibit the signaling pathway of CSCs. In tumor progression, up-regulation of HIF-1α plays a vital role in CSC-regulated cancer hallmarks by controlling different gene expressions involved in CSC maintenance. Irradiation-induced DNA damage exerts intense regulation of HIF-1α, which not only depends on the oxygenation status and aberrant stabilization by Nijmegen breakage syndrome protein 1 (NBS1), but also induces EMT, invasion, and other characteristics of the CSC phenotype [106,107]. Due to localization of cancer stem cells (CSCs) in hypoxic niches, head and neck squamous cell carcinoma (HNSCC) is resistant to standard treatments. The NF-κB/HIF-1α signaling pathway maintains cancer stemness and high radioresistance in CD133-positive CSCs, whereas the inhibition of this HIF-1α involving pathway reverses the EMT and reduces the radioresistance in a model with laryngeal squamous carcinoma CNE-2 stem cells. Hypoxia-inducible factor-1α (HIF-1α) is involved in the resistance to photons, but its role in response to carbon ions remains unclear. HIF-1α mainly describes the radioresistance of CSCs of head and neck squamous cell carcinoma to both photon and carbon ion irradiation, which makes the HIF-1α targeting an attractive therapeutic challenge [108].

In high-grade serous ovarian cancer (HGSOC), a novel cancer stem cell (CSC) marker, ZIP4, while it converts to cisplatin (CDDP), it has been found that ZIP4 induced sensitization of HGSOC cells to histone deacetylase inhibitors (HDACis). On the other hand, ZIP4 selectively up-regulates HDAC IIa HDACs, with little or no effect on HDACs in other classes. With endothelial growth factor A (VEGFA), functional downstream mediators of HDAC4, and hypoxia-inducible factor-1 alpha (HIF-1α), HDAC4 knockdown (KD) and LMK-235 inhibit spheroid formation in vitro and tumorigenesis in vivo. Moreover, Fan et al., reported that ZIP4, HDAC4, and HIF-1α are involved in regulating secreted VEGFA in HGSOC cells [109]. While many HDAC4 targets have been identified, they focused on HIF-1α, one of the central players of tumor progression and drug response and VEGFA, one of the best-characterized HIF-1α targets. Acriflavine and linifanib, selective inhibitors for HIF-1α and VEGFA respectively, inhibit cell proliferation and block spheroid formation in both PE04 and PEA2 cells in HGSOC. In PEA2 cells, ZIP4-KD and HDAC4-KD reduce the level of HIF-1α. It is related to HDAC4's effect on HIF-1α acetylation and stabilization. Additionally, ZIP4-KD and HDAC4-KD in PE04 and PEA2 extensively reduce VEGFA production/secretion in cell. LMK-235 and acriflavine considerably reduced VEGF production in PE04 and PEA2 cells [109].

Ammonia is a toxic by-product of metabolism that causes cellular stresses and it stabilizes and activates hypoxia-inducible factor-1α (HIF-1α). HIF-1α is also activated by ammonium chloride and compromises ammonia-induced apoptosis. Moreover, glutamine synthetase (GS), a key driver of cancer cell proliferation under ammonia stress and glutamine-dependent metabolism in ovarian cancer stem-like cells express CD90. Interestingly, activated HIF-1α counteracts the function of glutamine synthetase in glutamine metabolism by facilitating glycolysis and enriching glucose dependency. The functions of HIF-1α in a biphasic ammonia stress management in the cancer stem-like cells were unknown until now, where by GS facilitates cell propagation and HIF-1α contributes to the metabolic remodeling in energy fuel usage resulting in attenuated proliferation but conversely promoting cell survival [110]. In hypoxic conditions, some gastric CSCs improve the expression of hypoxia-inducible factor-1α (HIF-1α) and increase migration and invasion capabilities compared with the normoxic control. These CSCs are activated by mesenchymal cell marker Vimentin and by the inhibition of the epithelial cell marker E-cadherin. In this case, expressions of both HIF-1α and Snail increase, initiating a cascade of events that leads to the changes in characteristic of EMT, including decreased E-cadherin expression, increased Vimentin expression and enhanced invasion ability. Yang et al., reported that HIF-1α is responsible for activating EMT via increased expression of the transcription factor Snail in gastric CSCs [111]. Furthermore, the inhibition of Snail by shRNA reduces HIF-1α-induced EMT in gastric CSCs. As a result, hypoxia-induced EMT-like CSCs depend on HIF-1α to activate Snail, which may result in recurrence and metastasis of gastric cancer [111].

3.5. Role of HIF-1α in Cancer-Related Inflammatory Response

In the 19th century, Rudolf Virchow identified the presence of leucocyte within a tumor, indicating a role of inflammation in tumor progression and development [112]. Several studies reported that inflammation induced by many bacteria and viruses increases cancer risk [112,113]. As a result of this inflammatory response, various innate immune cells such as macrophages, neutrophils, mast cells, myeloid-derived suppressor cells, dendritic cells, and natural killer cells as well as adaptive immune cells such as T and B lymphocytes are recruited into tumor microenvironment. These cells communicate with each other using several autocrine and paracrine signaling methods mediated by cytokines and chemokines to regulate tumor growth [114]. HIF-1α plays a key role in generating an inflammatory response within the tumor microenvironment. Immune cells infiltrated at the tumor site also suffer from oxygen starvation resulting in activation of HIF-1α in them [57]. An important immune check point receptor, programmed death ligand 1 (PD-L1), is activated by HIF-1α resulting in myeloid-derived suppressor cells (MDSC)-mediated T cell activation [115]. Another important HIF-1α mediated inflammatory response occurs via activation of nuclear factor-kappa B (NF-κB) in tumor region. After dissociation from I-κB, NF-κB translocates into the nucleus and activates an array of proteins such as interleukin-6 (IL-6), cyclooxygenase 2 (COX-2), inducible nitric oxide synthase (NOS2), platelet endothelial cell adhesion molecule-1 (PECAM-1) and matrix metalloproteinase 9 (MMP9) [116,117]. Crosstalk between NF-κB and HIF-1α signaling cascades elevates inflammatory response in cancer by transcribing different downstream modulators such as IL-6, MMP9, and COX2 [118]. However, as reported by Uden et al., NF-κB binds to the promoter of HIF-1α at −197/188 bp from the initiation site and induces HIF-1α transcription [119]. Although HIF-1α is up-regulated by all the subunits of NF-κB, p50 and p52 show the highest and the lowest transcriptional efficacy [119]. In the case of translational regulation, NF-κB subunits such as RelA and c-Rel play the most important role in HIF-1α protein expression compared to other subunits such as p50, p52, and RelB [119]. Along with this, HIF-1α also regulates NF-κB activation in hypoxic conditions in vivo and in vitro in inflammatory conditions [120]. According to Han et al., HIF-1α alteration also occurs via TLR4 pathway which has a crucial role in inflammatory response at tumor site [121]. They proved that HIF-1α and its target gene VEGF become activated by the TLR activator, lipopolysaccharide (LPS) in HSC3 and SCC4 cells. Knockdown of TLR3 and TLR4 using siRNAs showed a significant reduction in Polyinosinic-polycytidylic acid (poly (I:C)) induced HIF-1α and VEGF mRNA expression [121]. They also showed that NF-κB up-regulates TNF-mediated HIF-1α and VEGF expression in oral squamous carcinoma cells. Moreover, inflammatory cytokines such as IL-1β, IL-6, IL-8, and IL-12p70 were also increased in hypoxic conditions, suggesting a role of HIF-1α in NF-κB mediated inflammatory response [121]. HIF-1α activates membrane receptors like RAGE and P2X7R which in turn induces NF-κB expression [97]. This inflammatory response includes chemokines such as CCL2, CCL5, and CXCR1/CXCL8, which have a crucial role in cell migration [122]. Ligands specific to RAGE and P2X7R receptors, such as HMGB1 and BzATP, are also induced inside the tumor due to the depletion of oxygen. This in turn helps in accumulation of NF-κB into tumor cells and exerts proinflammatory response [97].

4. Different Natural and Synthetic Compounds Targeting HIF-1α

Several studies revealed that HIF-1α could be a promising target for anticancer therapy. There are several phytochemicals and chemotherapeutic drugs leading in their anticancer effects by targeting HIF-1α and its related signaling pathways. Different types of cancers in the breast, colon, lung, prostate and ovary have regulation of HIF-1α; therefore, these cancers can be controlled by targeting this specific protein. Like other proteins, HIF-1α is synthesized into the ribosome and then it goes to its intended location for its specific activity and becomes degraded when its function passes. Therefore, the activity of HIF-1α may be inhibited by various means. It is possible to interfere with the mechanism of protein synthesis at the transcriptional or translational level. Other than these, the function of a protein can be impaired by inhibiting its DNA-binding or by proteasomal degradation of the protein directly. There is another aspect of HIF-1α inhibition by which one can restrict the activity of a protein indirectly by inhibiting its interacting subunits to be a potent inhibitor. Therefore, depending on the mode of action of HIF-1α targeting drugs, these could be classified in four major groups which include (i) HIF-1α synthesis blocker, (ii) HIF-1α activity blocker, (iii) HIF-1α degradation enhancer and (iv) Degrader of HIF-1α interacting HIF subunits. Based on the origin of the potent HIF-1α inhibitors, they can be broadly categorized in two major groups i.e., (a) natural compounds and (b) synthetic compounds. Here we are mentioning about the known promising inhibitors of HIF-1α which have potency to become useful antineoplastic therapeutics (Figure 3).

4.1. Natural Compounds as HIF-1α Inhibitors

There are several potential bioactive natural components exert their effects in different field of medicine including anticancer drug development. Various metabolites such as alkaloids, amines, alkamides, terpenes, steroids, saponins, flavonoids, tannins, phenylpropanoids, lignin, coumarins, lignans, polyacetylenes, fatty acids, and waxes are used for treatment of different type of cancers. Here, we discuss a few natural products which have an inhibitory role on HIF-1α and are able to regulate different cancer progressions (Table 1).

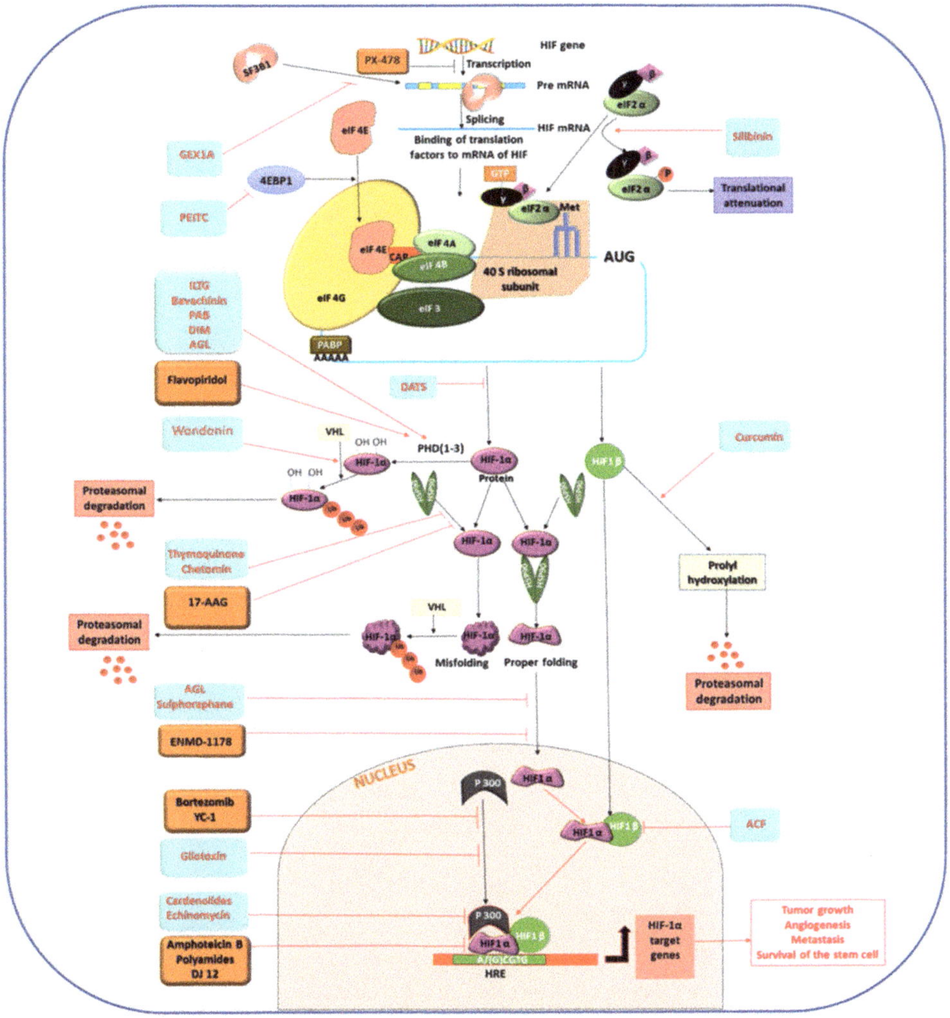

Figure 3. Molecular mechanism of different natural and synthetic compounds targeting HIF-1α and its pathways. The figure represents sky blue and brown boxes that symbolize natural and synthetic compounds, respectively. The illustration denotes GEX1A-mediated inhibition of HIF-1α mRNA splicing by inactivating spliceosome core protein SF3B1. The figure also shows translational attenuates of HIF-1α by silibinin via phosphorylation of α subunit of eIF2. ILTG, Bavachinin, PAB, DIM, AGL, and Flavopiridol-induced prolyl hydroxylation mediated proteasomal degradation of HIF-1α protein is also represented in this figure. Down-regulation of the activity of HIF-1α through proteasomal degradation of its interacting subunit αβ by Curcumin is also demonstrated here. This figure shows the mode of action of Chetomin and Thymoquinone by blocking HSP 90 mediated folding of HIF-1α protein which causes proteasomal degradation of it. Inhibition of nuclear localization of HIF-1α by AGL and Sulforaphane is shown in the figure. Suppression of HIF-1α and HIF-1β heterodimer formation by ACF is also represented. The figure denotes Bortezomib and Gliotoxin dependent inhibition of the interaction between p300 and HIF-1α. Cardenolides, Echinomycin, Amphotericin B, polyamides and DJ12 suppress the binding of HIF-1α/p300 complex to HRE, which are also represented in this figure.

Table 1. List of various potential natural HIF-1α inhibitors based on their mode of action in different cancers.

Oxygen Carbon Nitrogen Hydrogen Sulphur Bromine Chlorine

Serial No.	Name	Structure	Source	Mode of Action	Effective against Cancer Type	Clinical Trial	Reference
1	Silibinin	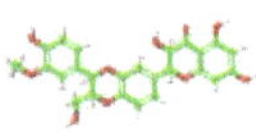	*Silybum marianum*	HIF-1α synthesis blocker	Prostate cancer, cervical cancer, hepatoma, colorectal cancer, nasopharyngeal cancer	Approved	[123–133]
2	Diallyl trisulfide (DATS)	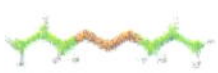	*Allium sativum*		Breast cancer	—	[127,133,134]
3	Herboxidiene (GEX1A)	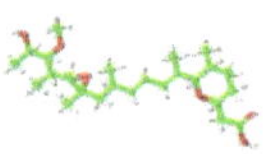	*Streptomyces chromofuscus*		Hepatoma	—	[127,133,135]
4	Celastrol	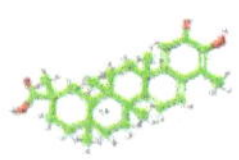	*Tripterygium wilfordii, Celastrus regelii*		Glioblastoma	—	[127,133,136–138]
5	PEITC	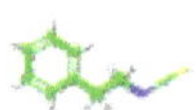	Cruciferous plants		Prostate cancer, human glioma cells, breast cancer	Phase II	[127,133,139,140]
6	Echinomycin	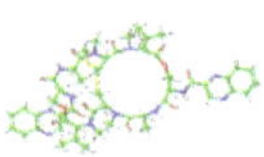	*Streptomyces echinatus*		Breast cancer, acute myeloid leukemia, uterine fibroids	Rejected after phase II trial	[127,133,141–145]
7	Chetomin	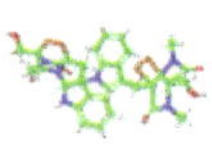	*Chaetomium globosum*	HIF-1α activity blocker	Lung cancer, multiple myeloma	—	[127,133,146–150]
8	Gliotoxin	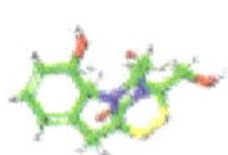	*Gliocladium fimbriatum*		Prostate cancer	—	[127,133,151–154]
9	Sulforaphane	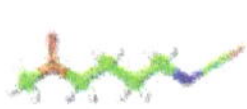	Cruciferous vegetables		Nonmuscle invasive bladder cancer, colon cancer and gastric cancer	Phase II	[127,133,155,156]
10	Acriflavin	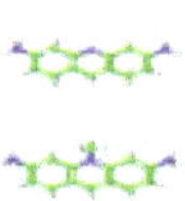	coal tar		Brain cancer, cholangiocarci-noma, ovarian and breast cancer	—	[127,133,157–161]
11	Emodin	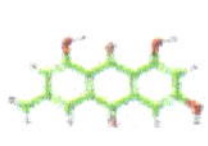	*Rheum palmatum, Polygo-nam multiflorum*		Prostate carcinoma	Rejected in clinical trial	[127,133,162,163]
12	Cardenolides	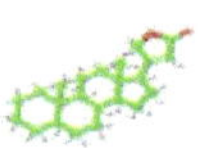	*Calotropis gigantea*		Breast cancer	Entered in clinical trial	[127,133,164,165]

Table 1. *Cont.*

Oxygen Carbon Nitrogen Hydrogen Sulphur Bromine Chlorine

Serial No.	Name	Structure	Source	Mode of Action	Effective against Cancer Type	Clinical Trial	Reference
13	DIM (3,3′-Diindolylmethane)	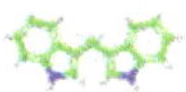	cruciferous vegetables such as broccoli (*Brassica oleracea*), Brussels sprouts, cabbage and kale.	HIF-1α degradation enhancer	Prostate, breast, colon, cervix and pancreas	Phase III	[127,133,166–168]
14	Pseudolaric acid B(PAB)	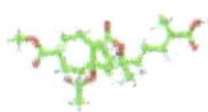	*Pseudolarix kaempferi*		Breast cancer	—	[127,133,169,170]
15	Bavachinin	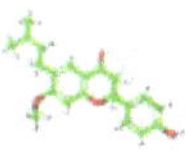	*Psoralea corylifolia*		Human KB carcinoma and HOS osteosarcoma	—	[127,133,171]
16	Andrographolide		*Andrographis paniculate*		Liver cancer, breast cancer	Phase III	[127,133,172,173]
17	Isoliquiritigenin (ILTG)	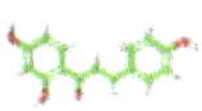	*Glycyrrhiza uralensis*, *Mongolian glycyrrhiza*, *Glycyrrhiza glabra*.		Breast cancer	—	[127,133,174,175]
18	Wondonin	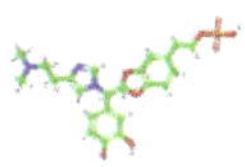	*Poecillastra wondoensis*		Keratinocyte	—	[127,133,176]
19	Thymoquinone	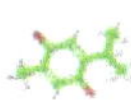	*Nigella sativa*		Renal cancer	Phase II	[127,133,177,178]
20	Curcumin		*Curcuma Longa*	Indirect inhibitors of HIF-1α	Breast cancer, pituitary adenoma	Phase II	[127,133,179,180]

4.1.1. HIF-1α Synthesis Blocker

First, we discuss the inhibition of de novo synthesis of HIF-1α by natural compounds. Silibinin (SL.1, Table 1) is a flavonolignan, found in fruit or seeds of milk thistle (*Silybum marianum*), has anti HIF-1α activity [123,124]. Silibinin is a clinically approved dietary compound against various liver diseases and cancers of the breast and prostate [125–127]. Dietary feeding of silibinin inhibits growth of advance human prostate carcinoma in athymic nude mice and increases plasma insulin-like growth factor-binding protein-3 levels [128–131]. Silibinin exerts its antitumor activity via inhibiting de novo synthesis of HIF-1α. Jung et al., suggested that Silibinin does not alter the transcription and degradation of HIF-1α, rather it inhibits translation of HIF-1α in LNCaP and PC-3 prostate cancer cells [132]. eIF4F complex is a key factor of HIF-1α translation which becomes altered by inhibition of phosphorylation of eIF-2a in Silibinin treated cells. Silibinin does not alter the mRNA level or half-life of HIF-1α rather it affects HIF-1α accumulation that promotes anticancer activity. A recent study by Deep et al., revealed that Silibinin inhibits HIF-1α synthesis by regulating its stimulator NOX [123]. In hypoxic conditions, NOX or NADPH oxidase generates ROS and promotes ROS mediated HIF-1α synthesis via PI3K/mTOR signaling pathways. Therefore, Silibinin has the potential to inhibit NOX mediated PI3K/mTOR signaling which in turn down-regulates HIF-1α synthesis [123]. Although Silibinin shows no significant toxicity, the prominent drawback of Silibinin is the inhibition of HIF-1α translation by it through targeting eIF4F complex which is a major factor of global protein synthesis. Considering the fact that eIF4F inhibition by Silibinin can cause normal cell death, researchers are interested to find other phytochemicals which can inhibit HIF-1α by exerting no toxicity or side effects.

In search of such effective bioactive components, researchers find DATS or Diallyl trisulfide (SL.2, Table 1), which is a water insoluble dietary organosulfur compound derived from the perennial plant *Allium sativum* (garlic). According to Wei et al., DATS is a natural Histone deacetylase that suppresses HIF-1α via inhibiting its upstream protein Trx-1 and attenuates metastasis of breast cancer [181]. However, DATS inhibits HIF-1α without promoting mRNA suppression or proteasomal degradation but it is able to inhibit HIF-1α synthesis at translational level [181]. Expression of L1CAM, VEGF-A, and EMT-related proteins (Slug, Snail, MMP-2, etc.) are repressed as a result of low HIF-1α level due to DATS treatment. Li et al., developed oil free microemulsion of DATS to reduce its toxicity and to improve its solubility and pharmacokinetics for making it a better anti HIF-1α compound for in vivo study [134]. Although these findings support DATS as a potent HIF-1α targeting anticancer agent, further investigation is needed into the detail mechanism of HIF-1α inhibition through Trx-1 suppression.

Other than these, Herboxidiene (GEX1A) (SL.3, Table 1) is an epoxide group containing polyketide molecule found to have anti HIF-1α activity. It is first isolated from the bacterium *Streptomyces chromofuscus* and known for its interference in the splicing of pre-mRNA of the gene that regulates cell cycle. According to Jung et al., in the case of HIF-1α, Herboxidiene inhibits splicing of it and decreases spliced HIF-1α mRNA level in hepatoma by targeting splicing factor 3B subunit 1 (SF3B1), the core spliceosome component [135]. Even though Herboxidiene intercedes in antitumor activity via inhibiting synthesis of HIF-1α, it has additive toxicity which needs to be reduced by further research.

Furthermore, Celastrol (tripterine) (SL.4, Table 1) is another phytochemical, isolated from the root extracts of *Tripterygium wilfordii* (Thunder god vine) and *Celastrus regelii*, used to improve apoptosis of cancer cells by targeting HIF-1α and inhibit HIF-1α mediated angiogenesis and metastasis. Growth, migration, and invasion of U87 and U251, human malignant glioblastoma cell lines, are inhibited by Celastrol [136]. Under hypoxic conditions, Celastrol inhibits HIF-1α mRNA levels and the hypoxia-induced accumulation of nuclear HIF-1α protein levels which subsequently resulted in the reduction of the transcriptional activities of HIF-1 target genes including VEGF. Celastrol also down-regulates the activities of PI3K, Akt, and mTOR signaling pathways which are the prominent regulators of HIF-1α synthesis. The expression of HIF-1 and vascular endothelial growth factor (VEGF) is promoted by LMP1 dependent JNK activation in Epstein-Barr virus (EBV)-associated nasopharyngeal carcinoma (NPC), which ultimately contributed toward radio-resistance in NPC patients. Phosphorylation of p38 MAPK and JNK1/2 in Cisplatin-resistant human NPC-039 and NPC-BM cells is increased by the treatment with Celastrol, increasing cytotoxicity by activation of caspase-mediated apoptotic pathways in these cells. The activation of JNK and p38 MAPK by LMP1 is related to the development of radio-resistance in NPC, but the mechanism remains to be unstated. A platelet-derived endothelial cell growth factor thymidine phosphorylase (TP) is related with poor prognosis in EBV associated NPC and expression of TP can be induced by triggering p38 MAPK pathway via the CTAR1 and CTAR2 domains of LMP1. In addition, anti-metastatic protein tissue inhibitor of metalloproteinase-3 (TIMP-3) through transcriptional repression via the p38 MAPK pathway is inhibited by LMP1 to promote metastasis [137]. As JNK and p38 MAPK pathway activation regulates HIF-1α expression, inhibition of these pathways by Celastrol may infer a negative effect on HIF-1α mediated cancer progression. To overcome the limitations of Celastrol, different derivatives of it were generated by Michael addition and ring-opening polymerization, which helps to make a compound with better solubility and pharmacological activities [138].

2-Phenethyl isothiocyanate (PEITC) (SL.5, Table 1) is another natural dietary HIF-1α synthesis blocker which is reported to have inhibitory effects on expression of HIF-1α target genes such as CAIX, GLUT1, BNIP3, and VEGF-A. PEITC does not depend on the activity of PHD and VHL, rather it inhibits phosphorylation of translational regulator 4E-BP1 resulting in attenuation of HIF-1α protein translation [139,140]. PEITC entered clinical trials for leukemia, whereas a phase II clinical trial has been completed for lung cancer [127]. PEITC has already completed phase I & II trials as a dietary supplement in oral cancer with mutant p53 [133]. However, it will be fascinating if PEITC could be developed as an anticancer drug by improving its narrow therapeutic window against a few types of cancer.

4.1.2. HIF-1α Activity Blocker

Other than inhibition of HIF-1α synthesis, its activity could also be regulated by transcriptional inhibition of HIF-1α target genes. There are several compounds which inhibit HIF-1α directly by interfering its binding to the DNA or its cofactor. After hypoxia, HIF-1α binds to its other subunits and forms HIF complex. There are some coactivators of HIF-1α like CBP/p300 helping transactivation of the genes. Some HIF-1α inhibitors target the binding of HIF-1α to the HRE element or they may hinder interaction of HIF-1α and CBP/p300.

Echinomycin (SL.6, Table 1) is such a cyclic peptide from the quinoxaline antibiotic family that inhibits HIF-1α activity by restricting HIF-1α-HRE or HIF-1α-CBP/p300 interaction. It is extracted from a bacterium *Streptomyces echinatus* [141]. Chromatin immunoprecipitation assay by Kong et al., showed that Echinomycin specifically hampers the HIF-1α binding to the promoter of VEGF in U251, human glioma and MCF-7, breast cancer cell lines. The main recognition sites of Echinomycin are 5′-ACGT-3′ and 5′-TCGT-3′ of VEGF promoter which possess common central 2-bp sequence 5′-CG-3′ that serves as strong binding sequence [142]. Echinomycin was introduced in phase I–II clinical trials for B16 melanoma and the P388 leukemia, but it was rejected because of side effects and low antitumor effects. A recent study by Vlaminck et al., explained the dual role of Echinomycin on HIF-1α activity [143]. They proposed that under normoxic conditions, it increases HIF-1α level. According to them, this drug is not efficient enough as an antitumor agent down-regulating HIF-1α. In the recent studies by Bailey et al., Echinomycin is used as liposomal nanoparticle against HIF-1α in triple negative breast cancer to improve its efficacy [144]. They proposed that liposomal Echinomycin shows more therapeutic efficacy and less toxicity than cremophor formulated Echinomycin. In another study, Wang et al., proved the therapeutic effects of Echinomycin by targeting HIF-1α in TP53 mutated acute myeloid leukemia (AML) [145]. TP53 mutation occurs in AML with 10% frequency and associated with poor prognosis. According to Wang et al., PEGylated liposomal Echinomycin prolongs the drug circulation time in the bloodstream [145]. Additionally, they showed that Echinomycin not only inhibits AML cell metastasis but also it targets CD34+CD38- stem cell population by inhibiting HIF-1α. These findings claim the reformulation and reintroduction of Echinomycin in clinical trial for cancer therapy.

Looking for an alternative, Min et al., discussed the anticancer role of Chetomin, (SL.7, Table 1), an antimicrobial metabolite extracted from the mesophilic saprophytic fungus *Chaetomium globosum* [146]. They reported the Chetomin-attenuated sphere forming ability of CSC and proliferation ability of non-CSC in Non-small cell lung cancer (NSCLC) [147]. The Co-Immunoprecipitation analysis by Min et al., demonstrated the interaction between the HIF-1α and HSP90 proteins in hypoxia mediated-activation of HIF-1α [147]. HSP90 (Heat shock protein 90) is a chaperone, frequently overexpressed in some cancers leading to chemoresistance and lower survivability of patients. Binding to the PAS-B domain of HIF-1α, HSP 90 stabilizes it and helps in the activity of HIFs [148,149]. Chetomin significantly affects the binding between HSP90 and HIF-1α without affecting the level of HSP90 or HSP70. They tested the inhibitory role of Chetomin on adhered monolayer cultures and spheroid cultures in a dose-dependent manner. Chetomin exposure inhibits survival promoting factors such as mitogen-activated protein kinase 1/2 (MEK1/2), insulin-like growth factor 1 (IGF1 R), epidermal growth factor receptor (EGFR), SRC, activation of protein kinase B (AKT) and mammalian target of rapamycin (mTOR), basic fibroblast growth factor (bFGF), platelet-derived growth factor (PDGF), etc. in monolayer cultures. Immunohistochemical analysis of lung revealed lower intensity of HIF-1 α and CD34 in Chetomin treated group. Thus, Chetomin alters the CSC mediated chemoresistance via attenuation of the activity of HIF-1α. In another study by Kung et al., it is shown that Chetomin constrains HIF-1α activity, precluding the binding of HIF-1α to p300 [150]. HIF-1α interacts with p300 through CH1(cysteine/histidine-rich 1) domain of p300. The CH1 domain is a zinc-containing transcription adapter zinc-binding (TAZ) domain. Chetomin disrupts this CH1 domain and prevents the interaction between HIF-1α and its coactivator p300. From these observations, it can be inferred that Chetomin inhibits HIF-1α directly and it can be a potent inhibitor of HIF-1α.

Apart from these, Gliotoxin (SL.8, Table 1) is a sulfur containing mycotoxin which mediates same kind of activity such as Chetomin. It is extracted from the marine Trichoderma species *Gliocladium fimbriatum*. It is the most effective member of epidithiodioxopiperazine fungal toxin family and exerts its anti-HIF-1α activity, down-regulating the transcription of VEGF-A, LDHA, and ENO1 genes. Reece et al., studied the antitumor activity of Gliotoxin in prostate cancer and proposed that it can disrupt the HIF-1α/p300 complex in vitro and in vivo [151]. Using a fluorescence binding assay, they confirmed that Gliotoxin disrupts the CTAD domain of HIF-1α and the CH1 domain of p300 which in turn interrupts their binding. Despite having such negative effects against HIF-1α, it has not started clinical trials because of its high toxicity. In earlier studies, it was observed that Gliotoxin treatment leads to death of experimental mice [152]. Therefore, further studies are going to need to reduce the toxic effects of Gliotoxin. In this context, Hubmann et al., proposed that lower dose of Gliotoxin shows no significant toxicity [153]. However, Comas et al., delivered Gliotoxin in cancer cell via nanoparticle and showed its antitumor activity with low toxicity [154]. The effects of Gliotoxin are also studied in combination with other chemotherapeutic agents in very low dose for reducing its toxic effects.

However, to overcome such toxic effects of therapeutics and to establish a new effective drug, Xia et al., introduced Sulforaphane (SL.9, Table 1), which is a nitrogen containing isothiocyanate compound that can be found in cruciferous vegetables such as broccoli, kale, cabbage, and watercress [155]. Xia et al., enlightened the anti HIF-1α activity of Sulforaphane in the case of non-muscle invasive bladder cancer cell line [155]. They explained that Sulforaphane blocks translocalization of HIF-1α to nucleus and suppresses hypoxia mediated glycolysis. In another study by Kim et al., Sulforaphane is introduced as an anti-HIF-1α component against human colon cancer and gastric cancer cells [156]. At first, Kim and his colleagues thought that inhibition of HIF-1α by Sulforaphane may be due to degradation by 26S proteasomal pathway. Further studies by Kim et al., confirmed that inhibitory effect of Sulforaphane against HIF-1α is neither 26S proteasomal pathway dependent nor Akt/mTOR signal mediated, rather they observed lysosomal degradation of HIF-1α [156]. Sulforaphane promotes lysosomal activity and degrades HIF-1α in hypoxia. There is no significant toxicity in Sulforaphane treatment and it is easily metabolized via the mercapturic acid pathway [156]. Although Sulforaphane undergoes phase IV clinical trials for treating Human immune deficiency virus (HIV) infection and phase II trial sfor combating autistic disorders, neurodevelopmental disorders, diabetes mellitus, and schizoaffective disorders, in the case of cancer there are only a few reports which mention phase II clinical trials for adenocarcinoma and recurrent prostate cancer [127,133]. Therefore, Sulforaphane could be inspected further as an anti-HIF-1α molecule.

One of the most potent inhibitors of HIF-1α activity is Acriflavin (ACF) (SL.10, Table 1). It is extracted from coal tar and was first introduced in medical science in 1972 by the German researcher Paul Ehrlich. ACF was generally used as antiseptic, trypanocides, anti-viral, and anti-bacterial agent. In recent years, ACF is repurposed in cancer to develop new therapeutic strategies [157]. Nehme et al., and Cheloni et al., demonstrated the anti-oncogenic properties of ACF against chronic myeloid leukemia [158,159]. Even though ACF is an FDA approved drug for urinary tract infections, recent research proved its anti-cancer efficacy which involves inhibition of HIF-1α activity [127,133,159]. Mangraviti et al., demonstrated that ACF down-regulates proangiogenic protein VEGF, PGK-1 by inhibiting the formation of HIF-1α and HIF-1β heterodimer in brain cancer [160]. According to Mangraviti and his colleagues, ACF can inhibit transcriptional activity of HIF-1α not only in brain cancer but also in cholangiocarcinoma (SK-ChA-1), ovarian (A2780), or breast cancer (MCF-7) cell lines [160]. They also explained that ACF has cytotoxic effects on glioma stem cells (GSCs) and reduces chemoresistance. ACF inhibits HIF-1α through binding to the PAS-B domain which in turn restricts its binding to HIF-1β subunit necessary for its transcriptional activity. The expression of hypoxia-induced GLUT1, a typical HIF-1α target gene, is reduced by the treatment with ACF [161]. Therefore, ACF could be used individually or in combination with other drugs as an antiangiogenic agent or a CSC inhibitor in cancer treatment.

In addition, Huang et al., proposed another phytochemical Emodin (SL.11, Table 1) as a HIF-1α inhibitor, enhancing antineoplastic effects of other chemotherapeutic drugs [162]. Emodin is an anthraquinone derivative extracted from the Chinese herbs such as *Rheum palmatum* and *Polygonam multiflorum*. Cancer cells explore multiple drug resistance against chemotherapeutic drug by MDR1 gene. MDR1 is a downstream targeted gene of HIF-1α. Generally, in cancer cells, ROS and HIF-1α maintain a balance. Increasing the ROS concentration causes the activation of HIF-1α and other members of the HIF family [163]. As a result, MDR1 gene becomes transcriptionally activated by HIF-1α and produces multidrug resistance. On the other hand, Huang and his colleagues showed that in combination therapy of Emodin with Cisplatin, ROS becomes increasingly relative to control and exerts ROS mediated HIF-1α suppression. By in vitro and in vivo study of Emodin cotreatment, it is documented that multiple drug resistance becomes altered as a result of the down-regulated MDR gene. As Emodin is a ROS generator, it suppresses transactivation of HIF-1α in combination therapy with Cisplatin to DU-145 cells [162]. Despite the fact that Emodin has failed to prove its clinical efficacy against polycystic kidney disease, its significant anti-neoplastic effects are encouraging researchers to include it in the preclinical study list of anticancer therapeutics.

Other than these phytochemicals, the lactone ring containing group of steroids, Cardenolides (SL.12, Table 1) also show attenuation of HIF-1α transcriptional activity. The latex containing Cardenolides are extracted from various medicinal plants belonging to the family *Asclepidaceae* and *Apocynaceae*, native to India, Africa, southern China, and southeast Asia. Parhira et al., isolated cardenolides from *Calotropis gigantea* and demonstrated their anticancer activity [164]. They isolated 20 Cardenolides, among which Digoxin has started clinical trials for breast cancer, colon cancer, rectal cancer, and sarcoma therapy. Although Digoxin has either completed or started phase IV clinical trials for arterial fibrillation, cardiac failure, Dilated Cardiomyopathy, and acute kidney injuries, the addition of these compounds to clinical trials is comparatively new [127,133]. Zheng et al., isolated six new non-classical Cardenolides, including 17 known ones [165]. 19-dihydrocalotoxinis one of those 23 compounds, which has more potential as an anti-HIF-1α component compared to Digotoxin. Further studies are necessary to verify its toxic effects and anti-HIF-1α activity in different cancers.

4.1.3. HIF-1α Degradation Enhancer

Degradation of an oncogenic protein by drug treatment could be a promising method for anticancer therapy. There are several HIF-1α inhibitors which can act by proteasomal degradation of the protein; thus, they represent a different approach compared with the other two mechanisms of HIF-1α inhibition stated earlier. 3,3′-Diindolylmethane or DIM (SL.13, Table 1) is one of those compounds, which exerts its anticancer effect by executing proteasomal degradation of HIF-1α. DIM is a phytochemical derived from Indole-3-carbinol (I3C), present in cruciferous vegetables [166]. Riby et al., demonstrated the inhibitory effects of DIM on HIF-1α in a dose-dependent manner [166]. They explained that DIM has a dual role in both synthesis and degradation of HIF-1α in MDAMB-231 cell line. To assure the effects of DIM on synthesis of HIF-1α they inhibited proteasomal degradation of HIF-1α using MG132, a proteasome inhibitor, and measured accumulation of HIF-1α protein level which became significantly reduced after DIM treatment. Furthermore, the rate of HIF-1α degradation was halted at a certain concentration of DIM which suggests the saturation of its activity. From this study, it was concluded that DIM significantly reduces the synthesis of HIF-1α and it mediates the degradation of HIF-1α with a concentration independent mechanism. Furin and VEGF are two HIF-1α regulated proangiogenic endogenous genes which are inhibited by DIM treatment in hypoxic tumor cells in concentration-dependent manner. Due to proven preclinical efficiency of DIM, it started different phases of clinical trials against different cancers such as cervical and prostate cancer [127,133]. As DIM is a natural compound and it was previously proved that DIM has no side effects in patients and it has a profound effect on HIF-1α, it could be one of the potent therapeutics against cancer targeting HIF-1α [167,168].

Additional studies of Li et al., displayed antiangiogenic role of pseudolaric acid B(PAB) (SL.14, Table 1) via proteasomal degradation of HIF-1α in MDAMB-468 cell line [169]. PAB is a natural diterpenoid antifungal compound occurring from *Pseudolarix kaempferi*. Li et al., stated that PAB inhibits paracrine secretion of VEGF via lowering the level of HIF-1α [169]. However, transcription of HIF-1α is not affected by PAB. On the contrary, HIF-1α level becomes decreased via ubiquitin proteasomal degradation mechanism. PAB has a dual effect as an antiangiogenic agent, i.e., not only has it an inhibitory effect on HIF-1α but also has a negative effect on the endothelial cell directly. PAB can also indirectly target HIF-1α stabilization by targeting c-Jun, cellular proliferation marker [170]. PAB phosphorylates c-Jun at Ser63/73 which causes destabilization of HIF-1α protein. This activity of PAB leads to premature degradation of HIF-1α protein and transcriptional attenuation of VEGF and MDR1 genes. For these antiangiogenic effects of PAB, it can be novel anticancer therapeutics for clinical studies.

Bavachinin (SL.15, Table 1) is another HIF-1α degrader and a natural prenylated flavanone, present in the Chinese herb *Psoralea corylifolia*. It has potent anti-angiogenic activity and an anti-inflammatory function. Under hypoxic conditions, Bavachinin decreases the activity of HIF-1α in a concentration-dependent manner in human KB carcinoma and HOS osteosarcoma cells, mainly by raising the interaction between VHL and HIF-1α followed by proteasomal degradation. Moreover, the downstream genes of HIF-1α (e.g., VEGF, GLUT-1, and hexokinase-2), which block angiogenesis and reduce energy metabolism, are suppressed by Bavachinin. Additionally, Bavachinin prohibits tube formation by HUVECs and also prevents the migration of KB cells. In KB cells, Bavachinin reduces the expression of CD31 [171]. As there is no significant toxicity is shown in Bavachinin treatment, it can be further investigated.

Aside from these, Ma et al., showed that Andrographolide (AGL) (SL. 16, Table 1) promotes proteasomal degradation of HIF-1α and hinders the nuclear localization of it in Hep3B and HepG2 liver cancer cell lines [136]. AGL is a bioactive diterpenoid isolated from *Andrographis paniculate* belonging to family *Acanthaceae* [136,172]. Inhibition of nuclear localization of HIF-1 α by AGL leads to transcriptional down-regulation of HIF-1α targeted genes. Cui et al., reported that the AGL is a HIF-1α and VEGF inhibitor targeting the PI3K/AKT signaling pathway and preventing choroidal neovasculature [173]. Therefore, despite having efficacy against acute exacerbation of chronic bronchitis and acute tonsillitis, these anti-oncogenic preclinical data support repurposing of AGL against different cancers [127,133]. Although the compound is in a phase III clinical study for squamous cell carcinoma, more research needs to be carried out to improve the activity of AGL [127,133].

Isoliquiritigenin or ILTG (SL.17,Table 1) facilitates its anticancer activity via the proteasomal degradation of HIF-1α and inhibits kinase activity of VEGFR-2(VEGF receptor 2) [174]. It is a chalcone type dietary flavonoid turned out from the roots of licorice plants such as *Glycyrrhiza uralensis*, *Mongolian glycyrrhiza*, and *Glycyrrhiza glabra* [175]. In vitro and in vivo study of Wang et al., validated that ILTG treatment inhibits cancer neo-angiogenesis in breast cancer and showed no significant toxicity. However, establishment of ILTG as anticancer therapeutics claims needs more research [174].

Wondonin (SL.18, Table 1) as a HIF-1α degrader, has significant potential in clinical applications against cancer. It is a bis-imidazole compound isolated from the sponge, *Poecillastra wondoensis*. Jun et al., proposed that Wondonin down-regulates the immunoreactivity of CD31 and VEGF, whereas it increases the affinity of binding of pVHL to HIF-1α, which leads to proteasomal degradation of HIF-1α [176]. Therefore, Wondonin has a prominent antiangiogenic role as it controls both HIF-1α and VEGF. Thus, Wondonin can be a novel anticancer therapeutic agent against HIF-1α.

Lee et al., introduced Thymoquinone (SL.19, Table 1) as another HIF-1α targeting antineoplastic benzoquinone phytochemical isolated from the seed of black cumin (*Nigella sativa*) [177]. Thymoquinone is a popular compound in phase II or phase III clinical trials for COVID-19 and polycystic ovarian syndrome [127,133]. In the case of cancer, researchers conducted phase II clinical trials of Thymoquinone against premalignant lesion [127,133]. However, according to the preclinical studies, Thymoquinone destabilizes the binding between HIF-1α and HSP90 and helps in pVHL-mediated polyubiquitination of HIF-1α protein, which is independent of prolyl hydroxylation in renal cancer cells [177]. Recent research by Homayoonfal et al., showed that targeting microRNAs with thymoquinone could be a new approach for cancer therapy [182]. PEGylated Thymoquinone up-regulates expression of miR-361 which suppresses proinflammatory and proangiogenic genes including HIF-1α. Thus, Thymoquinone could be a potent HIF-1α inhibitor which enhances proteasomal degradation of HIF-1α. Hence, it is an intriguing field of research where more drugs degrading the oncogenic HIF-1α are yet to be discovered.

4.1.4. Degrader of HIF-1α Interacting HIF Subunits

Earlier in this review, we discussed the natural compounds that target HIF-1α by blocking its synthesis, inhibiting its activity, and degrading the protein. Here another mechanism of HIF-1α inhibition is discussed, which is about down-regulation of its interacting subunits. As discussed in this review, HIF-2α and ARNT are the key interacting subunits of HIF-1α, expressing changes which can be noticed in hypoxia.

Strofer et al., suggested that at higher Curcumin (SL.20, Table 1) concentration, HIF-1α, HIF-2α and ARNT level decline in Hep3B, HepG2, and MCF-7 cancer cell lines [179]. Curcumin ((1,7-bis(4-hydroxy-3-methoxyphenyl)-1,6-heptadiene-3,5-dione) is a polyphenol compound extracted from rhizomatous herbaceous plant *Curcuma longa* (turmeric). For thousands of years, turmeric has been renowned for its medicinal value. At present, it is considered to be one of the FDA approved drugs in autosomal dominant polycystic kidney disease (ADPKD), chronic schizophrenia, major depressive disorder, periodontitis and type-2 diabetes mellitus [127,133]. Curcumin has been introduced into phase II and III clinical trials against various cancers such as breast, colorectal, and head and neck cancers [127,133]. Prolyl hydroxylase is an enzyme that hydroxylates HIF-1α and helps in proteasomal degradation in the presence of oxygen. Oxidation of iron is required for this oxygen mediated regulation of HIF-1α protein degradation. Curcumin acts as an iron chelator and disrupts enzymatic activity of the enzyme prolyl hydroxylase. Thus, Curcumin has miscellaneous effects on HIF-1α regulation [179]. In another study, the inhibitory role of Curcumin against HIFs was studied by Sarighieh et al., on MCF-7 cells and cancer stem- cells (CSCs) [180]. Presently, researchers focus on therapeutic efficacy of different synthetic drugs and natural compounds against CSCs (cancer stem cells). Sarighieh and his colleagues studied the role of curcumin on HIF-1α and HIF-2α in a dose-dependent manner. They revealed that Curcumin could not inhibit HIF-1α directly, rather it mediates antiangiogenic and antiproliferative effect via inhibiting HIF-2α and ARNT in both normoxic and hypoxic conditions, whereas HIF-1α level was not affected. Therefore, Curcumin interferes with metastasis by indirectly targeting HIF-1α by degrading HIF-2α and ARNT in CSCs of MCF7 and MDAMB-231 cell line [180]. Although, Curcumin is a well-known and effective natural compound with no side effects, it also has negative effects on CSCs, the momentous element of cancer cell metastasis, drug resistance and recurrence. Therefore, Curcumin could be a coercive element against HIF-1α mediated CSC progression.

The aforementioned natural compounds not only have fewer side effects, but they can also possess alternative anti-oncogenic response which involves targeting inflammatory marker such as NF-κB, a prominent pathway modulator of HIF-1α. Out of these natural compounds, Emodin has the ability to reduce the expression of NF-κB, whereas most of the natural compounds such as Silibinin, Celastrol, PEITC, DIM, Andrographolide, Thymoquinone and Curcumin can act as a inhibitor of IκBα degradation, which in turn cause suppression of phosphorylation, acylation, and nuclear translocation of NF-κB [178,183–192]. On the other hand, Gliotoxin is a natural compound that inhibits NF-κB activity by up-regulation of ROS, whereas, Sulforaphane decreases NF-κB activity by preventing the interaction between NF-κB and its consensus sequence [193,194].

Further research is going on for proving anticancer as well as anti-HIF-1α efficacy of different nontoxic and novel phytochemicals which need to be scrutinized for their better functionality.

4.2. *Synthetic Compounds as HIF-1α Inhibitors*

Conventional chemotherapy mostly includes synthetic drugs which have a specific effect on different cancers. Although synthetic drugs possess cytotoxic effects, they can interfere with cancer cell division, metastasis, proliferation, etc. Therefore, targeting oncogenic protein-like HIF-1α with synthetic drugs with low toxicity is a fascinating part of anticancer drug discovery. The common chemotherapeutic synthetic drugs able to regulate HIF-1α are listed in Table 2.

4.2.1. HIF-1α Synthesis Blocker

Some synthetic compounds also act as HIF-1α synthesis blockers that have different modes of action. 1-benzyl-3-(5′-hydroxymethyl-2′-furyl)-indazole or YC-1 (SL.1, Table 2) is such a compound which is introduced as HIF-1α inhibitor by Chun et al. [195]. Earlier studies demonstrated that YC-1 is a soluble guanylate cyclase inducer used in cardiopulmonary disease [196]. From the current research of Sun et al., it was proved that YC-1 can exert anticancer effect by targeting HIF-1α expression and activity through PI3K/AKT/mTOR/4E-BP axis regulation [197]. A recent study reported that YC-1 is able to block the induction of erythropoietin (EPO) and VEGF mRNA by inhibiting HIF-1α during hypoxia [198]. Along with these, several inflammation markers such as IL-6 and IL-8 are also significantly decreased by YC-1 treatment in vivo [199]. YC1 displayed excellent anti-cancer effects in different cancer cells including lung cancer, bladder cancer, breast cancer, canine lymphoma, colorectal cancer, gastric carcinoma, hepatocellular carcinoma cancer (HCC), preeclampsia, ovarian cancer, and non-small cell lung cancer (NSCLC) [200]. Another study stated that YC-1 can reverse the acquired resistance to gefitinib by inhibiting HIF-1α in HCC827 GR cells [201]. Additionally, Li et al., proposed that YC-1 exerts FIH dependent inhibition of p300 recruitment to HIF-1α and inhibits hypoxia mediated activation of NFκB, a downstream target of HIF-1α [202]. As NFκB is a key regulator of CSC maintenance and cancer cell proliferation, YC-1 could be a novel anticancer drug which may target cancer stem cell propagation as well as several other hallmarks of cancer such as angiogenesis and metastasis.

Another example of an HIF-1α inhibitor is PX-478 (S-2-amino-3- [40-N, N-bis(2-chloroethyl) amino] phenyl propionic acid N-oxide dihydrochloride) (SL.2, Table 2). PX-478 started clinical trials against lymphoma and other advanced metastatic cancer. VHL mediated ubiquitination has no role in PX-478 dependent down-regulation of HIF-1α expression, rather it was observed that it can cause a reduction of the HIF-1α m-RNA level. PX-478 also affects transcription of HIF-1α without altering eIF-4E/4E-BP axis [203]. According to Welsh et al., the transcription of VEGF and GLUT-1 are decreased in vitro and in vivo after PX-478 treatment in colon cancer [203].

Table 2. List of various potential synthetic HIF-1α inhibitors based on their mode of action in different cancers.

Oxygen | Carbon | Nitrogen | Hydrogen | Sulphur | Bromine | Chlorine

Serial No.	Name	Structure	Mode of Action	Effective against Cancer Type	Clinical Trial	Reference
1	YC-1	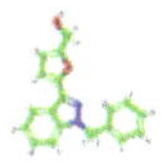	HIF-1α synthesis blocker	Hepatoma, gastric cancer, lung cancer, prostate cancer, pancreatic cancer, Human Bladder Transitional Carcinoma	—	[127,133,195–202]
2	PX-478	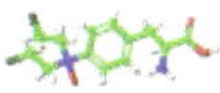		Colon carcinoma, Lung Adenocarcinoma, pancreatic ductal adenocarcinoma	Phase I	[127,133,203,204]
3	DJ12	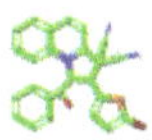	HIF-1α activity blocker	Breast cancer, Melanoma, Renal cancer	—	[127,133,205]
4	Bortezomib	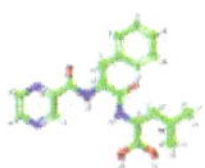		Human hepatoma, multiple myeloma, human embryonic kidney and human multiple myeloma	Approved	[127,133,206–208]
5	Amphotericin B (AmB)	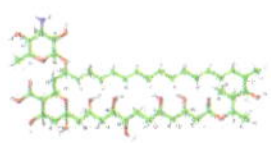		Hepatocellular carcinoma,	Approved	[127,133,209]
6	Polyamides			Adenocarcinoma	—	[127,133,210]
7	ENMD-1198 (Analog of 2ME2)	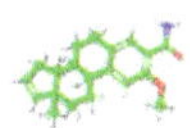	HIF-1α degradation enhancer	Prostate cancer, breast cancer, human hepatocellular carcinoma	—	[127,133,211–213]
8	Zebularine	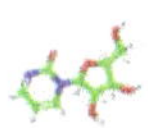		Oral squamous cell carcinoma	—	[127,133,214]
9	17-AAG (Analog of Gel-danamycin)	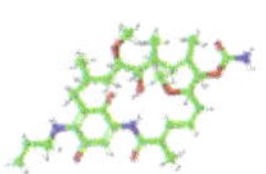		Prostate cancer, renal cell carcinoma, papillary thyroid carcinoma	Phase III	[127,133,215,216]
10	Flavopiridol	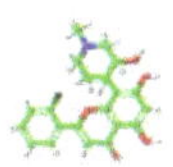		Leukemia, human glioma, neuroblastoma	Phase II	[127,133,217–222]

Furthermore, PX-478 helps in radiosensitization of hypoxic C6 glioma, HN5, and UMSCCa10 squamous cells and Panc-1 pancreatic adenocarcinoma cells by inhibiting HIF-1α in vitro. Along with these, PX-478 showed excellent radiosensitization of tumors by down-regulating post-radiation HIF-1α signaling pathways. PX-478 showed promising anti-cancer effects in phase I clinical trials with 41 patients at two sites in the USA. PX-478 is reported to be a well-tolerable drug in a relatively high proportion of patients. These pre-clinical and clinical studies decipher that PX-478 could be a potent anti-cancer agent by targeting HIF-1α in the near future [204].

4.2.2. HIF-1α Activity Blocker

Generally, the synthetic chemotherapeutic HIF-1α inhibitors may have specific and unique mechanism. For the development of an HIF-1α targeting cancer therapy, pathway-based regulation of HIF-1α with DNA-binding small molecules may represent an important approach.

Cisplatin is such a common synthetic anticancer agent that can down-regulate the expression of HIF-1 α in the Cisplatin-sensitive cell line by inducing apoptosis via elevating the expression of the cleaved PARP in dose and time dependent manner. Ai et al., tested the inhibitory role of Cisplatin against HIF-1α with two pairs of genetically matched Cisplatin-sensitive and Cisplatin-resistant ovarian cancer cell lines [223]. However, in normal conditions, the Cisplatin-resistant cell line showed inhibition in apoptosis, which was altered with low expression of HIF-1α. As discussed earlier, HIF-1α can regulate CSC properties that are responsible for drug resistance and cancer recurrence. For the researchers it is intriguing to find therapeutic modalities that can overcome HIF-1α mediated CSC dependent cancer resistance and recurrence. In this context, Zhihong and his colleagues showed that cancer recurrence and metastasis occurred in patient after Cisplatin treatment and PEO4 cells were derived from that patient after acquiring Cisplatin resistance, whereas PEO1 is the Cisplatin-sensitive primary ovarian cancer cells from the same patient before Cisplatin treatment [223]. A2780/CP is a Cisplatin-resistant ovarian cancer cell line which is derived from A2780, Cisplatin-sensitive ovarian cancer cell line., HIF-1α is down-regulated in A2780 and PEO1 cell line after Cisplatin treatment and the cancer cells undergo apoptosis, whereas Cisplatin sensitivity restores after HIF-1α knockdown and induces apoptosis in Cisplatin-resistant cell line A2780/CP and PEO4. From this study, it is confirmed that HIF-1α is an important factor in drug resistance and it can be targeted by Cisplatin [223].

Along with these, we focus on some other synthetic chemotherapeutic drugs which inhibit HIF-1α directly by interfering its binding to DNA. Jones and Harris explored the role of three suppositional molecules DJ12 (SL.3, Table 2), DJ15, and DJ30 against HIF-1α where DJ12 is more effective [205]. The effect of DJ15 and DJ30 is cell line specific. They studied the role of DJ12 in breast cancer cell lines MDA-MB-468 and ZR-75, melanoma cell line MDA-MB-435, and pVHL mutant renal cancer cell lines RCC4 and 786-O and concluded that DJ12 not only blocks the binding of HIF-1α to HRE region of DNA but also interferes with its transactivation. This leads to a reduction of mRNA expression of VEGF and BNIP3. With a HRE binding assay of nuclear and cytoplasmic extract of MDA-MB-468 cell line, it is confirmed that DJ12 did not affect DNA protein-binding directly, rather it inhibited the formation of HIF-1α, HIF-1β, and CBP/p300 transcription complex or folding of HIF-1α. Therefore, DJ12 can be used as a significant HIF-1α inhibitor in various cancer types.

Furthermore, Sung et al., studied the role of another synthetic drug Bortezomib (PS-341) (SL.4, Table 2) against HIF-1α in Hep3B (human hepatoma), ARH77 (human multiple myeloma), and U299 (human multiple myeloma) cell lines [206]. Bortezomib is a clinically approved proteasome inhibitor for myeloma and different solid tumors. Bortezomib mediates its effect on HIF-1α via interacting with CAD (C -terminal transactivation domain) and inhibits recruitment of p300 coactivator. In some cases, inhibition of HIF-1α by Bortezomib is improved by FIH (factor inhibiting HIF-1) hydroxylation. Bortezomib helps in prolyl hydroxylation at Pro402 and Pro564 by PHD (HIF-1 prolyl hydroxylases) and hydroxylation at Asn803 by FIH in hypoxic conditions [207]. These activities lead to repression of transcriptional activity of HIF-1α and inhibition of erythropoietin (EPO), vascular endothelial growth factor (VEGF), and carbonic anhydrase IX (CAIX) genes, downstream target of HIF-1α. Bortezomib is introduced to 15 clinical trials as monotherapy and 17 clinical trials as combinatorial therapy with other drugs. Bortezomib monotherapy showed poor and minimal effects in clinical studies. These results reported that Bortezomib offered no hope as a single agent as anti-cancer drug. Other clinical trials reported that Bortezomib with other drugs showed fewer anti-cancer activities with no statistical significance. In a phase I/II dose escalation study, it was observed that Bortezomib has anti-cancer effects and it has been tolerated by Advanced Androgen-Independent Prostate Cancer patients when combined with Docetaxel. However, further clinical trials using combinatorial therapy of Bortezomib with other drugs are needed for better understanding of its anti-cancer effects [208].

Additionally, several antimicrobial and antifungal agents play an antiangiogenic role targeting HIF-1α. Amphotericin B (AmB), (SL.5, Table 2) an FDA approved a polyene macrolide antifungal agent is one of them [127,133]. In the field of oncology, liposomal AmB is used in cancer patients with persistent unexplained fever [127,133]. Yeo et al., explained antitumor activity of AmB along with therapeutic efficacy against systemic mycoses [209]. AmB inhibits the transcriptional activity of HIF-1α, not the expression or nuclear translocation. From co-immunoprecipitation assay, it is concluded that after AmB treatment, FIH binding to CAD is increased and interaction between CAD and the C/H1 domain of p300 coactivator is blocked. Therefore, AmB treatment leads to anemia as a result of EPO (Erythropoietin) suppression. EPO is a glycoprotein, which stimulates red blood cell production from bone marrow and it is downstream target of HIF-1α. In the absence of EPO, patients suffer from chronic renal failure. Therefore, there might be nephrotoxicity in prolonged AmB treatment which may exhibit adverse side effects to cancer patients. Therefore, AmB is restricted for use as chemotherapeutic agents in medical oncology.

Polyamides (SL.6, Table 2) are another synthetic sequence specific DNA-binding oligomers, which exert restricted activity of HIF-1α. These are a new type of pyrrole imidazole compounds comprising of two and three aromatic N-methylpyrrole (Py) rings. The architecture of these Polyamides is inspired by the structure of Distamycin and Netropsin. Distamycin and Netropsin are two natural pyrrole-amidine antibiotic compounds originated from actinobacterium *Streptomyces netropsis*. Olenyuk et al., studied the antiangiogenic effects of Polyamides and the competitive inhibition of binding between HIF-1α/ARNT heterodimer to HRE [210]. Polyamides hinder HIF-1α binding to the promoter of proangiogenic protein VEGF and inhibit its transcription. As there are some sequence variations in VEGF splicing variants, Polyamides should be programmed for different sequence binding. Olenyuk and his colleagues designed two different Polyamides specific for 5′-WTWCGW-3′ and 5′-WGGWCW-3′ (where W = A or T) DNA sequences. Therefore, out of these different Polyamides, identification of the best one with prominent anti-cancer activity is dependent on further in vitro and in vivo investigations as well as clinical studies.

Other than these mechanisms, inhibition of nuclear translocation of HIF-1α is another approach for suppressing the transcription of HIF-1α target genes. A good example of an inhibitor that acts using this method is ENMD-1198 (SL.7, Table 2). ENMD-1198 is an analogue of 2-Methoxyestradiol (2ME2) synthesized by Lavallee et al. [211]. 2-Methoxyestradiol (2ME2) is a natural metabolite of estradiol compound with potential anti-HIF-1α activity [211]. Mabjeesh et al., explained that 2ME2 inhibits nuclear translocation of HIF-1α and promotes microtubule disruption [212]. However, according to Moser et al., ENMD-1198 is a tubulin binding agent and it reduces HIF-1α level [213]. Additionally, phosphorylation of MAPK/ERK, PI-3K/AKT and FAK are inhibited by ENMD-1198. Although, 2ME2 and ENMD-1198 have potential to inhibit the transcriptional ability of HIF-1α, further investigations into its anti-HIF-1α activity are required. Although, 2ME2 and ENMD-1198 are in phase II/phase I clinical trials, more investigations to identify their efficiency as HIF-1α inhibitor are needed.

In consideration of different mechanisms of targeting HIF-1α, another approach of HIF-1α regulation through destabilization or degradation by synthetic drugs is discussed in the following section.

4.2.3. HIF-1α Degradation Enhancer

Zeburaline (Zeb) (SL.8, Table 2) is a synthetic DNA methyl transferase inhibitor that also degrades HIF-1α by chemotherapeutic agent, which degrades HIF-1α by enhancing polyubiquitination. Suzuki et al., proved that nuclear localization of HIF-1α is interfered with in oral squamous cell carcinoma (OSCC) by Zeb [214]. It is a synthetic cytidine analogue with antitumor activity. Zeb mediates proteasomal degradation of HIF-1α in a dose-dependent manner whereas the HIF-1α level becomes increased after Zeb treatment in the presence of MG132, a potent proteasome inhibitor [214]. Zebularine affects angiogenesis through various mechanisms. In one case, Zebularine targets the HSP70/HSP90/HIF-1α axis and averts proper folding of HIF-1α leading to premature degradation of HIF-1α. On the other hand, Zebularine prevents nuclear localization of HIF-1α and represses its transcriptional activity. Moreover, Zebularine modulates epigenetic regulation of VEGF and acts as a DNA Methyl Transferase, leading to hypermethylation of proangiogenic genes. Despite having such targets, Zebularine cannot be a novel antiangiogenic drug in clinical use because of its miscellaneous targets and side effects.

Geldanamycin (GA), a natural benzoquinone ansamycin, represses transcription of VEGF, glucose transporters 1 and 3, most of the glycolytic enzymes, and erythropoietin inducing proteasomal degradation of HIF-1α. GA inhibits proper folding of HIF-1α by targeting HSP90. GA blocks ATPase activity of HSP90 by binding its N-terminal ATP binding pocket. Mabjeesh et al., proposed that in GA treatment transcriptional activity of HIF-1α cannot be restored even after use of LCN and MG-132, proteasome inhibitor [215]. In some cases, GA treatment leads to hepatotoxicity in cancer patients [216]. To overcome this toxicity, researchers have synthesized 17-allylamino-17-demethoxygeldanamycin (17-AAG) (SL.9, Table 2), a chemical derivative of Geldanamycin partaking anticancer role in blood, prostate, colonic, hepatic, ovarian, brain, skin, thyroid, renal, and breast cancers, which is supported by several clinical trial as well [127,133]. Geldanamycin and its derivatives are the novel inhibitor with promising pharmacology.

Another HIF-1α inhibitor, Flavopiridol (SL.10, Table 2) is the first semi-synthetic cyclin-dependent kinase (CDK) inhibitor to enter clinical trials [217]. Flavopiridol is a flavonoid compound which possess structural similarities with the alkaloid isolated from Indian endogenous plant *Dysoxylum binectariferum*. Newcomb et al., demonstrated that Flavopiridol mediates proteasomal degradation of HIF-1α [217]. Flavopiridol is used in combination therapy with bortezomib in human leukemia cell lines and induced apoptosis of cell. Proteasome inhibitors potentiate leukemic cell apoptosis induced by the cyclin-dependent kinase inhibitor Flavopiridol through a SAPK/JNK and NF-kβ-dependent process [218]. Along with these, Flavopiridol has a significant effect on regulating cell cytoskeleton, cell cycle components, and cell motility in CSCs. In the case of lung cancer, it was observed that Flavopiridol decreases $CD133^{high}/CD44^{high}$ population which is considered to be CSCs [219]. However, it will be fascinating to find the regulation of CD133 and CD44 by HIF-1α in the presence of Flavopiridol as HIF-1α is known to regulate these CD markers [220]. Interestingly, Flavopiridol passed a phase II clinical trial in acute myeloid leukemia, gastric cancer, pancreatic cancer, where it was proved that Flavopiridol is a safe drug with no acute toxicity [127,133,221]. Now, this promising synthetic anti-cancer compound is in phase III clinical trials [222].

Other than the aforesaid mode of action of the HIF-1α targeted synthetic compounds, an alternative mechanism of HIF-1α inhibition is present which targets NF-κB/HIF-1α axis. Figueroa et al., showed that YC-1 represses HIF-1α via inhibition of NF-κB which leads to a significant decrease of EPO expression [224]. Other than this, YC-1 inhibits NF-κB translocation and the mechanistic link between Akt/NF-κB and HIF-1α [225]. Out of the synthetic compounds mentioned in this review, Bortezomib is another one that restricts the degradation of NF-κB inhibitor, i.e., IκBα. Therefore, stabilization of IκBα causes cytoplasmic accumulation and degradation of NF-κB which controls a variety of subsequent oncogenic events, including angiogenesis [206,226–228]. Flavopiridol also has similar alternative mode of action such as Bortezomib, which comprises inflammatory response by inhibiting degradation of IκBα. It also inhibits tumor necrosis factor (TNF) mediated NF-κB activation and NF-κB downstream gene expression [229]. Apart from these, 17-AAG also has an alternative mode of action, which includes attenuation of the transcriptional activity of NF- κB by targeting HSP90 protein and preventing its coactivator recruitment [230]. Hence, identification of these alternative mode of actions of different synthetic compounds targeting oncogenic pathways such as NF-κB and/PI3K-Akt could be a promising approach for HIF-1α targeted cancer therapy.

Therefore, to follow the particular avenue of HIF-1α targeting, we believe more natural or synthetic compounds are worth exploring to enhance the possibility of successful clinical trial of better anticancer therapeutics. Further studies including combinatorial treatments with natural and synthetic compounds could provide an alternative platform to study HIF-1α targeting therapy.

5. Conclusions

Mounting data suggest that HIF-1α is one of the main culprits in the regulation of various hallmarks of cancer. Given the basic role of HIF-1α in the presence of hypoxia, which leads to tumorigenesis, several HIF-1α inhibitors were developed to efficiently treat cancer. Furthermore, the use of natural and synthetic products as HIF-1α inhibitors need more research to reduce their side effects and to decrease their toxicity. Although in clinical studies, until now only few HIF-1α inhibitors have been studied, failure in other cases exerts a major barrier in the development of HIF-1α-targeting anti-cancer therapeutics. As there are very few reports on efficient usage of some HIF-1α targeting drugs against certain type of cancers by clinicians, the pharmacological potential of common and newly proposed HIF-1α inhibitors should be properly documented. In addition, synergistic relationship between the phytochemicals and synthetic HIF-1α inhibitors could prove effective in managing different cancers. Therefore, studies on the synergistic mode of action of natural and synthetic compounds could pave a novel path in HIF-1α targeting anti-cancer therapy.

Author Contributions: Conceptualization, R.G., P.S. (Priya Samanta), R.S., S.B. and A.B.; validation, A.B., S.H. and P.S. (Prosenjit Saha); resources, P.S. (Prosenjit Saha) and S.H.; writing—original draft preparation, R.G., P.S. (Priya Samanta), R.S., S.B. and A.B.; writing—review and editing, A.B, S.H. and P.S. (Prosenjit Saha); supervision, A.B. and S.H.; project administration, A.B, S.H. and P.S. (Prosenjit Saha). All authors have read and agreed to the published version of the manuscript.

Funding: We are thankful to SERB sponsored projects (CRG/2021/007813 and EEQ/2020/000601) for their support.

Institutional Review Board Statement: Not applicable.

Informed Consent Statement: Not applicable.

Data Availability Statement: Not applicable.

Acknowledgments: We acknowledge Shampa Pakhira, Mrinmoyee Mondal, Sucheta Mondal, Shantam Ghosh, Soummadeep Sen for their help in review writing. We are also thankful to Jayanta Chakraborty, Chittaranjan National Cancer Institute.

Conflicts of Interest: The authors declare no conflict of interest.

Sample Availability: Samples of the compounds are not available from the authors.

Abbreviations

HIF-1α	hypoxia-inducible factor-1 α
CSCs	cancer stem cells
ROS	reactive oxygen species
HRE	hypoxia response element
VEGF	vascular endothelial growth factor
PDGF-β	platelet-derived growth factor-β
PAI-1	plasminogen activator-1
EPO	erythropoietin
ESA	epithelial-specific antigen
ALDH1	aldehyde dehydrogenase1
HSCs	hematopoietic stem cells
NOD	nonobese diabetic
SCID	severe combined immunodeficiency
OCT4	Octamer-binding transcription factor 4

BHLH	basic helix loop helix
PHD	Proline hydroxylase
pVHL	von Hippel-Lindau protein
CREB	cAMP-response-element-binding protein
CBP	CREB-binding protein
CTAD	carboxy-terminal transactivation domain
EGF	epidermal growth factor
ADM	adrenomedullin
eIF-4E	eukaryotic translation initiation factor 4E
4E-BP1	eIF-4E binding protein
PDK1	pyruvate dehydrogenase kinases 1
GLUT1	glucose transporter 1
PDH	pyruvate dehydrogenase
HK2	Hexokinase II
LDHA	lactate dehydrogenase A
MCT4	monocarboxylate transporter 4
TFAM	mitochondrial transcription factor A
MAP	mitogen-activated protein
NF-κβ	Nuclear Factor Kappa Beta
NOS	Nitric oxide synthases
NO	nitric oxide
PDGFR	platelet-derived growth factor receptor
EGFR	epidermal growth factor receptor
EMT	Epithelial to mesenchymal transitions
HDAC3	histone deacetylase 3
JUP	junction plakoglobin
LOX	Lysyl oxidase
MMPs	Matrix metalloproteinases
HGF	Hepatocyte growth factor; KLF4
KLF4	Krüppel-like factor 4
WNT	wingless and INT-1
ALDH1A1	Aldehyde Dehydrogenase 1 Family Member A1
HGSOC	high-grade serous ovarian cancer
HDACis	histone deacetylase inhibitors
GS	glutamine synthetase
NPC	nasopharyngeal carcinoma
NSCLC	Non-small cell lung cancer
HSP90	heat shock protein 90
I3C	Indole-3-carbinol
sGC	soluble guanylate cyclase
HCC	hepatocellular carcinoma cancer
AmB	Amphotericin B
2ME2	2-Methoxyestradiol
OSCC	oral squamous cell carcinoma
GA	Geldanamycin
17-AAG	17-allylamino-17-demethoxygeldanamycin

References

1. Müller, M.; Padberg, W.; Schindler, E.; Sticher, J.; Osmer, C.; Friemann, S.; Hempelmann, G. Renocortical tissue oxygen pressure measurements in patients undergoing living donor kidney transplantation. *Anesth. Analg.* **1998**, *87*, 474–476. [CrossRef] [PubMed]
2. Dings, J.; Meixensberger, J.; Jäger, A.; Roosen, K. Clinical experience with 118 brain tissue oxygen partial pressure catheter probes. *Neurosurgery* **1998**, *43*, 1082–1095. [CrossRef] [PubMed]
3. Vaupel, P.; Thews, O.; Hoeckel, M. Treatment resistance of solid tumors: Role of hypoxia and anemia. *Med. Oncol. Northwood Lond. Engl.* **2001**, *18*, 243–259. [CrossRef]
4. Vaupel, P. Tumor microenvironmental physiology and its implications for radiation oncology. *Semin. Radiat. Oncol.* **2004**, *14*, 198–206. [CrossRef] [PubMed]
5. Aggarwal, V.; Tuli, H.S.; Varol, A.; Thakral, F.; Yerer, M.B.; Sak, K.; Varol, M.; Jain, A.; Khan, M.A.; Sethi, G. Role of reactive oxygen species in cancer progression: Molecular mechanisms and recent advancements. *Biomolecules* **2019**, *9*, 735. [CrossRef]

6. Bridge, G.; Rashid, S.; Martin, S.A. DNA mismatch repair and oxidative DNA damage: Implications for cancer biology and treatment. *Cancers* **2014**, *6*, 1597–1614. [CrossRef]
7. Jing, X.; Yang, F.; Shao, C.; Wei, K.; Xie, M.; Shen, H.; Shu, Y. Role of hypoxia in cancer therapy by regulating the tumor microenvironment. *Mol. Cancer* **2019**, *18*, 157. [CrossRef]
8. Hirota, K.; Semenza, G.L. Regulation of angiogenesis by hypoxia-inducible factor 1. *Crit. Rev. Oncol. Hematol.* **2006**, *59*, 15–26. [CrossRef]
9. Masoud, G.N.; Li, W. HIF-1α Pathway: Role, regulation and intervention for cancer therapy. *Acta Pharm. Sin. B* **2015**, *5*, 378–389. [CrossRef]
10. Makino, Y.; Cao, R.; Svensson, K.; Bertilsson, G.; Asman, M.; Tanaka, H.; Cao, Y.; Berkenstam, A.; Poellinger, L. Inhibitory PAS domain protein is a negative regulator of hypoxia-inducible gene expression. *Nature* **2001**, *414*, 550–554. [CrossRef]
11. Jun, J.C.; Rathore, A.; Younas, H.; Gilkes, D.; Polotsky, V.Y. Hypoxia-inducible factors and cancer. *Curr. Sleep Med. Rep.* **2017**, *3*, 1–10. [CrossRef]
12. Viallard, C.; Larrivée, B. Tumor angiogenesis and vascular normalization: Alternative therapeutic targets. *Angiogenesis* **2017**, *20*, 409–426. [CrossRef]
13. Semenza, G.L.; Wang, G.L. A nuclear factor induced by hypoxia via de novo protein synthesis binds to the human erythropoietin gene enhancer at a site required for transcriptional activation. *Mol. Cell. Biol.* **1992**, *12*, 5447–5454.
14. Conway, E.M.; Collen, D.; Carmeliet, P. Molecular mechanisms of blood vessel growth. *Cardiovasc. Res.* **2001**, *49*, 507–521. [CrossRef]
15. Xing, F.; Okuda, H.; Watabe, M.; Kobayashi, A.; Pai, S.K.; Liu, W.; Pandey, P.R.; Fukuda, K.; Hirota, S.; Sugai, T.; et al. Hypoxia-induced jagged2 promotes breast cancer metastasis and self-renewal of cancer stem-like cells. *Oncogene* **2011**, *30*, 4075–4086. [CrossRef] [PubMed]
16. Bocca, C.; Ievolella, M.; Autelli, R.; Motta, M.; Mosso, L.; Torchio, B.; Bozzo, F.; Cannito, S.; Paternostro, C.; Colombatto, S.; et al. Expression of Cox-2 in human breast cancer cells as a critical determinant of epithelial-to-mesenchymal transition and invasiveness. *Expert Opin. Ther. Targets* **2014**, *18*, 121–135. [CrossRef] [PubMed]
17. Jo, M.; Lester, R.D.; Montel, V.; Eastman, B.; Takimoto, S.; Gonias, S.L. Reversibility of epithelial-mesenchymal transition (EMT) induced in breast cancer cells by activation of urokinase receptor-dependent cell signaling. *J. Biol. Chem.* **2009**, *284*, 22825–22833. [CrossRef] [PubMed]
18. Semenza, G.L. Hypoxia, clonal selection, and the Role of HIF-1 in tumor progression. *Crit. Rev. Biochem. Mol. Biol.* **2000**, *35*, 71–103. [CrossRef]
19. Zhang, Q.; Han, Z.; Zhu, Y.; Chen, J.; Li, W. Role of hypoxia inducible factor-1 in cancer stem cells (Review). *Mol. Med. Rep.* **2021**, *23*, 17. [CrossRef]
20. Yu, Z.; Pestell, T.G.; Lisanti, M.P.; Pestell, R.G. Cancer stem cells. *Int. J. Biochem. Cell Biol.* **2012**, *44*, 2144–2151. [CrossRef]
21. Al-Hajj, M.; Wicha, M.S.; Benito-Hernandez, A.; Morrison, S.J.; Clarke, M.F. Prospective identification of tumorigenic breast cancer cells. *Proc. Natl. Acad. Sci. USA* **2003**, *100*, 3983–3988. [CrossRef]
22. Prager, B.C.; Xie, Q.; Bao, S.; Rich, J.N. Cancer stem cells: The architects of the tumor ecosystem. *Cell Stem Cell* **2019**, *24*, 41–53. [CrossRef]
23. Danet, G.H.; Pan, Y.; Luongo, J.L.; Bonnet, D.A.; Simon, M.C. Expansion of human SCID-Repopulating cells under hypoxic conditions. *J. Clin. Investig.* **2003**, *112*, 126–135. [CrossRef]
24. Ezashi, T.; Das, P.; Roberts, R.M. Low O_2 Tensions and the prevention of differentiation of HES cells. *Proc. Natl. Acad. Sci. USA* **2005**, *102*, 4783–4788. [CrossRef]
25. Gustafsson, M.V.; Zheng, X.; Pereira, T.; Gradin, K.; Jin, S.; Lundkvist, J.; Ruas, J.L.; Poellinger, L.; Lendahl, U.; Bondesson, M. Hypoxia requires notch signaling to maintain the undifferentiated cell state. *Dev. Cell* **2005**, *9*, 617–628. [CrossRef]
26. Qiang, L.; Wu, T.; Zhang, H.-W.; Lu, N.; Hu, R.; Wang, Y.-J.; Zhao, L.; Chen, F.-H.; Wang, X.-T.; You, Q.-D.; et al. HIF-1α Is critical for hypoxia-mediated maintenance of glioblastoma stem cells by activating notch signaling pathway. *Cell Death Differ.* **2012**, *19*, 284–294. [CrossRef]
27. Wang, R.; Zhou, S.; Li, S. Cancer therapeutic agents targeting hypoxia-inducible factor-1. *Curr. Med. Chem.* **2011**, *18*, 3168–3189. [CrossRef]
28. Wang, G.L.; Semenza, G.L. Purification and characterization of hypoxia-inducible factor 1. *J. Biol. Chem.* **1995**, *270*, 1230–1237. [CrossRef]
29. Huang, L.E.; Bunn, H.F. Hypoxia-inducible factor and its biomedical relevance. *J. Biol. Chem.* **2003**, *278*, 19575–19578. [CrossRef]
30. Huang, L.E.; Gu, J.; Schau, M.; Bunn, H.F. Regulation of hypoxia-inducible factor 1alpha is mediated by an O2-dependent degradation domain via the ubiquitin-proteasome pathway. *Proc. Natl. Acad. Sci. USA* **1998**, *95*, 7987–7992. [CrossRef]
31. Wang, G.L.; Jiang, B.H.; Rue, E.A.; Semenza, G.L. Hypoxia-Inducible Factor 1 Is a Basic-Helix-Loop-Helix-PAS Heterodimer Regulated by Cellular O2 Tension. *Proc. Natl. Acad. Sci. USA* **1995**, *92*, 5510–5514. [CrossRef]
32. Huang, L.E.; Arany, Z.; Livingston, D.M.; Bunn, H.F. Activation of Hypoxia-Inducible Transcription Factor Depends Primarily upon Redox-Sensitive Stabilization of Its α Subunit. *J. Biol. Chem.* **1996**, *271*, 32253–32259. [CrossRef]
33. Kim, W.; Kaelin, W.G. The von Hippel-Lindau tumor suppressor protein: New insights into oxygen sensing and cancer. *Curr. Opin. Genet. Dev.* **2003**, *13*, 55–60. [CrossRef]

34. Kallio, P.J.; Wilson, W.J.; O'Brien, S.; Makino, Y.; Poellinger, L. Regulation of the hypoxia-inducible transcription factor 1alpha by the ubiquitin-proteasome pathway. *J. Biol. Chem.* **1999**, *274*, 6519–6525. [CrossRef]
35. Tanimoto, K.; Makino, Y.; Pereira, T.; Poellinger, L. Mechanism of Regulation of the hypoxia-inducible factor-1α by the von Hippel-Lindau tumor suppressor protein. *EMBO J.* **2000**, *19*, 4298–4309. [CrossRef]
36. Kaelin, W.G. Molecular Basis of the VHL hereditary cancer syndrome. *Nat. Rev. Cancer* **2002**, *2*, 673–682. [CrossRef]
37. Martin, A.R.; Ronco, C.; Demange, L.; Benhida, R. Hypoxia inducible factor down-regulation, cancer and cancer stem cells (CSCs): Ongoing success stories. *MedChemComm* **2017**, *8*, 21–52. [CrossRef] [PubMed]
38. Semenza, G.L. Regulation of mammalian O2 homeostasis by hypoxia-inducible factor 1. *Annu. Rev. Cell Dev. Biol.* **1999**, *15*, 551–578. [CrossRef] [PubMed]
39. Rankin, E.B.; Giaccia, A.J. The Role of Hypoxia-Inducible Factors in Tumorigenesis. *Cell Death Differ.* **2008**, *15*, 678–685. [CrossRef] [PubMed]
40. Keith, B.; Simon, M.C. Hypoxia-Inducible Factors, Stem Cells, and Cancer. *Cell* **2007**, *129*, 465–472. [CrossRef]
41. Vaupel, P.; Schmidberger, H.; Mayer, A. The Warburg effect: Essential Part of metabolic reprogramming and central contributor to cancer progression. *Int. J. Radiat. Biol.* **2019**, *95*, 912–919. [CrossRef]
42. Wang, Q.; Wang, P.; Qin, Z.; Yang, X.; Pan, B.; Nie, F.; Bi, H. Altered glucose metabolism and cell function in keloid fibroblasts under hypoxia. *Redox Biol.* **2021**, *38*, 101815. [CrossRef]
43. Zheng, F.; Chen, J.; Zhang, X.; Wang, Z.; Chen, J.; Lin, X.; Huang, H.; Fu, W.; Liang, J.; Wu, W.; et al. The HIF-1α Antisense Long Non-Coding RNA Drives a Positive Feedback Loop of HIF-1α Mediated Transactivation and Glycolysis. *Nat. Commun.* **2021**, *12*, 1341. [CrossRef]
44. Tang, K.; Zhu, L.; Chen, J.; Wang, D.; Zeng, L.; Chen, C.; Tang, L.; Zhou, L.; Wei, K.; Zhou, Y.; et al. Hypoxia promotes breast cancer cell growth by activating a glycogen metabolic program. *Cancer Res.* **2021**, *81*, 4949–4963. [CrossRef]
45. Grasmann, G.; Smolle, E.; Olschewski, H.; Leithner, K. Gluconeogenesis in cancer cells—Repurposing of a starvation-induced metabolic pathway? *Biochim. Biophys. Acta Rev. Cancer* **2019**, *1872*, 24–36. [CrossRef]
46. Mylonis, I.; Simos, G.; Paraskeva, E. Hypoxia-inducible factors and the regulation of lipid metabolism. *Cells* **2019**, *8*, 214. [CrossRef]
47. Gaspar, J.M.; Velloso, L.A. Hypoxia inducible factor as a central regulator of metabolism—Implications for the development of obesity. *Front. Neurosci.* **2018**, *12*, 813. [CrossRef]
48. Lee, Y.S.; Kim, J.-W.; Osborne, O.; Oh, D.Y.; Sasik, R.; Schenk, S.; Chen, A.; Chung, H.; Murphy, A.; Watkins, S.M.; et al. Increased Adipocyte O2 Consumption Triggers HIF-1α, Causing Inflammation and Insulin Resistance in Obesity. *Cell* **2014**, *157*, 1339–1352. [CrossRef]
49. Ziello, J.E.; Jovin, I.S.; Huang, Y. Hypoxia-Inducible Factor (HIF)-1 regulatory pathway and its potential for therapeutic intervention in malignancy and ischemia. *Yale J. Biol. Med.* **2007**, *80*, 51–60.
50. Papandreou, I.; Cairns, R.A.; Fontana, L.; Lim, A.L.; Denko, N.C. HIF-1 Mediates adaptation to hypoxia by actively downregulating mitochondrial oxygen consumption. *Cell Metab.* **2006**, *3*, 187–197. [CrossRef]
51. Gonzalez, F.J.; Xie, C.; Jiang, C. The Role of hypoxia-inducible factors in metabolic diseases. *Nat. Rev. Endocrinol.* **2019**, *15*, 21–32. [CrossRef]
52. Denko, N.C. Hypoxia, HIF1 and glucose metabolism in the solid tumour. *Nat. Rev. Cancer* **2008**, *8*, 705–713. [CrossRef]
53. Golias, T.; Papandreou, I.; Sun, R.; Kumar, B.; Brown, N.V.; Swanson, B.J.; Pai, R.; Jaitin, D.; Le, Q.-T.; Teknos, T.N.; et al. Hypoxic repression of pyruvate dehydrogenase activity is necessary for metabolic reprogramming and growth of model tumours. *Sci. Rep.* **2016**, *6*, 31146. [CrossRef]
54. Maxwell, P.H.; Dachs, G.U.; Gleadle, J.M.; Nicholls, L.G.; Harris, A.L.; Stratford, I.J.; Hankinson, O.; Pugh, C.W.; Ratcliffe, P.J. Hypoxia-inducible factor-1 modulates gene expression in solid tumors and influences both angiogenesis and tumor growth. *Proc. Natl. Acad. Sci. USA* **1997**, *94*, 8104–8109. [CrossRef]
55. Chen, C.; Pore, N.; Behrooz, A.; Ismail-Beigi, F.; Maity, A. Regulation of Glut1 MRNA by hypoxia-inducible factor-1. interaction between h-ras and hypoxia. *J. Biol. Chem.* **2001**, *276*, 9519–9525. [CrossRef]
56. Sakagami, H.; Makino, Y.; Mizumoto, K.; Isoe, T.; Takeda, Y.; Watanabe, J.; Fujita, Y.; Takiyama, Y.; Abiko, A.; Haneda, M. Loss of HIF-1α Impairs GLUT4 translocation and glucose uptake by the skeletal muscle cells. *Am. J. Physiol. Endocrinol. Metab.* **2014**, *306*, E1065–E1076. [CrossRef]
57. Balamurugan, K. HIF-1 at the crossroads of hypoxia, inflammation, and cancer. *Int. J. Cancer* **2016**, *138*, 1058–1066. [CrossRef]
58. Mathupala, S.P.; Rempel, A.; Pedersen, P.L. Glucose catabolism in cancer cells: Identification and characterization of a marked activation response of the type ii hexokinase gene to hypoxic conditions. *J. Biol. Chem.* **2001**, *276*, 43407–43412. [CrossRef]
59. Firth, J.D.; Ebert, B.L.; Ratcliffe, P.J. Hypoxic regulation of lactate dehydrogenase A. interaction between hypoxia-inducible factor 1 and CAMP response elements. *J. Biol. Chem.* **1995**, *270*, 21021–21027. [CrossRef]
60. Ullah, M.S.; Davies, A.J.; Halestrap, A.P. The Plasma Membrane Lactate Transporter MCT4, but Not MCT1, Is up-Regulated by Hypoxia through a HIF-1alpha-Dependent Mechanism. *J. Biol. Chem.* **2006**, *281*, 9030–9037. [CrossRef]
61. Pérez de Heredia, F.; Wood, I.S.; Trayhurn, P. Hypoxia stimulates lactate release and modulates monocarboxylate transporter (MCT1, MCT2, and MCT4) expression in human adipocytes. *Pflugers Arch.* **2010**, *459*, 509–518. [CrossRef] [PubMed]
62. Kim, J.; Tchernyshyov, I.; Semenza, G.L.; Dang, C.V. HIF-1-Mediated expression of pyruvate dehydrogenase kinase: A metabolic switch required for cellular adaptation to hypoxia. *Cell Metab.* **2006**, *3*, 177–185. [CrossRef] [PubMed]

63. Krishnan, J.; Suter, M.; Windak, R.; Krebs, T.; Felley, A.; Montessuit, C.; Tokarska-Schlattner, M.; Aasum, E.; Bogdanova, A.; Perriard, E.; et al. Activation of a HIF1alpha-PPARgamma axis underlies the integration of glycolytic and lipid anabolic pathways in pathologic cardiac hypertrophy. *Cell Metab.* **2009**, *9*, 512–524. [CrossRef] [PubMed]
64. Wise, D.R.; Ward, P.S.; Shay, J.E.S.; Cross, J.R.; Gruber, J.J.; Sachdeva, U.M.; Platt, J.M.; DeMatteo, R.G.; Simon, M.C.; Thompson, C.B. Hypoxia promotes isocitrate dehydrogenase-dependent carboxylation of α-ketoglutarate to citrate to support cell growth and viability. *Proc. Natl. Acad. Sci. USA* **2011**, *108*, 19611–19616. [CrossRef]
65. Xiang, L.; Mou, J.; Shao, B.; Wei, Y.; Liang, H.; Takano, N.; Semenza, G.L.; Xie, G. Glutaminase 1 expression in colorectal cancer cells is induced by hypoxia and required for tumor growth, invasion, and metastatic colonization. *Cell Death Dis.* **2019**, *10*, 40. [CrossRef]
66. Zhang, H.; Gao, P.; Fukuda, R.; Kumar, G.; Krishnamachary, B.; Zeller, K.I.; Dang, C.V.; Semenza, G.L. HIF-1 Inhibits mitochondrial biogenesis and cellular respiration in vhl-deficient renal cell carcinoma by repression of C-MYC activity. *Cancer Cell* **2007**, *11*, 407–420. [CrossRef]
67. Ben-Yosef, Y.; Lahat, N.; Shapiro, S.; Bitterman, H.; Miller, A. Regulation of Endothelial matrix metalloproteinase-2 by hypoxia/reoxygenation. *Circ. Res.* **2002**, *90*, 784–791. [CrossRef]
68. Ceradini, D.J.; Kulkarni, A.R.; Callaghan, M.J.; Tepper, O.M.; Bastidas, N.; Kleinman, M.E.; Capla, J.M.; Galiano, R.D.; Levine, J.P.; Gurtner, G.C. Progenitor cell trafficking is regulated by hypoxic gradients through hif-1 induction of SDF-1. *Nat. Med.* **2004**, *10*, 858–864. [CrossRef]
69. Dang, C.V.; Lewis, B.C.; Dolde, C.; Dang, G.; Shim, H. Oncogenes in tumor metabolism, tumorigenesis, and apoptosis. *J. Bioenerg. Biomembr.* **1997**, *29*, 345–354. [CrossRef]
70. Maxwell, P.H.; Wiesener, M.S.; Chang, G.W.; Clifford, S.C.; Vaux, E.C.; Cockman, M.E.; Wykoff, C.C.; Pugh, C.W.; Maher, E.R.; Ratcliffe, P.J. The tumour suppressor protein VHL targets hypoxia-inducible factors for oxygen-dependent proteolysis. *Nature* **1999**, *399*, 271–275. [CrossRef]
71. Ravi, R.; Mookerjee, B.; Bhujwalla, Z.M.; Sutter, C.H.; Artemov, D.; Zeng, Q.; Dillehay, L.E.; Madan, A.; Semenza, G.L.; Bedi, A. Regulation of tumor angiogenesis by P53-induced degradation of hypoxia-inducible factor 1alpha. *Genes Dev.* **2000**, *14*, 34–44. [CrossRef]
72. Zundel, W.; Schindler, C.; Haas-Kogan, D.; Koong, A.; Kaper, F.; Chen, E.; Gottschalk, A.R.; Ryan, H.E.; Johnson, R.S.; Jefferson, A.B.; et al. Loss of PTEN Facilitates HIF-1-mediated gene expression. *Genes Dev.* **2000**, *14*, 391–396. [CrossRef]
73. Harris, A.L. Hypoxia—A key regulatory factor in tumour growth. *Nat. Rev. Cancer* **2002**, *2*, 38–47. [CrossRef]
74. Semenza, G.L. HIF-1 and human disease: One highly involved factor. *Genes Dev.* **2000**, *14*, 1983–1991. [CrossRef]
75. Mack, F.A.; Rathmell, W.K.; Arsham, A.M.; Gnarra, J.; Keith, B.; Simon, M.C. Loss of PVHL is sufficient to cause HIF dysregulation in primary cells but does not promote tumor growth. *Cancer Cell* **2003**, *3*, 75–88. [CrossRef]
76. Mack, F.A.; Patel, J.H.; Biju, M.P.; Haase, V.H.; Simon, M.C. Decreased Growth of Vhl−/− Fibrosarcomas is associated with elevated levels of cyclin kinase inhibitors p21 and p27. *Mol. Cell. Biol.* **2005**, *25*, 4565–4578. [CrossRef]
77. Ryan, H.E.; Lo, J.; Johnson, R.S. HIF-1 alpha is required for solid tumor formation and embryonic vascularization. *EMBO J.* **1998**, *17*, 3005–3015. [CrossRef]
78. Tang, N.; Wang, L.; Esko, J.; Giordano, F.J.; Huang, Y.; Gerber, H.-P.; Ferrara, N.; Johnson, R.S. Loss of HIF-1α in Endothelial cells disrupts a hypoxia-driven VEGF autocrine loop necessary for tumorigenesis. *Cancer Cell* **2004**, *6*, 485–495. [CrossRef]
79. Das, B.; Yeger, H.; Tsuchida, R.; Torkin, R.; Gee, M.F.W.; Thorner, P.S.; Shibuya, M.; Malkin, D.; Baruchel, S. A Hypoxia-driven vascular endothelial growth factor/flt1 autocrine loop interacts with hypoxia-inducible factor-1α through mitogen-activated protein kinase/extracellular signal-regulated kinase 1/2 pathway in neuroblastoma. *Cancer Res.* **2005**, *65*, 7267–7275. [CrossRef]
80. Du, R.; Lu, K.V.; Petritsch, C.; Liu, P.; Ganss, R.; Passegué, E.; Song, H.; VandenBerg, S.; Johnson, R.S.; Werb, Z.; et al. HIF-1α Induces the recruitment of bone marrow-derived vascular modulatory cells to regulate tumor angiogenesis and invasion. *Cancer Cell* **2008**, *13*, 206–220. [CrossRef]
81. Quintero, M.; Brennan, P.A.; Thomas, G.J.; Moncada, S. Nitric oxide is a factor in the stabilization of hypoxia-inducible factor-1α in cancer: Role of free radical formation. *Cancer Res.* **2006**, *66*, 770–774. [CrossRef]
82. Hendrickson, M.D.; Poyton, R.O. Crosstalk between nitric oxide and hypoxia-inducible factor signaling pathways: An update. *Res. Rep. Biochem.* **2015**, *5*, 147–161. [CrossRef]
83. Pore, N.; Liu, S.; Haas-Kogan, D.A.; O'Rourke, D.M.; Maity, A. PTEN Mutation and epidermal growth factor receptor activation regulate vascular endothelial growth factor (VEGF) MRNA expression in human glioblastoma cells by transactivating the proximal VEGF Promoter1. *Cancer Res.* **2003**, *63*, 236–241.
84. Shi, Y.-H.; Wang, Y.-X.; Bingle, L.; Gong, L.-H.; Heng, W.-J.; Li, Y.; Fang, W.-G. In vitro study of HIF-1 activation and VEGF Release by BFGF in the T47D Breast cancer cell line under normoxic conditions: Involvement of PI-3K/Akt and MEK1/ERK Pathways. *J. Pathol.* **2005**, *205*, 530–536. [CrossRef]
85. Karar, J.; Maity, A. PI3K/AKT/MTOR pathway in angiogenesis. *Front. Mol. Neurosci.* **2011**, *4*, 51. [CrossRef]
86. Liao, D.; Johnson, R.S. Hypoxia: A key regulator of angiogenesis in cancer. *Cancer Metastasis Rev.* **2007**, *26*, 281–290. [CrossRef]
87. Zeeshan, R.; Mutahir, Z. Cancer Metastasis—Tricks of the Trade. *Bosn. J. Basic Med. Sci.* **2017**, *17*, 172–182. [CrossRef]
88. Yang, M.-H.; Wu, M.-Z.; Chiou, S.-H.; Chen, P.-M.; Chang, S.-Y.; Liu, C.-J.; Teng, S.-C.; Wu, K.-J. Direct Regulation of TWIST by HIF-1α Promotes Metastasis. *Nat. Cell Biol.* **2008**, *10*, 295–305. [CrossRef]

89. Tam, S.Y.; Wu, V.W.C.; Law, H.K.W. Hypoxia-induced epithelial-mesenchymal transition in cancers: HIF-1α and Beyond. *Front. Oncol.* **2020**, *10*, 486. [CrossRef]
90. Imai, T.; Horiuchi, A.; Wang, C.; Oka, K.; Ohira, S.; Nikaido, T.; Konishi, I. Hypoxia attenuates the expression of E-Cadherin via Up-Regulation of SNAIL in ovarian carcinoma cells. *Am. J. Pathol.* **2003**, *163*, 1437–1447. [CrossRef]
91. Erler, J.T.; Bennewith, K.L.; Nicolau, M.; Dornhöfer, N.; Kong, C.; Le, Q.-T.; Chi, J.-T.A.; Jeffrey, S.S.; Giaccia, A.J. Lysyl oxidase is essential for hypoxia-induced metastasis. *Nature* **2006**, *440*, 1222–1226. [CrossRef] [PubMed]
92. Krishnamachary, B.; Zagzag, D.; Nagasawa, H.; Rainey, K.; Okuyama, H.; Baek, J.H.; Semenza, G.L. Hypoxia-Inducible Factor-1-Dependent Repression of *E-Cadherin* in von hippel-lindau tumor suppressor–null renal cell carcinoma mediated by TCF3, ZFHX1A, and ZFHX1B. *Cancer Res.* **2006**, *66*, 2725–2731. [CrossRef] [PubMed]
93. Muñoz-Nájar, U.M.; Neurath, K.M.; Vumbaca, F.; Claffey, K.P. Hypoxia Stimulates Breast Carcinoma Cell Invasion through MT1-MMP and MMP-2 activation. *Oncogene* **2006**, *25*, 2379–2392. [CrossRef] [PubMed]
94. Azab, A.K.; Hu, J.; Quang, P.; Azab, F.; Pitsillides, C.; Awwad, R.; Thompson, B.; Maiso, P.; Sun, J.D.; Hart, C.P.; et al. Hypoxia promotes dissemination of multiple myeloma through acquisition of epithelial to mesenchymal transition-like features. *Blood* **2012**, *119*, 5782–5794. [CrossRef]
95. Muz, B.; de la Puente, P.; Azab, F.; Azab, A.K. The Role of hypoxia in cancer progression, angiogenesis, metastasis, and resistance to therapy. *Hypoxia* **2015**, *3*, 83–92. [CrossRef]
96. Pennacchietti, S.; Michieli, P.; Galluzzo, M.; Mazzone, M.; Giordano, S.; Comoglio, P.M. Hypoxia promotes invasive growth by transcriptional activation of the met protooncogene. *Cancer Cell* **2003**, *3*, 347–361. [CrossRef]
97. Tafani, M.; Schito, L.; Pellegrini, L.; Villanova, L.; Marfe, G.; Anwar, T.; Rosa, R.; Indelicato, M.; Fini, M.; Pucci, B.; et al. Hypoxia-Increased RAGE and P2X7R expression regulates tumor cell invasion through phosphorylation of Erk1/2 and akt and nuclear translocation of NF-KB. *Carcinogenesis* **2011**, *32*, 1167–1175. [CrossRef]
98. Tafani, M.; Sansone, L.; Limana, F.; Arcangeli, T.; De Santis, E.; Polese, M.; Fini, M.; Russo, M.A. The interplay of reactive oxygen species, hypoxia, inflammation, and sirtuins in cancer initiation and progression. *Oxid. Med. Cell. Longev.* **2015**, *2016*, e3907147. [CrossRef]
99. Helbig, G.; Christopherson, K.W.; Bhat-Nakshatri, P.; Kumar, S.; Kishimoto, H.; Miller, K.D.; Broxmeyer, H.E.; Nakshatri, H. NF-KappaB Promotes breast cancer cell migration and metastasis by inducing the expression of the chemokine receptor CXCR4. *J. Biol. Chem.* **2003**, *278*, 21631–21638. [CrossRef]
100. Cheng, J.-C.; Klausen, C.; Leung, P.C.K. Hypoxia-inducible factor 1 alpha mediates epidermal growth factor-induced down-regulation of e-cadherin expression and cell invasion in human ovarian cancer cells. *Cancer Lett.* **2013**, *329*, 197–206. [CrossRef]
101. Liu, Z.; Semenza, G.L.; Zhang, H. Hypoxia-inducible Factor 1 and breast cancer metastasis. *J. Zhejiang Univ.-Sci. B* **2015**, *16*, 32–43. [CrossRef]
102. Chua, H.L.; Bhat-Nakshatri, P.; Clare, S.E.; Morimiya, A.; Badve, S.; Nakshatri, H. NF-KappaB Represses E-cadherin expression and enhances epithelial to mesenchymal transition of mammary epithelial cells: Potential involvement of ZEB-1 and ZEB-2. *Oncogene* **2007**, *26*, 711–724. [CrossRef]
103. Huang, G.; Chen, J.; Liu, J.; Zhang, X.; Duan, H.; Fang, Q. MiR-935/HIF-1α feedback loop inhibits the proliferation and invasiveness of glioma. *OncoTargets Ther.* **2020**, *13*, 10817–10828. [CrossRef]
104. Arnold, C.R.; Mangesius, J.; Skvortsova, I.-I.; Ganswindt, U. The role of cancer stem cells in radiation resistance. *Front. Oncol.* **2020**, *10*, 164. [CrossRef]
105. Zhang, H.; Lu, H.; Xiang, L.; Bullen, J.W.; Zhang, C.; Samanta, D.; Gilkes, D.M.; He, J.; Semenza, G.L. HIF-1 Regulates CD47 Expression in breast cancer cells to promote evasion of phagocytosis and maintenance of cancer stem cells. *Proc. Natl. Acad. Sci. USA* **2015**, *112*, E6215–E6223. [CrossRef]
106. Bian, L.; Meng, Y.; Zhang, M.; Li, D. MRE11-RAD50-NBS1 Complex alterations and DNA damage response: Implications for cancer treatment. *Mol. Cancer* **2019**, *18*, 169. [CrossRef]
107. Syed, A.; Tainer, J.A. The MRE11–RAD50–NBS1 Complex conducts the orchestration of damage signaling and outcomes to stress in DNA replication and repair. *Annu. Rev. Biochem.* **2018**, *87*, 263–294. [CrossRef]
108. Kabakov, A.E.; Yakimova, A.O. Hypoxia-induced cancer cell responses driving radioresistance of hypoxic tumors: Approaches to targeting and radiosensitizing. *Cancers* **2021**, *13*, 1102. [CrossRef]
109. Fan, Q.; Li, L.; Wang, T.-L.; Emerson, R.E.; Xu, Y. A Novel ZIP4-HDAC4-VEGFA axis in high-grade serous ovarian cancer. *Cancers* **2021**, *13*, 3821. [CrossRef]
110. Kitajima, S.; Lee, K.L.; Hikasa, H.; Sun, W.; Huang, R.Y.-J.; Yang, H.; Matsunaga, S.; Yamaguchi, T.; Araki, M.; Kato, H.; et al. Hypoxia-inducible factor-1α promotes cell survival during ammonia stress response in ovarian cancer stem-like cells. *Oncotarget* **2017**, *8*, 114481–114494. [CrossRef]
111. Yang, S.; Zhang, Z.; Hao, Y.; Zhao, Y.; Qian, F.; Shi, Y.; Li, P.; Liu, C.; Yu, P. HIF-1α Induces the epithelial-mesenchymal transition in gastric cancer stem cells through the snail pathway. *Oncotarget* **2017**, *8*, 9535–9545. [CrossRef] [PubMed]
112. Grivennikov, S.I.; Greten, F.R.; Karin, M. Immunity, inflammation, and cancer. *Cell* **2010**, *140*, 883–899. [CrossRef] [PubMed]
113. de Martel, C.; Franceschi, S. Infections and cancer: Established associations and new hypotheses. *Crit. Rev. Oncol. Hematol.* **2009**, *70*, 183–194. [CrossRef] [PubMed]
114. de Visser, K.E.; Eichten, A.; Coussens, L.M. Paradoxical roles of the immune system during cancer development. *Nat. Rev. Cancer* **2006**, *6*, 24–37. [CrossRef]

115. Noman, M.Z.; Desantis, G.; Janji, B.; Hasmim, M.; Karray, S.; Dessen, P.; Bronte, V.; Chouaib, S. PD-L1 Is a Novel Direct Target of HIF-1α, and Its Blockade under Hypoxia Enhanced MDSC-Mediated T Cell Activation. *J. Exp. Med.* **2014**, *211*, 781–790. [CrossRef]
116. Hoesel, B.; Schmid, J.A. The Complexity of NF-KB Signaling in inflammation and cancer. *Mol. Cancer* **2013**, *12*, 86. [CrossRef]
117. Bruning, U.; Fitzpatrick, S.F.; Frank, T.; Birtwistle, M.; Taylor, C.T.; Cheong, A. NFκB and HIF display synergistic behaviour during hypoxic inflammation. *Cell. Mol. Life Sci. CMLS* **2012**, *69*, 1319–1329. [CrossRef]
118. D'Ignazio, L.; Batie, M.; Rocha, S. Hypoxia and inflammation in cancer, focus on HIF and NF-KB. *Biomedicines* **2017**, *5*, 21. [CrossRef]
119. van Uden, P.; Kenneth, N.S.; Rocha, S. Regulation of hypoxia-inducible factor-1alpha by NF-KappaB. *Biochem. J.* **2008**, *412*, 477–484. [CrossRef]
120. Bandarra, D.; Biddlestone, J.; Mudie, S.; Müller, H.-A.J.; Rocha, S. HIF-1α Restricts NF-KB-dependent gene expression to control innate immunity signals. *Dis. Model. Mech.* **2015**, *8*, 169–181. [CrossRef]
121. Han, S.; Xu, W.; Wang, Z.; Qi, X.; Wang, Y.; Ni, Y.; Shen, H.; Hu, Q.; Han, W. Crosstalk between the HIF-1 and Toll-like receptor/nuclear factor-κb pathways in the oral squamous cell carcinoma microenvironment. *Oncotarget* **2016**, *7*, 37773–37789. [CrossRef]
122. Malkov, M.I.; Lee, C.T.; Taylor, C.T. Regulation of the hypoxia-inducible factor (HIF) by pro-inflammatory cytokines. *Cells* **2021**, *10*, 2340. [CrossRef]
123. Deep, G.; Kumar, R.; Nambiar, D.K.; Jain, A.K.; Ramteke, A.M.; Serkova, N.J.; Agarwal, C.; Agarwal, R. Silibinin inhibits hypoxia-induced hif-1α-mediated signaling, angiogenesis and lipogenesis in prostate cancer cells: In vitro evidence and in vivo functional imaging and metabolomics: *Silibinin inhibits* hypoxia effects in PCa. *Mol. Carcinog.* **2017**, *56*, 833–848. [CrossRef]
124. Rugamba, A.; Kang, D.Y.; Sp, N.; Jo, E.S.; Lee, J.-M.; Bae, S.W.; Jang, K.-J. Silibinin regulates tumor progression and tumorsphere formation by suppressing pd-l1 expression in non-small cell lung cancer (NSCLC) cells. *Cells* **2021**, *10*, 1632. [CrossRef]
125. Gu, M.; Singh, R.P.; Dhanalakshmi, S.; Agarwal, C.; Agarwal, R. Silibinin Inhibits inflammatory and angiogenic attributes in photocarcinogenesis in SKH-1 hairless mice. *Cancer Res.* **2007**, *67*, 3483–3491. [CrossRef]
126. Wellington, K.; Jarvis, B. Silymarin: A review of its clinical properties in the management of hepatic disorders. *BioDrugs* **2001**, *15*, 465–489. [CrossRef]
127. Clinical Trials. Available online: https://www.who.int/observatories/global-observatory-on-health-research-and-development/resources/databases/databases-on-processes-for-r-d/clinical-trials (accessed on 20 March 2022).
128. Singh, R.P.; Deep, G.; Chittezhath, M.; Kaur, M.; Dwyer-Nield, L.D.; Malkinson, A.M.; Agarwal, R. Effect of silibinin on the growth and progression of primary lung tumors in mice. *JNCI J. Natl. Cancer Inst.* **2006**, *98*, 846–855. [CrossRef]
129. Roy, S.; Kaur, M.; Agarwal, C.; Tecklenburg, M.; Sclafani, R.A.; Agarwal, R. P21 and P27 Induction by silibinin is essential for its cell cycle arrest effect in prostate carcinoma cells. *Mol. Cancer Ther.* **2007**, *6*, 2696–2707. [CrossRef]
130. Singh, R.P.; Agarwal, R. Prostate cancer chemoprevention by silibinin: Bench to bedside. *Mol. Carcinog.* **2006**, *45*, 436–442. [CrossRef]
131. Silibinin Decreases Prostate-Specific Antigen with Cell Growth Inhibition via G1 Arrest, Leading to Differentiation of Prostate Carcinoma Cells: Implications for Prostate Cancer Intervention | PNAS. Available online: https://www.pnas.org/content/96/13/7490.short (accessed on 15 February 2022).
132. Jung, H.-J.; Park, J.-W.; Lee, J.S.; Lee, S.-R.; Jang, B.-C.; Suh, S.-I.; Suh, M.-H.; Baek, W.-K. Silibinin Inhibits Expression of HIF-1α through Suppression of Protein Translation in Prostate Cancer Cells. *Biochem. Biophys. Res. Commun.* **2009**, *390*, 71–76. [CrossRef]
133. DrugBank Online | Database for Drug and Drug Target Info. Available online: https://go.drugbank.com/ (accessed on 20 March 2022).
134. Li, X.; Yue, Y.; Zhou, Y.; Fan, Y.; Fan, C.; Huang, Y.; Wu, F.; Liu, Y. An Oil-free microemulsion for intravenous delivery of diallyl trisulfide: Formulation and evaluation. *Int. J. Pharm.* **2011**, *407*, 158–166. [CrossRef]
135. Jung, H.J.; Kim, Y.; Shin, J.Y.; Sohng, J.K.; Kwon, H.J. Antiangiogenic activity of herboxidiene via downregulation of vascular endothelial growth factor receptor-2 and hypoxia-inducible factor-1α. *Arch. Pharm. Res.* **2015**, *38*, 1728–1735. [CrossRef]
136. Ma, Z.; Xiang, X.; Li, S.; Xie, P.; Gong, Q.; Goh, B.-C.; Wang, L. Targeting hypoxia-inducible factor-1, for cancer treatment: Recent advances in developing small-molecule inhibitors from natural compounds. *Semin. Cancer Biol.* **2020**, *80*, 379–390. [CrossRef]
137. Pua, L.J.W.; Mai, C.-W.; Chung, F.F.-L.; Khoo, A.S.-B.; Leong, C.-O.; Lim, W.-M.; Hii, L.-W. Functional roles of jnk and p38 mapk signaling in nasopharyngeal carcinoma. *Int. J. Mol. Sci.* **2022**, *23*, 1108. [CrossRef]
138. Chen, S.-R.; Dai, Y.; Zhao, J.; Lin, L.; Wang, Y.; Wang, Y. A Mechanistic Overview of triptolide and celastrol, natural products from tripterygium wilfordii hook F. *Front. Pharmacol.* **2018**, *9*, 104. [CrossRef]
139. Cavell, B.E.; Syed Alwi, S.S.; Donlevy, A.M.; Proud, C.G.; Packham, G. Natural Product-derived antitumor compound phenethyl isothiocyanate inhibits mtorc1 activity via TSC2. *J. Nat. Prod.* **2012**, *75*, 1051–1057. [CrossRef]
140. Qin, X.; Jiang, B.; Zhang, Y. 4E-BP1, a multifactor regulated multifunctional protein. *Cell Cycle* **2016**, *15*, 781–786. [CrossRef]
141. Sharbeen, G.; McCarroll, J.A.; Akerman, A.; Kopecky, C.; Youkhana, J.; Kokkinos, J.; Holst, J.; Boyer, C.; Erkan, M.; Goldstein, D.; et al. Cancer-associated fibroblasts in pancreatic ductal adenocarcinoma determine response to slc7a11 inhibition. *Cancer Res.* **2021**, *81*, 3461–3479. [CrossRef]
142. Van Dyke, M.M.; Dervan, P.B. Echinomycin binding sites on DNA. *Science* **1984**, *225*, 1122–1127. [CrossRef]
143. Vlaminck, B.; Toffoli, S.; Ghislain, B.; Demazy, C.; Raes, M.; Michiels, C. Dual effect of echinomycin on hypoxia-inducible factor-1 activity under normoxic and hypoxic conditions. *FEBS J.* **2007**, *274*, 5533–5542. [CrossRef] [PubMed]

144. Bailey, C.M.; Liu, Y.; Peng, G.; Zhang, H.; He, M.; Sun, D.; Zheng, P.; Liu, Y.; Wang, Y. Liposomal formulation of HIF-1α Inhibitor echinomycin eliminates established metastases of triple-negative breast cancer. *Nanomed. Nanotechnol. Biol. Med.* **2020**, *29*, 102278. [CrossRef] [PubMed]
145. Wang, Y.; Liu, Y.; Bailey, C.; Zhang, H.; He, M.; Sun, D.; Zhang, P.; Parkin, B.; Baer, M.R.; Zheng, P.; et al. Therapeutic targeting of TP53-Mutated acute myeloid leukemia by inhibiting HIF-1α with echinomycin. *Oncogene* **2020**, *39*, 3015–3027. [CrossRef]
146. Xu, G.-B.; He, G.; Bai, H.-H.; Yang, T.; Zhang, G.-L.; Wu, L.-W.; Li, G.-Y. Indole alkaloids from *Chaetomium globosum*. *J. Nat. Prod.* **2015**, *78*, 1479–1485. [CrossRef] [PubMed]
147. Min, S.; Wang, X.; Du, Q.; Gong, H.; Yang, Y.; Wang, T.; Wu, N.; Liu, X.; Li, W.; Zhao, C.; et al. Chetomin, a Hsp90/HIF-1α pathway inhibitor, effectively targets lung cancer stem cells and non-stem cells. *Cancer Biol. Ther.* **2020**, *21*, 698–708. [CrossRef]
148. Katschinski, D.; Le, L.; Schindler, S.; Thomas, T.; Voss, A.; Wenger, R. Interaction of the PAS B Domain with HSP90 accelerates hypoxia-inducible factor-1α stabilization. *Cell. Physiol. Biochem.* **2004**, *14*, 351–360. [CrossRef]
149. Minet, E.; Mottet, D.; Michel, G.; Roland, I.; Raes, M.; Remacle, J.; Michiels, C. Hypoxia-Induced Activation of HIF-1: Role of HIF-1alpha-Hsp90 interaction. *FEBS Lett.* **1999**, *460*, 251–256. [CrossRef]
150. Kung, A.L.; Zabludoff, S.D.; France, D.S.; Freedman, S.J.; Tanner, E.A.; Vieira, A.; Cornell-Kennon, S.; Lee, J.; Wang, B.; Wang, J.; et al. Small molecule blockade of transcriptional coactivation of the hypoxia-inducible factor pathway. *Cancer Cell* **2004**, *1*, 33–43. [CrossRef]
151. Reece, K.M.; Richardson, E.D.; Cook, K.M.; Campbell, T.J.; Pisle, S.T.; Holly, A.J.; Venzon, D.J.; Liewehr, D.J.; Chau, C.H.; Price, D.K.; et al. Epidithiodiketopiperazines (ETPs) Exhibit in vitro antiangiogenic and in vivo antitumor activity by disrupting the hif-1α/p300 complex in a preclinical model of prostate cancer. *Mol. Cancer* **2014**, *13*, 91. [CrossRef]
152. Waring, P.; Beaver, J. Gliotoxin and related epipolythiodioxopiperazines. *Gen. Pharmacol.* **1996**, *27*, 1311–1316. [CrossRef]
153. Hubmann, R.; Sieghart, W.; Schnabl, S.; Araghi, M.; Hilgarth, M.; Reiter, M.; Demirtas, D.; Valent, P.; Zielinski, C.; Jäger, U.; et al. Gliotoxin targets nuclear NOTCH2 in human solid tumor derived cell lines in vitro and inhibits melanoma growth in xenograft mouse model. *Front. Pharmacol.* **2017**, *8*, 319. [CrossRef]
154. Comas, L.; Polo, E.; Domingo, M.P.; Hernández, Y.; Arias, M.; Esteban, P.; Martínez-Lostao, L.; Pardo, J.; Martínez de la Fuente, J.; Gálvez, E.M. Intracellular delivery of biologically-active fungal metabolite gliotoxin using magnetic nanoparticles. *Materials* **2019**, *12*, 1092. [CrossRef]
155. Xia, Y.; Kang, T.W.; Jung, Y.D.; Zhang, C.; Lian, S. Sulforaphane inhibits nonmuscle invasive bladder cancer cells proliferation through suppression of hif-1α-mediated glycolysis in hypoxia. *J. Agric. Food Chem.* **2019**, *67*, 7844–7854. [CrossRef]
156. Kim, D.H.; Sung, B.; Kang, Y.J.; Hwang, S.Y.; Kim, M.J.; Yoon, J.-H.; Im, E.; Kim, N.D. Sulforaphane inhibits hypoxia-induced hif-1α and vegf expression and migration of human colon cancer cells. *Int. J. Oncol.* **2015**, *47*, 2226–2232. [CrossRef]
157. Nehme, R.; Hallal, R.; El Dor, M.; Kobeissy, F.; Gouilleux, F.; Mazurier, F.; Zibara, K. Repurposing of acriflavine to target chronic myeloid leukemia treatment. *Curr. Med. Chem.* **2021**, *28*, 2218–2233. [CrossRef]
158. Hallal, R.; Nehme, R.; Brachet-Botineau, M.; Nehme, A.; Dakik, H.; Deynoux, M.; Dello Sbarba, P.; Levern, Y.; Zibara, K.; Gouilleux, F.; et al. Acriflavine targets oncogenic stat5 signaling in myeloid leukemia cells. *J. Cell. Mol. Med.* **2020**, *24*, 10052–10062. [CrossRef]
159. Cheloni, G.; Tanturli, M.; Tusa, I.; Ho DeSouza, N.; Shan, Y.; Gozzini, A.; Mazurier, F.; Rovida, E.; Li, S.; Dello Sbarba, P. Targeting chronic myeloid leukemia stem cells with the hypoxia-inducible factor inhibitor acriflavine. *Blood* **2017**, *130*, 655–665. [CrossRef]
160. Mangraviti, A.; Raghavan, T.; Volpin, F.; Skuli, N.; Gullotti, D.; Zhou, J.; Asnaghi, L.; Sankey, E.; Liu, A.; Wang, Y.; et al. HIF-1α-Targeting acriflavine provides long term survival and radiological tumor response in brain cancer therapy. *Sci. Rep.* **2017**, *7*, 14978. [CrossRef]
161. Cowman, S.; Pizer, B.; Sée, V. Downregulation of both mismatch repair and non-homologous end-joining pathways in hypoxic brain tumour cell lines. *PeerJ* **2021**, *9*, e11275. [CrossRef]
162. Huang, X.; Wang, J.; Huang, C.; Chen, Y.; Shi, G.; Hu, Q.; Yi, J. Emodin Enhances cytotoxicity of chemotherapeutic drugs in prostate cancer cells: The mechanisms involve ROS-Mediated suppression of multidrug resistance and hypoxia inducible factor-1. *Cancer Biol. Ther.* **2008**, *7*, 468–475. [CrossRef]
163. Castelli, S.; Ciccarone, F.; Tavian, D.; Ciriolo, M.R. ROS-dependent HIF-1α activation under Forced lipid catabolism entails glycolysis and mitophagy as mediators of higher proliferation rate in cervical cancer cells. *J. Exp. Clin. Cancer Res. CR* **2021**, *40*, 94. [CrossRef]
164. Parhira, S.; Zhu, G.-Y.; Chen, M.; Bai, L.-P.; Jiang, Z.-H. Cardenolides from Calotropis gigantea as potent inhibitors of hypoxia-inducible factor-1 transcriptional activity. *J. Ethnopharmacol.* **2016**, *194*, 930–936. [CrossRef] [PubMed]
165. Zheng, Z.; Zhou, Z.; Zhang, Q.; Zhou, X.; Yang, J.; Yang, M.-R.; Zhu, G.-Y.; Jiang, Z.-H.; Li, T.; Lin, Q.; et al. Non-classical cardenolides from calotropis gigantea exhibit anticancer effect as hif-1 inhibitors. *Bioorg. Chem.* **2021**, *109*, 104740. [CrossRef] [PubMed]
166. Riby, J.E.; Firestone, G.L.; Bjeldanes, L.F. 3,3′-Diindolylmethane Reduces Levels of HIF-1alpha and HIF-1 activity in hypoxic cultured human cancer cells. *Biochem. Pharmacol.* **2008**, *75*, 1858–1867. [CrossRef] [PubMed]
167. Bhowmik, A.; Chakravarti, S.; Ghosh, A.; Shaw, R.; Bhandary, S.; Bhattacharyya, S.; Sen, P.C.; Ghosh, M.K. Anti-SSTR2 Peptide based targeted delivery of potent plga encapsulated 3,3′-diindolylmethane nanoparticles through blood brain barrier prevents glioma progression. *Oncotarget* **2017**, *8*, 65339–65358. [CrossRef]

168. Hajra, S.; Basu, A.; Singha Roy, S.; Patra, A.R.; Bhattacharya, S. Attenuation of doxorubicin-induced cardiotoxicity and genotoxicity by an indole-based natural compound 3,3′-diindolylmethane (dim) through activation of nrf2/are signaling pathways and inhibiting apoptosis. *Free Radic. Res.* **2017**, *51*, 812–827. [CrossRef]
169. Li, M.-H.; Miao, Z.-H.; Tan, W.-F.; Yue, J.-M.; Zhang, C.; Lin, L.-P.; Zhang, X.-W.; Ding, J. Pseudolaric Acid B inhibits angiogenesis and reduces hypoxia-inducible factor 1α by promoting proteasome-mediated degradation. *Clin. Cancer Res.* **2004**, *10*, 8266–8274. [CrossRef]
170. Yu, B.; Li, M.-H.; Wang, W.; Wang, Y.-Q.; Jiang, Y.; Yang, S.-P.; Yue, J.-M.; Ding, J.; Miao, Z.-H. Pseudolaric Acid B-Driven phosphorylation of c-jun impairs its role in stabilizing HIF-1alpha: A novel function-converter model. *J. Mol. Med.* **2012**, *90*, 971–981. [CrossRef]
171. Nepal, M.; Jung Choi, H.; Choi, B.-Y.; Lim Kim, S.; Ryu, J.-H.; Hee Kim, D.; Lee, Y.-H.; Soh, Y. Anti-angiogenic and anti-tumor activity of bavachinin by targeting hypoxia-inducible Factor-1α. *Eur. J. Pharmacol.* **2012**, *691*, 28–37. [CrossRef]
172. Islam, M.T. Andrographolide, a new hope in the prevention and treatment of metabolic syndrome. *Front. Pharmacol.* **2017**, *8*, 571. [CrossRef]
173. Cui, K.; Liu, J.; Huang, L.; Qin, B.; Yang, X.; Li, L.; Liu, Y.; Gu, J.; Wu, W.; Yu, Y.; et al. Andrographolide attenuates choroidal neovascularization by inhibiting the HIF-1α/VEGF signaling pathway. *Biochem. Biophys. Res. Commun.* **2020**, *530*, 60–66. [CrossRef]
174. Wang, Z.; Wang, N.; Han, S.; Wang, D.; Mo, S.; Yu, L.; Huang, H.; Tsui, K.; Shen, J.; Chen, J. Dietary compound isoliquiritigenin inhibits breast cancer neoangiogenesis via VEGF/VEGFR-2 signaling pathway. *PLoS ONE* **2013**, *8*, e68566. [CrossRef]
175. Peng, F.; Du, Q.; Peng, C.; Wang, N.; Tang, H.; Xie, X.; Shen, J.; Chen, J. A review: The pharmacology of isoliquiritigenin: A Review: The pharmacology of ISL. *Phytother. Res.* **2015**, *29*, 969–977. [CrossRef]
176. Jun, H.-O.; Kim, Y.; Kwon, Y.-W.; Hong, S.-S.; Kim, K.-W.; Shin, J.; Kim, T.-Y. Wondonin, a novel compound, inhibits hypoxia-induced angiogenesis through hypoxia-inducible factor 1 alpha. *FEBS Lett.* **2007**, *581*, 4977–4982. [CrossRef]
177. Lee, Y.-M.; Kim, G.-H.; Park, E.-J.; Oh, T.-I.; Lee, S.; Kan, S.-Y.; Kang, H.; Kim, B.M.; Kim, J.H.; Lim, J.-H. Thymoquinone selectively kills hypoxic renal cancer cells by suppressing HIF-1α-Mediated glycolysis. *Int. J. Mol. Sci.* **2019**, *20*, 1092. [CrossRef]
178. Beesetti, S.L.; Jayadev, M.; Subhashini, G.V.; Mansour, L.; Alwasel, S.; Harrath, A.H. Andrographolide as a therapeutic agent against breast and ovarian cancers. *Open Life Sci.* **2019**, *14*, 462–469. [CrossRef]
179. Ströfer, M.; Jelkmann, W.; Depping, R. Curcumin decreases survival of Hep3B Liver and MCF-7 breast cancer cells: The role of HIF. *Strahlenther. Onkol.* **2011**, *187*, 393–400. [CrossRef]
180. Sarighieh, M.A.; Montazeri, V.; Shadboorestan, A.; Ghahremani, M.H.; Ostad, S.N. The inhibitory effect of curcumin on hypoxia inducer factors (Hifs) as a regulatory factor in the growth of tumor cells in breast cancer stem-like cells. *Drug Res.* **2020**, *70*, 512–518. [CrossRef]
181. Wei, Z.; Shan, Y.; Tao, L.; Liu, Y.; Zhu, Z.; Liu, Z.; Wu, Y.; Chen, W.; Wang, A.; Lu, Y. Diallyl Trisulfides, a natural histone deacetylase inhibitor, attenuate hif-1α synthesis, and decreases breast cancer metastasis. *Mol. Carcinog.* **2017**, *56*, 2317–2331. [CrossRef]
182. Homayoonfal, M.; Asemi, Z.; Yousefi, B. Targeting MicroRNAs with thymoquinone: A new approach for cancer therapy. *Cell. Mol. Biol. Lett.* **2021**, *26*, 43. [CrossRef]
183. Rasmi, R.R.; Sakthivel, K.M.; Guruvayoorappan, C. NF-KB Inhibitors in treatment and prevention of lung cancer. *Biomed. Pharmacother.* **2020**, *130*, 110569. [CrossRef]
184. Dhanalakshmi, S.; Singh, R.; Agarwal, C.; Agarwal, R. Silibinin Inhibits Constitutive and TNFa-Induced activation of NF-KB and sensitizes human prostate carcinoma DU145 Cells to TNFa-Induced Apoptosis. *Oncogene* **2002**, *21*, 1759–1767. [CrossRef] [PubMed]
185. Raina, K.; Agarwal, C.; Agarwal, R. Effect of silibinin in human colorectal cancer cells: Targeting the activation of NF-KB signaling. *Mol. Carcinog.* **2013**, *52*, 195–206. [CrossRef] [PubMed]
186. Ni, H.; Zhao, W.; Kong, X.; Li, H.; Ouyang, J. NF-Kappa B modulation is involved in celastrol induced human multiple myeloma cell apoptosis. *PLoS ONE* **2014**, *9*, e95846. [CrossRef] [PubMed]
187. Xu, C.; Shen, G.; Chen, C.; Gélinas, C.; Kong, A.-N.T. Suppression of NF-KappaB and NF-KappaB-regulated gene expression by sulforaphane and PEITC through IkappaBalpha, IKK pathway in human prostate cancer PC-3 Cells. *Oncogene* **2005**, *24*, 4486–4495. [CrossRef]
188. Abdellatef, A.A.; Fathy, M.; Mohammed, A.E.-S.I.; Bakr, M.S.A.; Ahmed, A.H.; Abbass, H.S.; El-Desoky, A.H.; Morita, H.; Nikaido, T.; Hayakawa, Y. Inhibition of cell-intrinsic NF-KB Activity and metastatic abilities of breast cancer by aloe-emodin and emodic-acid isolated from *Asphodelus microcarpus*. *J. Nat. Med.* **2021**, *75*, 840–853. [CrossRef]
189. Rahman, K.W.; Sarkar, F.H. Inhibition of nuclear translocation of nuclear factor—KB Contributes to 3,3V-diindolylmethane-induced apoptosis in breast cancer cells. *Cancer Res* **2005**, *65*, 364–371. [CrossRef]
190. Zhang, L.; Bai, Y.; Yang, Y. Thymoquinone chemosensitizes colon cancer cells through inhibition of NF-KB. *Oncol. Lett.* **2016**, *12*, 2840–2845. [CrossRef]
191. Ghasemi, F.; Shafiee, M.; Banikazemi, Z.; Pourhanifeh, M.H.; Khanbabaei, H.; Shamshirian, A.; Amiri Moghadam, S.; ArefNezhad, R.; Sahebkar, A.; Avan, A.; et al. Curcumin Inhibits NF-KB and Wnt/β-catenin pathways in cervical cancer cells. *Pathol. Res. Pract.* **2019**, *215*, 152556. [CrossRef]

192. Marquardt, J.U.; Gomez-Quiroz, L.; Arreguin Camacho, L.O.; Pinna, F.; Lee, Y.-H.; Kitade, M.; Domínguez, M.P.; Castven, D.; Breuhahn, K.; Conner, E.A.; et al. Curcumin effectively inhibits oncogenic NF-KB signaling and restrains stemness features in liver cancer. *J. Hepatol.* **2015**, *63*, 661–669. [CrossRef]
193. Kim, Y.-S.; Park, S.J. Gliotoxin from the marine fungus aspergillus fumigatus induces apoptosis in HT1080 fibrosarcoma cells by downregulating NF-KB. *Fish. Aquat. Sci.* **2016**, *19*, 35. [CrossRef]
194. Checker, R.; Gambhir, L.; Thoh, M.; Sharma, D.; Sandur, S.K. Sulforaphane, a naturally occurring isothiocyanate, exhibits anti-inflammatory effects by targeting GSK3β/Nrf-2 and NF-KB pathways in T Cells. *J. Funct. Foods* **2015**, *19*, 426–438. [CrossRef]
195. Chun, Y.S.; Yeo, E.J.; Choi, E.; Teng, C.M.; Bae, J.M.; Kim, M.S.; Park, J.W. Inhibitory Effect of YC-1 on the hypoxic induction of erythropoietin and vascular endothelial growth factor in Hep3B cells. *Biochem. Pharmacol.* **2001**, *61*, 947–954. [CrossRef]
196. Stasch, J.-P.; Pacher, P.; Evgenov, O.V. Soluble guanylate cyclase as an emerging therapeutic target in cardiopulmonary disease. *Circulation* **2011**, *123*, 2263–2273. [CrossRef]
197. Sun, H.-L.; Liu, Y.-N.; Huang, Y.-T.; Pan, S.-L.; Huang, D.-Y.; Guh, J.-H.; Lee, F.-Y.; Kuo, S.-C.; Teng, C.-M. YC-1 Inhibits HIF-1 expression in prostate cancer cells: Contribution of Akt/NF-KappaB Signaling to HIF-1alpha accumulation during hypoxia. *Oncogene* **2007**, *26*, 3941–3951. [CrossRef]
198. Zhao, Q.; Du, J.; Gu, H.; Teng, X.; Zhang, Q.; Qin, H.; Liu, N. Effects of YC-1 on hypoxia-inducible factor 1-driven transcription activity, cell proliferative vitality, and apoptosis in hypoxic human pancreatic cancer cells. *Pancreas* **2007**, *34*, 242–247. [CrossRef]
199. Yan, Z.; An, J.; Shang, Q.; Zhou, N.; Ma, J. YC-1 Inhibits VEGF and Inflammatory mediators expression on experimental central retinal vein occlusion in rhesus monkey. *Curr. Eye Res.* **2018**, *43*, 526–533. [CrossRef]
200. Yu, K.-H.; Hung, H.-Y. synthetic strategy and structure–activity relationship (SAR) studies of 3-(5′-Hydroxymethyl-2′-Furyl)-1-benzyl indazole (YC-1, Lificiguat): A review. *RSC Adv.* **2022**, *12*, 251–264. [CrossRef]
201. Jin, Q.; Zheng, J.; Chen, M.; Jiang, N.; Xu, X.; Huang, F. HIF-1 Inhibitor YC-1 reverses the acquired resistance of EGFR-mutant HCC827 cell line with MET amplification to gefitinib. *Oxid. Med. Cell. Longev.* **2021**, *2021*, e6633867. [CrossRef]
202. Li, S.H.; Shin, D.H.; Chun, Y.-S.; Lee, M.K.; Kim, M.-S.; Park, J.-W. A novel mode of action of YC-1 in HIF inhibition: Stimulation of FIH-dependent P300 dissociation from HIF-1{alpha}. *Mol. Cancer Ther.* **2008**, *7*, 3729–3738. [CrossRef]
203. Welsh, S.; Williams, R.; Kirkpatrick, L.; Paine-Murrieta, G.; Powis, G. Antitumor Activity and pharmacodynamic properties of PX-478, an inhibitor of hypoxia-inducible factor-1alpha. *Mol. Cancer Ther.* **2004**, *3*, 233–244.
204. Lee, K.; Kim, H.M. A novel approach to cancer therapy using PX-478 as a HIF-1α inhibitor. *Arch. Pharm. Res.* **2011**, *34*, 1583–1585. [CrossRef] [PubMed]
205. Jones, D.T.; Harris, A.L. Identification of novel small-molecule inhibitors of hypoxia-inducible factor-1 transactivation and DNA binding. *Mol. Cancer Ther.* **2006**, *5*, 2193–2202. [CrossRef] [PubMed]
206. Sung, M.-H.; Bagain, L.; Chen, Z.; Karpova, T.; Yang, X.; Silvin, C.; Voss, T.; McNally, J.; Waes, C.V.; Hager, G.L. Dynamic effect of bortezomib on NF-KB activity and gene expression in tumor cells. *Mol. Pharmacol.* **2008**, *74*, 1215–1222. [CrossRef] [PubMed]
207. Lando, D.; Peet, D.J.; Whelan, D.A.; Gorman, J.J.; Whitelaw, M.L. Asparagine Hydroxylation of the HIF transactivation domain: A hypoxic switch. *Science* **2002**, *295*, 858–861. [CrossRef] [PubMed]
208. Huang, Z.; Wu, Y.; Zhou, X.; Xu, J.; Zhu, W.; Shu, Y.; Liu, P. Efficacy of therapy with bortezomib in solid tumors: A review based on 32 clinical trials. *Future Oncol.* **2014**, *10*, 1795–1807. [CrossRef]
209. Yeo, E.-J.; Ryu, J.-H.; Cho, Y.-S.; Chun, Y.-S.; Huang, L.E.; Kim, M.-S.; Park, J.-W. Amphotericin B Blunts erythropoietin response to hypoxia by reinforcing fih-mediated repression of HIF-1. *Blood* **2006**, *107*, 916–923. [CrossRef]
210. Olenyuk, B.Z.; Zhang, G.-J.; Klco, J.M.; Nickols, N.G.; Kaelin, W.G.; Dervan, P.B. Inhibition of vascular endothelial growth factor with a sequence-specific hypoxia response element antagonist. *Proc. Natl. Acad. Sci. USA* **2004**, *101*, 16768–16773. [CrossRef]
211. LaVallee, T.M.; Burke, P.A.; Swartz, G.M.; Hamel, E.; Agoston, G.E.; Shah, J.; Suwandi, L.; Hanson, A.D.; Fogler, W.E.; Sidor, C.F.; et al. Significant antitumor activity in vivo following treatment with the microtubule agent ENMD-1198. *Mol. Cancer Ther.* **2008**, *7*, 1472–1482. [CrossRef]
212. Mabjeesh, N.J.; Escuin, D.; LaVallee, T.M.; Pribluda, V.S.; Swartz, G.M.; Johnson, M.S.; Willard, M.T.; Zhong, H.; Simons, J.W.; Giannakakou, P. 2ME2 inhibits tumor growth and angiogenesis by disrupting microtubules and dysregulating HIF. *Cancer Cell* **2003**, *3*, 363–375. [CrossRef]
213. Moser, C.; Lang, S.A.; Mori, A.; Hellerbrand, C.; Schlitt, H.J.; Geissler, E.K.; Fogler, W.E.; Stoeltzing, O. ENMD-1198, a novel tubulin-binding agent reduces HIF-1alpha and STAT3 Activity in Human Hepatocellular Carcinoma(HCC) cells, and inhibits growth and vascularization in vivo. *BMC Cancer* **2008**, *8*, 206. [CrossRef]
214. Suzuki, M.; Shinohara, F.; Rikiishi, H. Zebularine-induced reduction in VEGF secretion by HIF-1α degradation in oral squamous cell carcinoma. *Mol. Med. Rep.* **2008**, *1*, 465–471. [CrossRef]
215. Mabjeesh, N.J.; Post, D.E.; Willard, M.T.; Kaur, B.; Van Meir, E.G.; Simons, J.W.; Zhong, H. Geldanamycin induces degradation of hypoxia-inducible factor 1alpha protein via the proteosome pathway in prostate cancer cells. *Cancer Res.* **2002**, *62*, 2478–2482.
216. Li, Z.; Jia, L.; Tang, H.; Shen, Y.; Shen, C. Synthesis and biological evaluation of geldanamycin–ferulic acid conjugate as a potent Hsp90 inhibitor. *RSC Adv.* **2019**, *9*, 42509–42515. [CrossRef]
217. Newcomb, E.W.; Tamasdan, C.; Entzminger, Y.; Alonso, J.; Friedlander, D.; Crisan, D.; Miller, D.C.; Zagzag, D. Flavopiridol Induces mitochondrial-mediated apoptosis in murine glioma GL261 cells via release of cytochrome c and apoptosis inducing factor. *Cell Cycle Georget. Tex* **2003**, *2*, 243–250. [CrossRef]

218. Dai, Y.; Rahmani, M.; Grant, S. Proteasome inhibitors potentiate leukemic cell apoptosis induced by the cyclin-dependent kinase inhibitor flavopiridol through a SAPK/JNK- and NF-KappaB-dependent process. *Oncogene* **2003**, *22*, 7108–7122. [CrossRef]
219. Bozok Cetintas, V.; Acikgoz, E.; Yigitturk, G.; Demir, K.; Oktem, G.; Tezcanli Kaymaz, B.; Oltulu, F.; Aktug, H. Effects of flavopiridol on critical regulation pathways of CD133high/CD44high lung cancer stem cells. *Medicine* **2016**, *95*, e5150. [CrossRef]
220. Qin, J.; Liu, Y.; Lu, Y.; Liu, M.; Li, M.; Li, J.; Wu, L. Hypoxia-inducible Factor 1 alpha promotes cancer stem cells-like properties in human ovarian cancer cells by upregulating SIRT1 expression. *Sci. Rep.* **2017**, *7*, 10592. [CrossRef]
221. Awan, F.T.; Jones, J.A.; Maddocks, K.; Poi, M.; Grever, M.R.; Johnson, A.; Byrd, J.C.; Andritsos, L.A. A Phase 1 clinical trial of flavopiridol consolidation in chronic lymphocytic leukemia patients following chemoimmunotherapy. *Ann. Hematol.* **2016**, *95*, 1137–1143. [CrossRef]
222. Flich-Carbonell, J.; Alegre Martinez, A.; Alfonso-Sanchez, J.; Torres-Sanchez, M.; Gomez-Abril, S.; Martínez-Martínez, M.; Martin-Moreno, J.M. The potential link between episodes of diverticulitis or hemorrhoidal proctitis and diets with selected plant foods: A case–control study. *Nutrients* **2021**, *13*, 1791. [CrossRef]
223. Ai, Z.; Lu, Y.; Qiu, S.; Fan, Z. Overcoming cisplatin resistance of ovarian cancer cells by targeting HIF-1-regulated cancer metabolism. *Cancer Lett.* **2016**, *373*, 36–44. [CrossRef]
224. Figueroa, Y.G.; Chan, A.K.; Ibrahim, R.; Tang, Y.; Burow, M.E.; Alam, J.; Scandurro, A.B.; Beckman, B.S. NF-KappaB Plays a key role in hypoxia-inducible factor-1-regulated erythropoietin gene expression. *Exp. Hematol.* **2002**, *30*, 1419–1427. [CrossRef]
225. Lee, W.-T.; Tai, S.-H.; Lin, Y.-W.; Wu, T.-S.; Lee, E.-J. YC-1 Reduces inflammatory responses by inhibiting nuclear factor-κB translocation in mice subjected to transient focal cerebral ischemia. *Mol. Med. Rep.* **2018**, *18*, 2043–2051. [CrossRef]
226. Van Waes, C. Nuclear factor-kappab in development, prevention, and therapy of cancer. *Clin. Cancer Res. Off. J. Am. Assoc. Cancer Res.* **2007**, *13*, 1076–1082. [CrossRef]
227. Richardson, P.G.; Mitsiades, C.; Hideshima, T.; Anderson, K.C. Proteasome inhibition in the treatment of cancer. *Cell Cycle Georget. Tex* **2005**, *4*, 290–296. [CrossRef]
228. Zavrski, I.; Jakob, C.; Schmid, P.; Krebbel, H.; Kaiser, M.; Fleissner, C.; Rosche, M.; Possinger, K.; Sezer, O. Proteasome: An emerging target for cancer therapy. *Anticancer. Drugs* **2005**, *16*, 475–481. [CrossRef]
229. Takada, Y.; Aggarwal, B.B. Flavopiridol inhibits NF-KB activation induced by various carcinogens and inflammatory agents through inhibition of IκBα Kinase and P65 Phosphorylation: Abrogation of cyclin d1, cyclooxygenase-2, and matrix metalloprotease-9. *J. Biol. Chem.* **2004**, *279*, 4750–4759. [CrossRef]
230. Thangjam, G.S.; Dimitropoulou, C.; Joshi, A.D.; Barabutis, N.; Shaw, M.C.; Kovalenkov, Y.; Wallace, C.M.; Fulton, D.J.; Patel, V.; Catravas, J.D. Novel mechanism of attenuation of lps-induced nf-κb activation by the heat shock protein 90 inhibitor, 17-n-allylamino-17-demethoxygeldanamycin, in human lung microvascular endothelial cells. *Am. J. Respir. Cell Mol. Biol.* **2014**, *50*, 942–952. [CrossRef]

molecules

MDPI

Article

Bursatella leachii Purple Ink Secretion Concentrate Exerts Cytotoxic Properties against Human Hepatocarcinoma Cell Line (HepG2): In Vitro and In Silico Studies

Zeyad I. Alehaideb [1,†], Anuradha Venkatraman [2,†], Mahadev Kokane [2,3,†], Syed Ali Mohamed [4], Saranya Rameshbabu [4], Rasha S. Suliman [1,5], Sahar S. Alghamdi [1,5], Hamad Al-Eidi [1], Bandar Alghanem [1], Maha-Hamadien Abdulla [6] and Sabine Matou-Nasri [1,*]

1 King Abdullah International Medical Research Center, King Saud bin Abdulaziz University for Health Sciences, Riyadh 11481, Saudi Arabia; alehaidebze1@NGHA.MED.SA (Z.I.A.); sulimanr@ksau-hs.edu.sa (R.S.S.); ghamdisa@ksau-hs.edu.sa (S.S.A.); aleidiha@NGHA.MED.SA (H.A.-E.); GhanemBa@NGHA.MED.SA (B.A.)
2 PG & Research Department of Biochemistry, Mohamed Sathak Arts and Science College, Tamil Nadu 600119, India; vanuradha.2003@gmail.com (A.V.); mahadevciba@gmail.com (M.K.)
3 Central Institutes of Fisheries Nautical and Engineering Training, Tamil Nadu 600013, India
4 PG & Research Department of Biotechnology, Mohamed Sathak Arts and Science College, Tamil Nadu 600119, India; syedmicro555@gmail.com (S.A.M.); saranyasundar2012@gmail.com (S.R.)
5 College of Pharmacy, King Saud bin Abdulaziz University for Health Sciences, Riyadh 11481, Saudi Arabia
6 Colorectal Research Chair, Department of Surgery, College of Medicine, King Khalid University Hospital, King Saud University, Riyadh 11472, Saudi Arabia; mabdulla@ksu.edu.sa
* Correspondence: matouepnasrisa@ngha.med.sa; Tel.: +966-(11)-429-4535; Fax: +966-(11)-429-4440
† These authors contributed equally to this work.

Citation: Alehaideb, Z.I.; Venkatraman, A.; Kokane, M.; Mohamed, S.A.; Rameshbabu, S.; Suliman, R.S.; Alghamdi, S.S.; Al-Eidi, H.; Alghanem, B.; Abdulla, M.-H.; et al. *Bursatella leachii* Purple Ink Secretion Concentrate Exerts Cytotoxic Properties against Human Hepatocarcinoma Cell Line (HepG2): In Vitro and In Silico Studies. *Molecules* **2022**, *27*, 826. https://doi.org/10.3390/molecules27030826

Academic Editors: Manuela Pintado, Ezequiel Coscueta and María Emilia Brassesco

Received: 6 December 2021
Accepted: 24 January 2022
Published: 26 January 2022

Abstract: Liver cancer is a leading cause of cancer death globally. Marine mollusc-derived drugs have gained attention as potential natural-based anti-cancer agents to overcome the side effects caused by conventional chemotherapeutic drugs during cancer therapy. Using liquid chromatography-mass spectrometry, the main biomolecules in the purple ink secretion released by the sea hare, named *Bursatella leachii* (*B. leachii*), were identified as hectochlorin, malyngamide X, malyngolide S, bursatellin and lyngbyatoxin A. The cytotoxic effects of *B. leachii* ink concentrate against human hepatocarcinoma (HepG2) cells were determined to be dose- and time-dependent, and further exploration of the underlying mechanisms causing the programmed cell death (apoptosis) were performed. The expression of cleaved-caspase-8 and cleaved-caspase-3, key cysteine-aspartic proteases involved in the initiation and completion of the apoptosis process, appeared after HepG2 cell exposure to the *B. leachii* ink concentrate. The gene expression levels of pro-apoptotic *BAX*, *TP53* and *Cyclin D1* were increased after treatment with the *B. leachii* ink concentrate. Applying in silico approaches, the high scores predicted that bioactivities for the five compounds were protease and kinase inhibitors. The ADME and cytochrome profiles for the compounds were also predicted. Altogether, the *B. leachii* ink concentrate has high pro-apoptotic potentials, suggesting it as a promising safe natural product-based drug for the treatment of liver cancer.

Keywords: ADME; apoptosis; *Bursatella leachii* ink; caspase; target prediction; TP53; liver cancer

1. Introduction

Liver cancer is the sixth most frequently diagnosed cancer and third leading cause of cancer death globally, with a high incidence observed in Asian and African countries [1]. The variable geographical distribution of liver cancer overlaps with the geographic incidence of viral hepatitis (i.e., hepatitis B and C viruses) and of the human immunodeficiency virus (HIV) [2]. This viral infection results in the onset of liver cancer and the progression from chronic hepatitis, liver cirrhosis to heterogeneous hepatocellular carcinoma (HCC) [3].

The onset of liver cancer can also be due to aging, exposure to toxic compounds, autoimmunity and metabolic diseases [4]. Conventional treatments such as surgery, radiotherapy and chemotherapy, as well as gene- and immune-based therapeutic drugs are currently used [5]. Cytotoxic chemotherapy is not the first-line treatment (i.e., protein kinase inhibitor, Sorafenib) for HCC, the main type of primary liver cancer, which is a chemotherapy-refractory tumour [6]. However, there is ongoing discovery of natural bioactive compounds as neoadjuvant agents, which inhibit liver cancer cell growth and enhance liver cancer prevention, as well as overcome hepatotoxicity side effects from conventional therapy and liver cancer recurrence [7].

An alternative treatment for liver cancer prior to transplantation is urgently required. Natural product extracts have been investigated for their bioactive compounds with anti-proliferative activity and pro-apoptotic effects, revealed by tyrosine kinase inhibition [8] and by caspase activation, cell cycle- and apoptosis-related gene up-regulation [9]. The main advantage of the induction of apoptosis is the absence of the inflammatory reaction triggered by necrotic cells [10]. Some natural bioactive compounds with anti-inflammatory and anti-angiogenic effects, which prevent cancer progression, have been discovered [11].

Molluscs are the second largest animal phylum on earth and provide a rich source of medicinal natural bioactive molecules [12]. The opisthobranch molluscs are a subclass of the *Gastropoda* family *Aplysiidae*, order *Anaspide*, genus and species *Bursatella (B.) leachii*, commonly known as sea hares [13]. Sea hare-derived bioactive compounds with anti-cancer activity, including soblidotin (dolastatin 10 derivative), synthadotin/ILX_{651}, cemadotin and kahalalide F, have been in clinical trials, and brentuximab vedotin Adcetris® (dolastatin 10), an antibody drug conjugate, has been approved by the Food and Drug Administration for the treatment of Hodgkin lymphoma and systemic anaplastic large cell lymphoma [14–16]. Similar to a squid, sea hares release a purple ink to fend off predators which contains secondary metabolites with potential cytotoxicity [16]. We previously reported *B. leachii* purple ink-derived anti-HIV protein [17], 7,9-di-tert-butyl-1-oxaspiro [4,5], deca-6,9-diene-2,8-dione and digoxigenin acetate as potent anti-inflammatory compounds [18]. However, *B. leachii* purple ink-derived concentrate, including the anti-inflammatory compounds identified, has not been studied for its potential anti-cancer activity. We chemically analysed *B. leachii* ink concentrate to identify secondary metabolites and evaluated the potential cytotoxic effects of a crude ink concentrate of *B. leachii* against the growth of the human hepatocellular carcinoma (HCC) cell line HepG2. Protein and gene expression levels of apoptosis and cell cycle regulatory markers in the *B. leachii* ink concentrate-treated HepG2 cells were assessed and several biological target predictions were performed.

2. Results

2.1. Chemical Identification of B. leachii Ink Concentrate Using Liquid Chromatography-Quadrupole Time of Flight (LC-QTOF)

The crude ink concentrate of the *B. leachii* was subjected to total ion current spectra raw data (See Figure 1). Qualitative and quantitative analysis software from the data analysis program MassHunter (Agilent Technologies) were also used. After conducting a mass screening on the spectra (Figure 1), the chemical features were extracted from the LC-QTOF data using the Molecular Features Extraction algorithm and the recursive analysis workflow. Features were extracted by screening the detected nodes at various retention times per minute, with a minimum intensity of 6000 counts, and aligned with previously detected compounds considering adducts ($[M + K]^+$ and $[M - H]^-$). The tentatively identified compounds were hectochlorin (A), malyngamide X (B), bursatellin (C), malyngamide S (D) and lyngbyatoxin A (E).

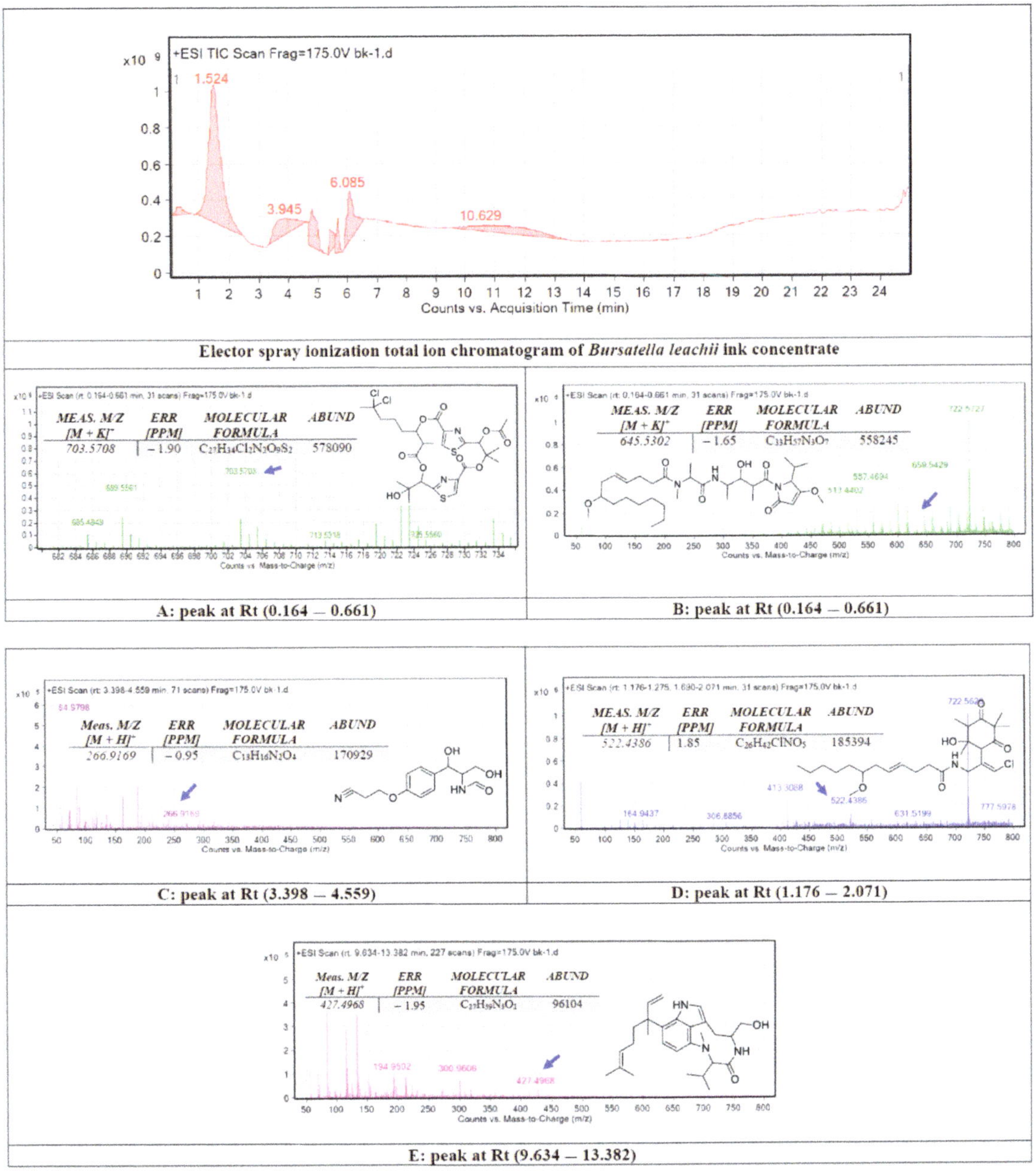

Figure 1. Base peak chromatogram of *Bursatella leachii* ink concentrate. Tentatively identified secondary metabolites are hectochlorin (**A**) [19], malyngamide X (**B**) [20], bursatellin (**C**) [21], malyngamide S (**D**) [22] and lyngbyatoxin A (**E**) [23]. Means m/z implies measured m/z.

2.2. Cytotoxic Effect of the B. leachii Ink Concentrate

The effect of *B. leachii* ink concentrate on HepG2 cell proliferation was tested at different concentrations and time periods. The increasing concentrations (from 10 to 1000.0 µg/mL) of the *B. leachii* ink concentrate and the lengthening of the exposure times (from 24 to 72 h) resulted in a decrease in the viability of the HepG2 cells, based on the ATP generated by the living cells, compared to the control cells, which described a dose- and time-dependent effect (Figure 2). The half-maximal inhibitory concentration (IC_{50}) values of *B. leachii* ink concentrate required in order to inhibit 50% of HepG2 cell growth were determined at each exposure time. The treatment of HepG2 cells with *B. leachii* ink concentrate for 72 h of exposure displayed the lowest IC_{50} value of 242.9 μg/mL, followed by 48 h exposure with an IC_{50} value of 447.5 μg/mL, and the IC_{50} value for 24 h was more than 1000 μg/mL of *B. leachii* ink concentrate.

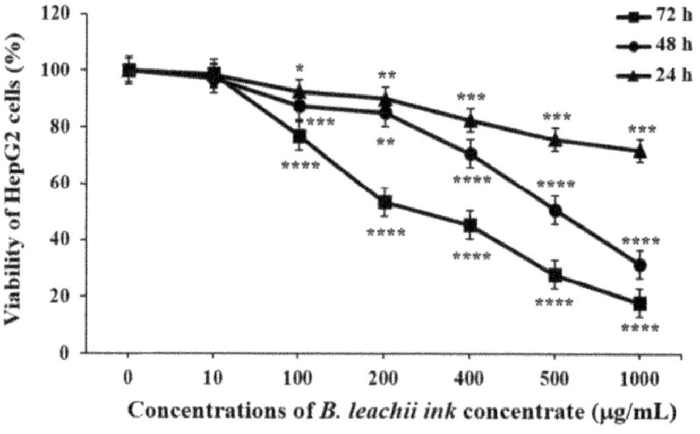

Figure 2. Effects of B. leachii concentrate on HepG2 cell viability. The cell viability was determined using CellTiter-Glo® kit and expressed as percentage of the control, the untreated cell viability, corresponding to 100%. * $p < 0.05$, ** $p < 0.01$, *** $p < 0.001$, and **** $p < 0.0001$ signify a statistically significant difference compared with the control, from three independent experiments.

2.3. Induction of Apoptosis by B. leachii Ink Concentrate

As one of the cell death mechanisms, the potential induction of apoptosis in HepG2 cells treated with 100 μg/mL and 400 μg/mL of the *B. leachii* ink concentrate was investigated using Western blot analysis. After 24 h of exposure, the expression levels of pro-apoptotic proteins, such as the cleaved caspase-8 (key enzyme prompting extrinsic apoptotic pathway) and cleaved-caspase-3 (key enzyme resulting in apoptosis), were evaluated in the *B. leachii* ink concentrate-treated cells compared with the untreated cells. Used as a positive control, staurosporine (STS) led to a cleavage of both caspase-8 and caspase-3 in HepG2 cells, and, as expected, a quasi-absence of cleaved-caspase-8 and cleaved-caspase-3 expression was detected in the untreated cells (Figure 3). A cleavage of both caspase-8 and caspase-3 was observed in *B. leachii* ink concentrate-treated HepG2 cells (Figure 3). Higher levels of expression of the cleaved-caspase-8 were observed in the HepG2 cells treated with 400 μg/mL of *B. leachii* ink concentrate compared with the cells treated with 100 μg/mL of *B. leachii* ink concentrate (Figure 3). A similar expression level of cleaved-caspase-3 was noticed in the HepG2 cells exposed to the two tested concentrations of *B. leachii* ink concentrate (Figure 3).

2.4. Modulation of Gene Expression Levels of Apoptotic and Cell Cycle Regulatory Genes by B. leachii Ink Concentrate

The pro-apoptotic effect of *B. leachii* ink concentrate on the HepG2 cells was evaluated by monitoring the expression levels of apoptosis and cell cycle-related genes, including *BCL-2*, *BCL-xL*, *TP53*, *BAX*, *CDKN1A*, *CCNA* (Cyclin A), *CCND1* (Cyclin D1) and *Survivin* using reverse transcription-quantitative polymerase chain reaction (RT-qPCR). The treatment of the HepG2 cells with 400 μg/mL of *B. leachii* ink concentrate significantly enhanced the expression levels of the pro-apoptotic genes *BAX* (2.6-fold, p = 0.000012), *TP53* (2.3-fold,

$p = 0.00023$) and *CCND1* (2.1-fold, $p = 0.0012$), compared with the basal gene expression level detected in the untreated cells (Figure 4). A significant up-regulation of the gene expression level of the anti-apoptotic *BCL-xL* (2.0-fold, $p = 0.0043$) was noticed in the *B. leachii* ink concentrate-treated HepG2 cells (Figure 4). Of note, the up-regulation of the pro-apoptotic *CCNA* (1.77-fold) and anti-apoptotic *Survivin* (1.64-fold) gene expression levels remained non-significant (Figure 4).

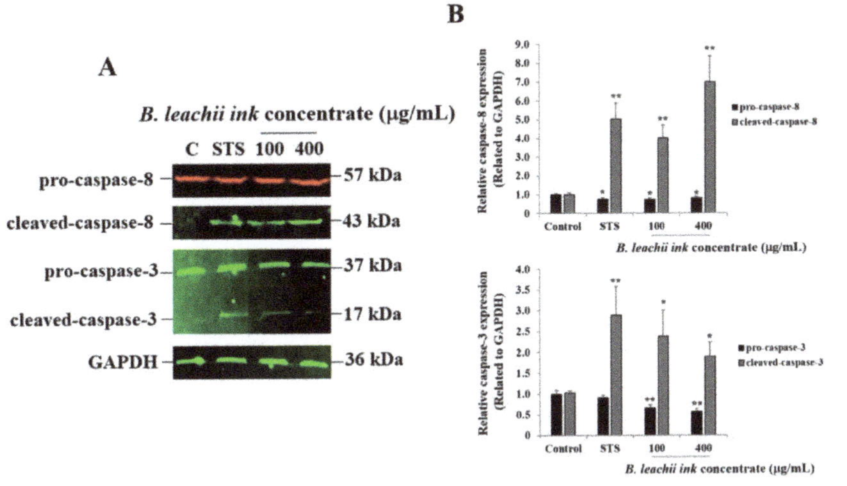

Figure 3. *B. leachii* ink concentrate triggers caspase-8 and caspase-3 cleavage in HepG2 cells. (**A**) Representative Western blot gels showing the detection of cleaved-caspase-8 and of cleaved-caspase-3 in HepG2 cells after 24 h of treatment with 100 and 400 g/mL of *B. leachii* ink concentrate along with 1 M STS, used as a positive control. Quasi-absence of the cleavage of caspase was observed in untreated HepG2 cells, the control (C). (**B**) Bar graphs indicating the relative expression levels of caspase-8 (top) and of caspase-3 (bottom), calculated as a ratio of the expression to GAPDH, used as a loading control. * $p < 0.05$ and ** $p < 0.01$ signify a statistically significant difference compared with the control, from three independent experiments.

2.5. B. leachii Ink Concentrate Bioactivity Predictions

In this study, we sought to determine the bioactivity score of each bioactive metabolite identified in the *B. leachii* ink concentrate. These bioactivity predictions provided more information about which molecule in the ink concentrate could contribute to the observed anti-cancer activity. The bioactivity score of the five bioactive molecules was investigated using the PASS online webserver. Our results showed that hectochlorin and malyngamide S had the highest bioactivity scores (Table 1), with Pa 0.933 and Pa 0.747, respectively, suggesting a promising anti-neoplastic activity for these two molecules. Additionally, the anti-cancer activity of *B. leachii* ink concentrate could be attributed to the presence of hectochlorin and malyngamide S metabolites. The remaining molecules exhibited lower bioactivity scores, and no predicted score was identified for bursatellin.

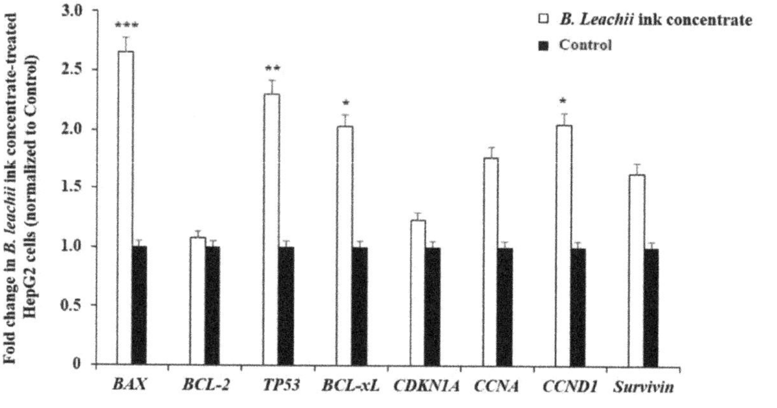

Figure 4. Effects of *B. leachii* ink concentrate on apoptosis and cell cycle regulatory gene expression detected in HepG2 cells. The bar graph shows the relative gene expression of apoptotic (Bax, Bcl2, Bcl-xL, Survivin), tumour suppressor TP53 and cell cycle (cyclin A, cyclin D1, cyclin-dependent kinase inhibitor CDKN1A) regulatory transcripts determined by RT-qPCR analysis in HepG2 cells treated with 400 g/mL of *B. leachii* ink concentrate, as compared with the basal level of gene expression monitored in untreated HepG2 cells, the control. * $p < 0.05$, ** $p < 0.01$, and *** $p < 0.001$ signify a statistically significant difference compared with the control, from three independent experiments.

Table 1. The bioactivity scores of identified metabolites from *B. leachii* ink concentrate using the PASS online webserver.

Anti-Neoplastic Activity	Probability of Being Active (Pa)	Probability of Being Inactive (Pi)
Hectochlorin	0.933	0.002
Malyngamide X	0.295	0.231
Malyngamide S	0.747	0.019
Bursatellin	not applicable	not applicable
Lyngbyatoxin A	0.169	0.075

2.6. Molecular Target Predictions of B. leachii Ink Concentrate

Molinspiration was used to investigate the possible molecular targets that could mediate the observed and predicted anti-cancer activity. Each bioactive molecule was evaluated as a G protein-coupled receptor (GPCR) ligand, ion channel modulator, kinase inhibitor, nuclear receptor ligand and protease and enzyme inhibitor. Interestingly, almost all of the five molecules exhibited a positive bioactivity score as protease inhibitors, and malyngamide X, malyngamide S and lyngbyatoxin A had the highest scores (0.46, 0.32 and 0.36, respectively). Malyngamide X, malyngamide S and lyngbyatoxin A had high bioactivity scores as enzyme inhibitors with values of 0.32, 0.32 and 0.35, respectively. Other possible targets such as the ion channels, GPCR and kinase inhibition were seen with malyngamide X, malyngamide S and lyngbyatoxin A, suggesting that these bioactive molecules could regulate several molecular targets. Additional target mapping was conducted with SwissTargetPrediction and the results were comparable with the Molinspiration webserver with a high probability of targeting proteases and kinases (Table 2).

Table 2. Molecular Target Predictions for bioactive molecules identified from *B. leachii* ink concentrate using Molinspiration and SwissTargetPrediction.

Name	Target Prediction (Molinspiration)		Target Prediction (SwissTargetPrediction)
Hectochlorin	GPCR ligand Ion channel modulator Kinase inhibitor Nuclear receptor-ligand Protease inhibitor Enzyme inhibitor	−0.09 −0.46 −0.46 −0.18 0.1 −0.08	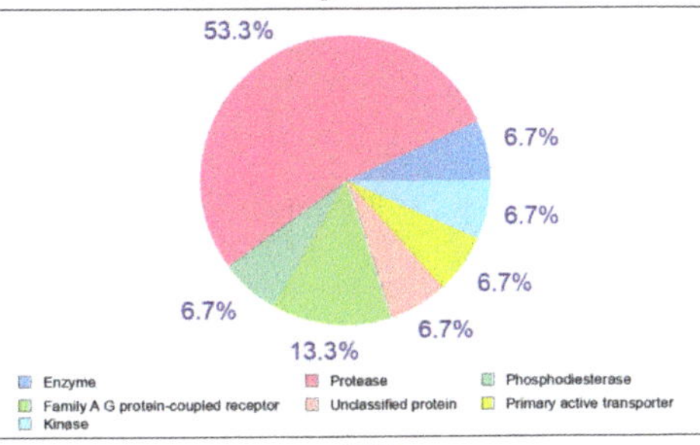
Malyngamide X	GPCR ligand Ion channel modulator Kinase inhibitor Nuclear receptor-ligand Protease inhibitor Enzyme inhibitor	0.19 −0.44 −0.42 −0.29 0.46 0.01	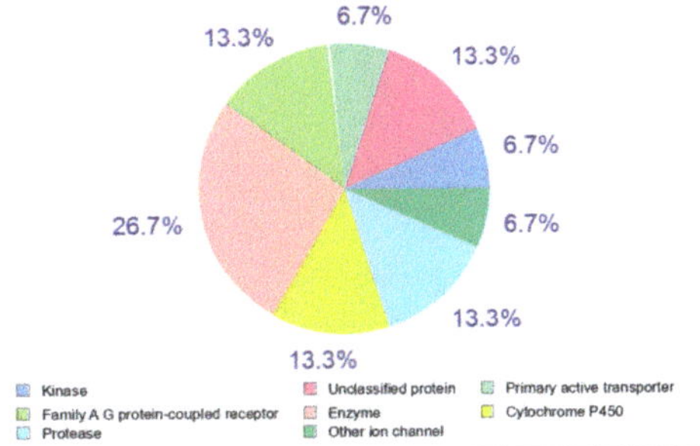
Malyngamide S	GPCR ligand Ion channel modulator Kinase inhibitor Nuclear receptor ligand Protease inhibitor Enzyme inhibitor	0.17 0.24 −0.28 0.23 0.32 0.32	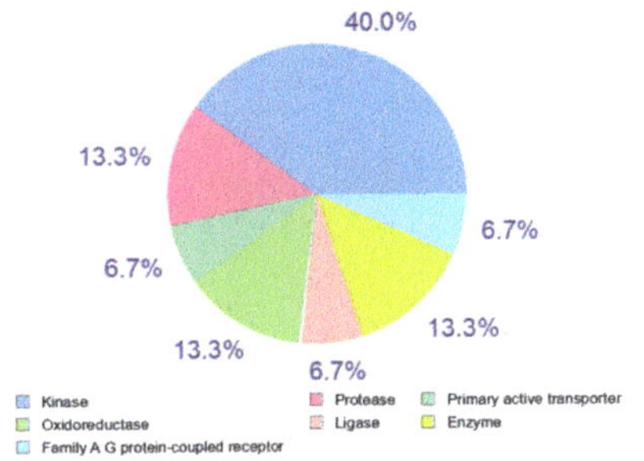
Bursatellin	GPCR ligand Ion channel modulator Kinase inhibitor Nuclear receptor-ligand Protease inhibitor Enzyme inhibitor	0.04 −0.22 −0.36 −0.21 0.01 0.32	NA
Lyngbyatoxin A	GPCR ligand Ion channel modulator Kinase inhibitor Nuclear receptor ligand Protease inhibitor Enzyme inhibitor	0.49 0.2 0.42 0.06 0.36 0.35	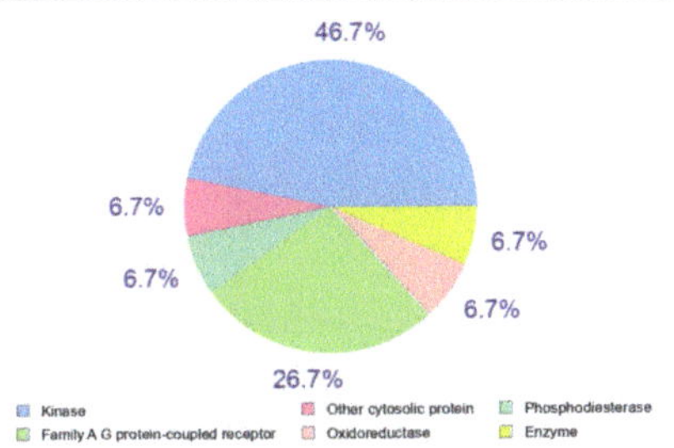

2.7. Pharmacokinetics Absorption, Distribution, Metabolism and Excretion (ADME) Predictions and Cytochrome (CYP) P450 Enzyme Inhibition Profiling

To assess the potential pharmaceutical properties of the *B. leachii* ink concentrate-derived bioactive molecules, SwissADME (Swiss Institute of Bioinformatics, Lausanne, Switzerland) was used to evaluate several parameters important for drug discovery. Of the five bioactive metabolites, malyngamide S, bursatellin and lyngbyatoxin A demonstrated a molecular weight of fewer than 500 daltons. All the compounds that demonstrated a high lipophilicity, except for bursatellin, exhibited a low Log *p* value. The solubility of the compounds was poor, with the exception of bursatellin, which demonstrated excellent solubility (Log S −2.39). Most of the compounds were predicted to have a peripheral effect (no blood–brain barrier (BBB) penetration except for lyngbyatoxin A) and high gastrointestinal (GI) absorption (Table 3).

Table 3. The predicted ADME properties for the bioactive molecules identified in *B. leachii* ink concentrate.

Compound Name	Molecular Weight	Log Po/w (WLOGP)	Log S (SILICOS- IT)	BBB Permeant	GI Absorption	Rule of Five (ROF)
Hectochlorin	665.60 g/mol	5.09	−6.86 Poorly soluble	No	Low	No; 2 violations: MW > 500, NorO > 10
Malyngamide X	607.82 g/mol	3.98	−5.07 Moderately soluble	No	High	Yes; 1 violation: MW > 500
Malyngamide S	484.07 g/mol	4.73	−6.25 Poorly soluble	No	High	Yes; 0 violations
Bursatellin	264.28 g/mol	−0.21	−2.39 Soluble	No	High	Yes; 0 violations
Lyngbyatoxin A	437.62 g/mol	4.09	−6.69 Poorly soluble	Yes	High	Yes; 0 violations

We determined and qualitatively predicted the possibility of CYP enzyme inhibition that could be associated with these bioactive molecules using the SWISS webserver. Our results showed that hectochlorin and bursatellin did not exhibit any CYP enzyme inhibition. Malyngamide X demonstrated only two inhibitions for CYP2C19 and CYP3A4, and the malyngamide S was predicted to inhibit CYP2C19, CYP2D6 and CYP3A4 enzymes. Lyngbyatoxin A inhibited three CYP enzymes, including CYP2C19, CYP2C9 and CYP3A4 (Table 4).

Table 4. The CYP enzyme inhibition profile of the bioactive molecules identified in *B. leachii* ink concentrate.

Compound Name	CYP1A2 Inhibitor	CYP2C19 Inhibitor	CYP2C9 Inhibitor	CYP2D6 Inhibitor	CYP3A4 Inhibitor
Hectochlorin	No	No	No	No	No
Malyngamide X	No	Yes	No	No	Yes
Malyngamide S	No	Yes	No	Yes	Yes
Bursatellin	No	No	No	No	No
Lyngbyatoxin A	No	Yes	Yes	No	Yes

3. Discussion

Intensive exploration of the marine ecosystem has provided a valuable source of diverse bioactive compounds. Recently, the purple ink concentrate released by the sea hare *B. leachii* was studied, and anti-HIV and anti-inflammatory activities were identified [17,18], two prominent properties required for a potential liver cancer treatment. To widen the biological activities of this *B. leachii* purple ink-derived concentrate as a promising natural neo-adjuvant for the treatment of liver cancer, we investigated its potential cytotoxic effects against HCC HepG2 cells and established molecular target and pharmacokinetic predictions of the identified metabolites-derived *B. leachii* purple ink concentrate.

In this study, the chemical analysis was performed using high-resolution Q-TOF analysis, which supports the tentative identification of the chemicals more accurately and the comparison with previous structure identification studies for the *B. leachii* ink concentrate biomolecules. For example, the m/z value at retention time (0.164–0.661) was correlated with the parent compound hectochlorin [19], with m/z $[M + K]^+$ 703.5708 daltons and a molecular formula of $[C_{27}H_{34}Cl_2N_2O_9S_2]^+$, in the positive ion mode $[M + H]^+$ m/z 666.236 and $[M - H]^-$ with m/z 664.163 daltons in the negative mode, indicating that the compound had a molecular weight of 665.603 $g{\cdot}mol^{-1}$. The m/z value at retention time (0.164–0.661) was correlated with the parent compound malyngamide X [20], with m/z $[M + K]^+$ 646.5302 daltons and a molecular formula of $[C_{33}H_{57}N_3O_7]^+$, in the positive ion mode $[M + H]^+$ m/z 607.420 and $[M - H]^-$ with m/z 606.822 daltons in the negative mode, indicating that the compound had a molecular weight of 607.420 $g{\cdot}mol^{-1}$. The m/z value at retention time (3.398–4.559) was correlated with the parent compound bursatellin [21], with m/z $[M + 2H]^+$ 266.9169 daltons and a molecular formula of $[C_{13}H_{16}N_2O_4]^+$, in the positive ion mode $[M + H]^+$ m/z 264.111 and $[M - H]^-$ with m/z 263.210 daltons in the negative mode, indicating that the compound had a molecular weight of 264.277 $g{\cdot}mol^{-1}$. The m/z value at retention time (1.176–2.071) was correlated with the parent compound malyngamide S [22], with m/z $[M + K]^+$ 522.4386 daltons and a molecular formula of $[C_{26}H_{42}ClNO_5]^+$, in the positive ion mode $[M + H]^+$ m/z 483.279 and $[M - H]^-$ with m/z 482.109 daltons in the negative mode, indicating that the compound had a molecular weight of 484.069 $g{\cdot}mol^{-1}$. The m/z value at retention time (9.634–13.382) was correlated with the parent compound lyngbyatoxin A [23], with m/z $[M - CH_3]^+$ 426.4968 daltons and a molecular formula of $[C_{27}H_{39}N_3O_2]^+$, in the positive ion mode 437.304 and $[M - H]^-$ with m/z 437.304 daltons in the negative mode, indicating that the compound had a molecular weight of 437.617 $g{\cdot}mol^{-1}$.

The human HCC cell line HepG2 exposed to the *B. leachii* purple ink secretion concentrate led to an inhibition of the cell proliferation in a dose- and time-dependent manner. An induction of apoptosis in *B. leachii* ink concentrate-treated HepG2 cells was observed at intermediate concentrations (100 and 400 μg/mL) of *B. leachii* ink concentrate after 24 h exposure. *B. leachii* ink concentrate added to the HepG2 cells for 72 h of treatment resulted in the lowest IC_{50} value of 242.9 μg/mL compared with IC_{50} values determined after 24 and 48 h of exposure. A study conducted by Suntornchashwej and colleagues [19] reported the cytotoxicity exhibited by the ethyl acetate-derived ink extract of *B. leachii* against the human small cell lung cancer (NCI-H187), oral human epidermoid carcinoma (KB) and breast cancer (BC) cell lines with half-maximal effective dose (ED_{50}) values of 16.2, 7.2 and 6.6 μg/mL, respectively. The protective effect of *B. leachii* ink extract against neuroblastoma cell line SH-SY5Y pre-treated with hydrogen peroxide and against microglia cells stimulated by a bacterial lipopolysaccharide was studied [24]. *B. leachii* ink extract was effective against microglia cells by decreasing the intracellular nitric oxide production with an IC_{50} value of 5.74 μg/mL; however, the SH-SY5Y cells had no cell response to the concentration of the *B. leachii* ink extract studied. The determination of different IC_{50} values reflecting the anti-proliferative activity of *B. leachii* ink extract indicate the specificity of the ink extract to exert cytotoxicity against various cancer cell lines.

Similarly, the anti-cancer potential of *B. leachii* ink concentrate at 100 and 400 μg/mL through apoptosis induction in HepG2 cells treated for 24 h was confirmed using Western blot technology by quantitatively detecting the expression of the most important pro-apoptotic proteins, namely the initiator caspase-8 and the effector/executioner caspase 3. Apoptosis is initiated and achieved by the cleavage of the initiator and executioner pro-caspases into cleaved-caspases, the activated form of the enzymes. In the present study, the cleavage of caspase-8 and caspase-3 was observed in the HepG2 cells after 24 h of cell exposure to *B. leachii* ink concentrate. An equal level of cleaved-caspase-3 was observed at both concentrations (100 and 400 μg/mL) of *B. leachii* ink concentrate, confirming the slight decrease in HepG2 cell growth noticed at similar conditions. Tested at the same intermediate concentrations, a concomitant cleavage of pro-caspase-3 would be expected after 72 h of exposure, due to the decrease in the *B. leachii* ink concentrate-treated HepG2 cell growth. However, the degree of pro-caspase-8 cleavage increased with the concentrations of *B. leachii* ink concentrate, confirming the role of caspase-8 as the initiator of apoptosis. In addition, the expression of cleaved-caspase-8 and cleaved-caspase-3 detected in *B. leachii* ink concentrate-treated HepG2 cells may reveal that the induction of apoptosis occurs via the activation of the extrinsic death receptor pathway by transmitting the death signal from the cell surface to the intracellular signalling pathways through the tumour necrosis factor receptor gene family [25]. Active caspase-8 either initiates apoptosis directly by cleaving pro-caspase-3 into activated cleaved-caspase-3 or through the mitochondria by the cleavage of BID to induce cell death [26]. An investigation of the involvement of the mitochondrial-dependent intrinsic apoptosis pathway, such as monitoring the mitochondrial membrane potential or release of cytochrome and other apoptotic proteins, in *B. leachii* ink concentrate-treated HepG2 cells would be of interest.

The quantitative expression of cell cycle and apoptotic regulatory genes in the HepG2 cells treated with 400 μg/mL of *B. leachii* after 48 h of exposure was analysed using RT-qPCR. The expressions of the predominant pro-apoptotic *BAX*, *TP53* and *CCND1* (Cyclin D1) genes were significantly up-regulated over the increased expression of the anti-apoptotic *BCL-xL* gene. The gene expression of the anti-apoptotic *BCL-2*, belonging to the main Bcl-2 family protein regulators of apoptosis that are endowed with pro- and anti-apoptotic activities, was not modulated even after the HepG2 cell exposure to *B. leachii* ink concentrate. The tumour suppressor p53 is a transcriptional protein activated by a variety of oncogenic/hyperproliferative stimuli, including DNA damage or chemotherapeutic drugs, and can regulate downstream pro-apoptotic (i.e., *BAX* and p53 upregulated modulator of apoptosis *PUMA* up-regulation) and anti-apoptotic genes (i.e., *BCL-2* repression) [27,28]. Of note, concerning the observed up-regulation of *BCL-xL* and *CCND1* gene expression, both the *BCL-xL* and *CCND1* promoters contain signal transducer and activators of transcription (STAT) binding sites [29], suggesting a possible STAT transcription factor activation in *B. leachii* ink concentrate-treated HepG2 cells. The protein p53 located in the cytosol induces the activation of pro-apoptotic Bax by protein–protein interactions and with Bcl-xL and Bcl-2 by p300/CBP binding [30]. Bax protein homodimerization results in pore formation in the outer mitochondrial membrane, facilitating the release of pro-apoptotic proteins, and Bax-Bcl-2 heterodimerization results in the neutralization of Bcl-2 anti-apoptotic activity [25]. The elevated gene expression levels of *TP53* and *BAX* in *B. leachii* ink concentrate-treated HepG2 cells endorses that cell death might be due to p53-dependent apoptosis.

The analysis of the bioactivity predictions indicates that of the five identified compounds, hectochlorin and malyngamide S had the highest scores as anti-neoplastic agents. This is important for lead optimization and development. The prediction supports the observed anti-cancer activity of the *B. leachii* ink concentrate that could be attributed to the presence of hectochlorin and malyngamide S. The molecular target predictions established from the *B. leachii* ink concentrate suggest the involvement of proteases and kinase inhibitors as potential targets that could explain the observed up-regulation of several apoptotic markers. Several studies reported the modulation of proteases and kinases by marine bioactive molecules, which induced apoptosis in cancer cells [23,31–33].

The in silico ADME predictions for the identified molecules are useful for the potential use of these compounds as a lead for the discovery of a novel anti-cancer therapy. The ADME properties data showed that most of the bioactive molecules have acceptable pharmaceutical properties and follow Lipinski's rule-of-five for drugability related to absorption/permeation, molecular weight and solubility [34,35], except for hectochlorin, which had two violations of this rule. Moreover, the CYP enzyme inhibition profile suggests that some of the compounds could inhibit CYP2C19 and CYP3A4 activities. However, this inhibition profile could be overcome with future optimization of the lead compound.

4. Materials and Methods

4.1. Collection of *B. leachii* Ink Concentrate

The adult *B. leachii* sea hares were collected from intertidal waters of Pulicat lake, position Lat. 13.452523° N Long. 80.319133° E +/− 0.03° N/E, and brought to the laboratory in live condition. The accession number was M-1697, obtained from the Zoological Survey of India, Marine Biology Regional Centre (MBRC), Chennai, India and dated 23 July 2015. Obtained by disturbing the *B. leachii*, the purple fluid ink was filtered through Whatman® filter paper (Sigma-Aldrich, St. Louis, MO, USA). All the aqueous ink-derived samples were centrifuged at 15,000× *g* for 15 min as described by Vennila and colleagues [36] and the supernatant was kept and lyophilized to purple ink residue using a freeze dryer and stored at 4 °C for further use.

4.2. Chemicals and Reagents

Dulbecco's Modified Eagles Medium (DMEM), foetal bovine serum (FBS), penicillin, streptomycin and L-glutamine were obtained from Gibco™ (Waltham, MA, USA). Staurosporine (STS) (>99%) was obtained from Santa Cruz biotechnology (Dallas, TX, USA). High-purity methanol (99.9%) was procured from Honeywell (Charlotte, NC, USA). Formic acid (>95.0%) was purchased from Sigma-Aldrich. Ultrapure water was produced using a Millipore (Billerica, MA, USA) system with a resistivity reading of 18.2 MΩ·cm at 25 °C.

4.3. Chemical Analysis Using LC-QTOF

The fingerprinting of *B. leachii* secretion of aqueous ink concentrate was performed using the Agilent (Santa Clara, CA, USA) 1260 Infinity high performance liquid chromatography system coupled to Agilent 6530 Q-TOF. The analysis was performed using an Agilent SB-C18 column (4.6 mm × 150 mm, 1.8 μm) with the following elution gradient: 0–2 min, 5% B; 2–17 min, 5–100% B; 17–21 min, 95% B; 21–25 min, 5% B, using mobile phase A (0.1% formic acid in water) and mobile phase B (0.1% formic acid in methanol). The injection volume was 10 μL and the flow rate was set at 250 μL/min. The scanning range was set at 50–800 (m/z) and the remaining parameters were set as follows: gas temperature at 300 °C, gas flow at 8 L/min, nebulizer pressure at 35 psi, sheath gas temperature at 350 °C and sheath gas flow rate at 11 L/min. The data were generated by the Agilent MassHunter qualitative analysis software (version B.06.00).

4.4. Cell Line and Culture Medium

The human HCC cell line HepG2 (#HB-8065, American Type Culture Collections, Manassas, VA, USA) was cultured in DMEM, supplemented with 10% FBS, 100 IU/mL of penicillin, 100 μg/mL of streptomycin and 2 mM of L-glutamine. The cells were maintained at 37 °C in a 5% humidified CO_2 incubator.

4.5. Cell Proliferation Assay

The HepG2 cells (5000/0.1 mL) were seeded in white flat-bottom 96-well plates (Costar®, Thermo Fisher Scientific, Waltham, MA, USA). After 24 h of incubation, the cells were independently treated 3 times in triplicate with various concentrations of *B. leachii* ink concentrate (10, 100, 200, 400, 500 and 1000 μg/mL). The wells containing the culture media and the cells without treatment served as blank and control, respectively. The cell proliferation was measured after 24, 48 and 72 h using the CellTiter-Glo® Luminescent Cell Viability Assay (Promega Corporation Inc., Fitchburg, WI, USA) and we determined the half-maximal inhibitory concentration (IC_{50}) value as described in [37].

4.6. Western Blot Analysis

The HepG2 cells (5×10^5/mL) were seeded in Nunc™ 12-well plates (Thermo Fisher Scientific, Inc.). After 24 h of incubation, the cells were treated in triplicate with *B. leachii* ink concentrate at 100 and 400 μg/mL along with 1 μM STS, used as a positive control. Western blot technology and image analysis were employed as described in [38]. Polyvinylidene difluoride membranes (Millipore, Thermo Fisher Scientific) were probed with (1:1000 dilution) mouse anti-pro/cleaved-caspase-3, rabbit anti-pro-caspase-8 and mouse anti-cleaved-caspase-8 antibodies (Cell Signalling Technology, Danvers, MA, USA) and mouse anti-GAPDH antibody (Abcam, Cambridge, UK).

4.7. Gene Expression Analysis

The HepG2 cells (1.5×10^6) were seeded in Nunc™ 6-well plates. After 24 h of incubation, the cells were treated in triplicate with or without 400 µg/mL of *B. leachii* ink concentrate and incubated for 48 h. From the total RNA ink concentrate ion to reverse-transcribed cDNA, RT-qPCR was performed as described in [39]. The relative quantifications of the mRNA expression level for the target genes are listed in the Table 1.

4.8. Activity Prediction Using PASS Online Webserver

The anti-neoplastic activity of the bioactive metabolites identified from *B. leachii* ink concentrate was assessed using Pass online webserver (Way2Drug, Moscow, Russia Version 2.0) [40]. For each compound, the SMILES (Simplified Molecular Input Line Entry System) was generated and entered in the webserver to perform the assessment. The results were classified based on the compound probability of being active (Pa) and inactive (Pi) for the specified activity.

4.9. Target Predictions Using Molinspiration and SwissTargetPrediction Tools

To investigate the possible molecular targets for these metabolites identified from *B. leachii* ink concentrate, Molinspiration (Molinspiration Cheminformatics, Slovenský Grob, Slovakia) [41] and SwissTargetPrediction (Swiss Institute of Bioinformatics, Lausanne, Switzerland) were used [42]. For both webservers, SMILES was applied to generate the data. The Molinspiration web server produced a score that reflected the bioactivity of the compound. Positive values indicated the highest probability that the compound was active at the molecular target. For the SWISS target predictions, a general mapping of the possible molecular targets was provided for any compound, which facilitated the identification of the biological targets of uncharacterized molecules [39–42].

4.10. Pharmacokinetic ADME Predictions and Cytochrome P450 Profiling Using SWISS Tool

The pharmacokinetics concerning the ADME of the identified bioactive metabolites from *B. leachii* ink concentrate was explored using the SwissADME web server (Swiss Institute of Bioinformatics, Lausanne, Switzerland), which provided detailed, fast, in silico predictions of the pharmaceutical profiles of the bioactive compounds [43]. The selected ADME parameters for the analysis were molecular weight, lipid solubility (Log P), water solubility (Log S), BBB penetration and GI absorption. After data generation, the results were compared with the established drug-likeness properties (rule-of-five, ROF) important for drug discovery [44].

Additional investigations were conducted to assess the CYP inhibition profile of the bioactive molecules using the SWISS web server. Each compound was evaluated against several CYP enzymes, including CYP1A2, CYP2C19, CYP2C9, CYP2D6 and CYP3A4. The CYP enzyme inhibition profile was important for the early identification of possible significant drug interactions.

4.11. Statistical Analysis

All the data are expressed as mean ± SD of three independent experiments. The IC_{50} values were calculated by a nonlinear dose/response regression model using GraphPad Prism software version 6 for Windows (San Diego, CA, USA, http://www.graphpad.com/, accessed on 10 February 2020). The relative changes in the expression of the gene were analysed by the $2^{-\Delta\Delta Ct}$ method [37]. The Student's paired *t*-test was used to calculate the *p* value and the significance was considered if $p < 0.05$.

5. Conclusions

The ink concentrate of *B. leachii* exerts anti-proliferative and pro-apoptotic activities in the human liver cancer HepG2 cell line, suggesting *B. leachii* ink concentrate as a promising, safe, natural-based, neo-adjuvant drug for liver cancer treatment. Our computational predictions for the *B. leachii* ink concentrate-derived identified bioactive molecules suggest that these compounds have promising anti-cancer properties with acceptable drug-likeness profiles and minimal CYP enzyme inhibitions, which warrants further optimization and development to discover novel drug entities from marine-derived natural resources. Additional chemical isolation and in vivo studies are still required.

Author Contributions: Conceptualization, Z.I.A., A.V., R.S.S. and S.S.A.; data curation, M.K., S.R. and M.-H.A.; formal analysis, M.-H.A.; funding acquisition, Z.I.A.; investigation, Z.I.A., A.V. and S.M.-N.; methodology, Z.I.A., M.K., S.A.M., S.R., R.S.S., S.S.A., H.A.-E., B.A. and M.-H.A.; resources, S.R., H.A.-E. and B.A.; software, R.S.S., S.A.M. and B.A.; supervision, Z.I.A., A.V. and S.M.-N.; validation, A.V., M.K., S.A.M., S.R., R.S.S., S.S.A., H.A.-E., B.A., M.-H.A. and S.M.-N.; visualization, M.K., S.A.M., S.R., R.S.S., S.S.A., H.A.-E., B.A., M.-H.A. and S.M.-N.; writing—original draft, A.V. and S.M.-N.; writing—review and editing, Z.I.A., M.K., S.A.M., S.R., R.S.S., S.S.A., H.A.-E., B.A., M.-H.A. and S.M.-N. All authors have read and agreed to the published version of the manuscript.

Funding: All the experimental work was partially funded by King Abdullah International Medical Research Center under grant number RC17/093/R. The funder had no role in study design, data collection and analysis, decision to publish or preparation of the manuscript.

Institutional Review Board Statement: The study was conducted in accordance with the Declaration of Helsinki, and approved by the Institutional Review Board of King Abdullah International Medical Research Center (IRB NCBE Registration no. H-01-R-005, approved on 06/05/2018).

Data Availability Statement: The data presented in this study are available on request from the corresponding author.

Conflicts of Interest: The authors have declared that there are no conflicts of interest.

Abbreviations

ADME	absorption distribution metabolism excretion
ATP	adenosine triphosphate
BBB	blood–brain barrier
cDNA	complementary deoxyribonucleic acid
CO_2	carbon dioxide
DMEM	Dulbecco's Modified Eagles Medium
FBS	foetal bovine serum
GAPDH	glyceraldehyde 3-phosphate dehydrogenase
GI	gastrointestinal
HCC	hepatocellular carcinoma
HIV	human immunodeficiency virus
IC_{50}	half-maximal inhibitory concentration
Log P	lipid solubility
Log S	water solubility
mRNA	messenger ribonucleic acid
QTOF	quadrupole time of flight
ROF	rule-of-five
RT-qPCR	reverse transcription-quantitative polymerase chain reaction
STS	staurosporine

References

1. Sung, H.; Ferlay, J.; Siegel, R.L.; Laversanne, M.; Soerjomataram, I.; Jemal, A.; Bray, F. Global Cancer Statistics 2020: GLOBOCAN Estimates of Incidence and Mortality Worldwide for 36 Cancers in 185 Countries. *CA Cancer J. Clin.* **2021**, *71*, 209–249. [CrossRef] [PubMed]
2. Okeke, E.; Davwar, P.M.; Roberts, L.; Sartorius, K.; Spearman, W.; Malu, A.; Duguru, M. Epidemiology of Liver Cancer in Africa: Current and Future Trends. *Semin. Liver Dis.* **2020**, *40*, 111–123. [CrossRef] [PubMed]
3. Al-Anazi, M.R.; Matou-Nasri, S.; Al-Qahtani, A.A.; Alghamdi, J.; Abdo, A.A.; Sanai, F.M.; Al-Hamoudi, W.K.; Alswat, K.A.; Al-Ashgar, H.I.; Khan, M.Q.; et al. Association between IL-37 gene polymorphisms and risk of HBV-related liver disease in a Saudi Arabian population. *Sci. Rep.* **2019**, *9*, 7123. [CrossRef] [PubMed]
4. Nishida, N. Metabolic disease as a risk of hepatocellular carcinoma. *Clin. Mol. Hepatol.* **2021**, *27*, 87–90. [CrossRef]
5. Liu, C.Y.; Chen, K.F.; Chen, P.J. Treatment of liver cancer. *Cold Spring Harb. Perspect. Med.* **2015**, *5*, a021535. [CrossRef]

6. Shaaban, S.; Negm, A.; Ibrahim, E.E.; Elrazak, A.A. Chemotherapeutic agents for the treatment of hepatocellular carcinoma: Efficacy and mode of action. *Oncol. Rev.* **2014**, *8*, 25–35. [CrossRef]
7. Zhou, Y.; Li, Y.; Zhou, T.; Zheng, J.; Li, S.; Li, H.-B. Dietary Natural Products for Prevention and Treatment of Liver Cancer. *Nutrients* **2016**, *8*, 156. [CrossRef]
8. Yin, B.; Fang, D.-M.; Zhou, X.-L.; Gao, F. Natural products as important tyrosine kinase inhibitors. *Eur. J. Med. Chem.* **2019**, *182*, 111664. [CrossRef]
9. Rajabi, S.; Maresca, M.; Yumashev, A.; Choopani, R.; Hajimehdipoor, H. The Most Competent Plant-Derived Natural Products for Targeting Apoptosis in Cancer Therapy. *Biomolecules* **2021**, *11*, 534. [CrossRef]
10. Szondy, Z.; Sarang, Z.; Kiss, B.; Garabuczi, É.; Köröskényi, K. Anti-inflammatory Mechanisms Triggered by Apoptotic Cells during Their Clearance. *Front. Immunol.* **2017**, *8*, 909. [CrossRef]
11. Strathearn, L.S.; Stepanov, A.I.; Font-Burgada, J. Inflammation in Primary and Metastatic Liver Tumorigenesis–Under the Influence of Alcohol and High-Fat Diets. *Nutrients* **2020**, *12*, 933. [CrossRef]
12. Benkendorff, K. Molluscan biological and chemical diversity: Secondary metabolites and medicinal resources produced by marine molluscs. *Biol. Rev. Camb. Philos. Soc.* **2010**, *85*, 757–775. [CrossRef]
13. Sethi, S.; Kokane, M.R.; Otta, S.K.; Sethi, G. First record of Ragged Sea Hare Bursatella leachii Blainville, 1817 (Opisthobranchia: Euopisthobranchia: Aplysiidae) in Pulicat Lake, east coast of India. *Mar. Biodivers. Rec.* **2015**, *8*, e34. [CrossRef]
14. Zhang, Q.T.; Liu, Z.D.; Wang, Z.; Wang, T.; Wang, N.; Wang, N.; Zhang, B.; Zhao, Y.F. Recent Advances in Small Peptides of Marine Origin in Cancer Therapy. *Mar. Drugs* **2021**, *19*, 115. [CrossRef]
15. Pereira, R.B.; Andrade, P.B.; Valentão, P. Chemical Diversity and Biological Properties of Secondary Metabolites from Sea Hares of Aplysia Genus. *Mar. Drugs* **2016**, *14*, 39. [CrossRef]
16. Pereira, R.B.; Evdokimov, N.M.; Lefranc, F.; Valentão, P.; Kornienko, A.; Pereira, D.M.; Andrade, P.B.; Gomes, N.G.M. Marine-Derived Anticancer Agents: Clinical Benefits, Innovative Mechanisms, and New Targets. *Mar. Drugs* **2019**, *17*, 329. [CrossRef]
17. Rajaganapathi, J.; Kathiresan, K.; Singh, T.P. Purification of Anti-HIV Protein from Purple Fluid of the Sea Hare Bursatella leachii de Blainville. *Mar. Biotechnol.* **2002**, *4*, 447–453. [CrossRef]
18. Kokane, M.R.; Anuradha, V.; Revathi, K. GC MS analysis and in vitro anti-inflammatory activity of purple ink secreted by Bursatella leachii. *Eur. J. Biomed. Pharm. Sci.* **2019**, *6*, 212–216.
19. Suntornchashwej, S.; Chaichit, N.; Isobe, M.; Suwanborirux, K. Hectochlorin and Morpholine Derivatives from the Thai Sea Hare, *Bursatella leachii*. *J. Nat. Prod.* **2005**, *68*, 951–955. [CrossRef]
20. Suntornchashwej, S.; Suwanborirux, K.; Koga, K.; Isobe, M. Malyngamide X: The First (7R)-Lyngbic Acid that Connects to a New Tripeptide Backbone from the Thai Sea Hare *Bursatella leachii*. *Chem. Asian J.* **2007**, *2*, 114–122. [CrossRef]
21. Gopichand, Y.; Schmitz, F.J. Bursatellin: A new diol dinitrile from the sea hare *Bursatella leachii* pleii. *J. Org. Chem.* **1980**, *45*, 5383–5385. [CrossRef]
22. Appleton, D.R.; Sewell, M.A.; Berridge, M.V.; Copp, B.R. A new biologically active malyngamide from a New Zealand collection of the sea hare *Bursatella leach*. *J. Nat. Prod.* **2002**, *65*, 630–631. [CrossRef]
23. Capper, A.; Tibbetts, I.R.; O'Neil, J.M.; Shaw, G.R. The Fate of Lyngbya majuscula Toxins in Three Potential Consumers. *J. Chem. Ecol.* **2005**, *31*, 1595–1606. [CrossRef]
24. Braga, T.; Rodrigues, M.J.; Pereira, H.; Varela, J.; Barreira, L.; González-Wangüemert, M.; Custódio, L. *Bursatella leachiifrom* Mar Menor as a Source of Bioactive Molecules: Preliminary Evaluation of the Nutritional Profile, In Vitro Biological Activities, and Fatty Acids Contents. *J. Aquat. Food Prod. Technol.* **2017**, *26*, 1337–1350. [CrossRef]
25. Guicciardi, M.E.; Gores, G.J. Life and death by death receptors. *FASEB J.* **2009**, *23*, 1625–1637. [CrossRef]
26. Fulda, S.; Debatin, K.M. Extrinsic versus intrinsic apoptosis pathways in anticancer chemotherapy. *Oncogene* **2006**, *25*, 4798–4811. [CrossRef]
27. Mantovani, F.; Collavin, L.; Del Sal, G. Mutant p53 as a guardian of the cancer cell. *Cell Death Differ.* **2019**, *26*, 199–212. [CrossRef]
28. Feroz, W.; Sheikh, A.M.A. Exploring the multiple roles of guardian of the genome: P53. *Egypt. J. Med. Hum. Genet.* **2020**, *21*, 49. [CrossRef]
29. de Groot, R.P.; Raaijmakers, J.A.; Lammers, J.W.J.; Koenderman, L. STAT5-dependent cyclin D1 and Bcl-xL expression in Bcr-Abl-transformed cells. *Mol. Cell Res. Commun.* **2000**, *3*, 299–305. [CrossRef] [PubMed]
30. Ha, J.H.; Shin, J.S.; Yoon, M.K.; Lee, M.S.; He, F.; Bae, K.H.; Yoon, H.S.; Lee, C.K.; Park, S.G.; Muto, Y.; et al. Dual-site Interactions of p53 Protein Transactivation Domain with Anti-apoptotic Bcl-2 Family Proteins Reveal a Highly Convergent Mechanism of Divergent p53 Pathways. *J. Biol. Chem.* **2013**, *288*, 7387–7398. [CrossRef] [PubMed]
31. Teruya, T.; Sasaki, H.; Fukazawa, H.; Suenaga, K. Bisebromoamide, a Potent Cytotoxic Peptide from the Marine Cyanobacterium *Lyngbya* sp.: Isolation, Stereostructure, and Biological Activity. *Org. Lett.* **2009**, *11*, 5062–5065. [CrossRef]
32. Rubio, B.K.; Parrish, S.M.; Yoshida, W.; Schupp, P.J.; Schils, T.; Williams, P.G. Depsipeptides from a Guamanian marine cyanobacterium, *Lyngbya bouillonii*, with selective inhibition of serine proteases. *Tetrahedron Lett.* **2010**, *51*, 6718–6721. [CrossRef]
33. Taori, K.; Paul, V.J.; Luesch, H. Kempopeptins A and B, Serine Protease Inhibitors with Different Selectivity Profiles from a Marine Cyanobacterium, *Lyngbya* sp. *J. Nat. Prod.* **2008**, *71*, 1625–1629. [CrossRef]
34. Matthew, S.; Paul, V.J.; Luesch, H. Largamides A–C, Tiglic Acid-Containing Cyclodepsipeptides with Elastase-Inhibitory Activity from the Marine Cyanobacterium *Lyngbya confervoides*. *Planta Med.* **2009**, *75*, 528–533. [CrossRef]

35. Benet, L.Z.; Hosey, C.M.; Ursu, O.; Oprea, T. BDDCS, the Rule of 5 and drugability. *Adv. Drug Deliv. Rev.* **2016**, *101*, 89–98. [CrossRef]
36. Vennila, R.; Kanchana, S.; Arumugam, M.; Balasubramanian, T. Investigation of antimicrobial and plasma coagulation property of some molluscan ink extracts: Gastropods and Cephalopods. *Afr. J. Biochem. Res.* **2011**, *5*, 14–21.
37. Rameshbabu, S.; Messaoudi, S.A.; Alehaideb, Z.I.; Ali, M.S.; Venktraman, A.; Alajmi, H.; Al-Eidi, H.; Matou-Nasri, S. *Anastatica hierochuntica* (L.) methanolic and aqueous extracts exert antiproliferative effects through the induction of apoptosis in MCF-7 breast cancer cells. *Saudi Pharm. J.* **2020**, *28*, 985–993. [CrossRef]
38. Alehaideb, Z.; AlGhamdi, S.; Bin Yahya, W.; Al-Eidi, H.; Alharbi, M.; Alaujan, M.; Albaz, A.; Tukruni, M.; Nehdi, A.; Abdulla, M.-H.; et al. Anti-proliferative and pro-apoptotic effects of *Calligonum comosum* (L'Her.) methanolic extract in human triple-negative MDA-MB-231 breast cancer cells. *J. Evid. Based Integr. Med.* **2020**, *25*, 2515690X20978391. [CrossRef]
39. Livak, K.J.; Schmittgen, T.D. Analysis of relative gene expression data using real-time quantitative PCR and the 2[-Delta Delta C(t)] method. *Methods* **2001**, *25*, 402–408. [CrossRef]
40. Filimonov, D.; Lagunin, A.A.; Gloriozova, T.A.; Rudik, A.; Druzhilovskii, D.S.; Pogodin, P.V.; Poroikov, V.V. Prediction of the Biological Activity Spectra of Organic Compounds Using the Pass Online Web Resource. *Chem. Heterocycl. Compd.* **2014**, *50*, 444–457. [CrossRef]
41. Molinspiration, Cheminformatics. Calculation of Molecular Properties and Bioactivity Score. Available online: http://www.molinspiration.com/cgi-bin/properties (accessed on 11 November 2021).
42. Gfeller, D.; Grosdidier, A.; Wirth, M.; Daina, A.; Michielin, O.; Zoete, V. SwissTargetPrediction: A web server for target prediction of bioactive small molecules. *Nucleic Acids Res.* **2014**, *42*, W32–W38. [CrossRef]
43. Daina, A.; Michielin, O.; Zoete, V. SwissADME: A free web tool to evaluate pharmacokinetics, drug-likeness and medicinal chemistry friendliness of small molecules. *Sci. Rep.* **2017**, *7*, 42717. [CrossRef]
44. Lipinski, C.A. Lead- and drug-like compounds: The rule-of-five revolution. *Drug Discov. Today Technol.* **2004**, *1*, 337–341. [CrossRef]

molecules

Review

Phenylethyl Isothiocyanate: A Bioactive Agent for Gastrointestinal Health

Ezequiel R. Coscueta [1,*], Ana Sofia Sousa [1], Celso A. Reis [2,3,4] and Maria Manuela Pintado [1]

[1] CBQF—Centro de Biotecnologia e Química Fina-Laboratório Associado, Escola Superior de Biotecnologia, Universidade Católica Portuguesa, Rua Diogo Botelho 1327, 4169-005 Porto, Portugal; assousa@ucp.pt (A.S.S.); mpintado@ucp.pt (M.M.P.)
[2] i3S—Instituto de Investigação e Inovação em Saúde, University of Porto, 4169-005 Porto, Portugal; celsor@ipatimup.pt
[3] IPATIMUP—Institute of Molecular Pathology and Immunology, University of Porto, 4169-005 Porto, Portugal
[4] Medical Faculty, University of Porto, Al. Prof. Hernâni Monteiro, 4169-005 Porto, Portugal
* Correspondence: ecoscueta@ucp.pt; Tel.: +351-225-580-001 (ext. 8047)

Abstract: The incidence of gastrointestinal pathologies (cancer in particular) has increased progressively, with considerable morbidity and mortality, and a high economic impact on the healthcare system. The dietary intake of natural phytochemicals with certain bioactive properties has shown therapeutic and preventive effects on these pathologies. This includes the cruciferous vegetable derivative phenylethyl isothiocyanate (PEITC), a bioactive compound present in some vegetables, such as watercress. Notably, PEITC has antioxidant, anti-inflammatory, bactericidal, and anticarcinogenic properties. This review summarized the current knowledge on the role of PEITC as a potential natural nutraceutical or an adjuvant against oxidative/inflammatory-related disorders in the gastrointestinal tract. We also discussed the safe and recommended dose of PEITC. In addition, we established a framework to guide the research and development of sustainable methodologies for obtaining and stabilizing this natural molecule for industrial use. With PEITC, there is great potential to develop a viable strategy for preventing cancer and other associated diseases of the gastrointestinal tract. However, this topic still needs more scientific studies to help develop new PEITC products for the nutraceutical, pharmaceutical, or food industries.

Keywords: gastrointestinal health; natural anti-inflammatory; natural antioxidant; natural anti-cancer; nutraceutical; watercress

Citation: Coscueta, E.R.; Sousa, A.S.; Reis, C.A.; Pintado, M.M. Phenylethyl Isothiocyanate: A Bioactive Agent for Gastrointestinal Health. *Molecules* **2022**, *27*, 794. https://doi.org/10.3390/molecules27030794

Academic Editor: George Grant

Received: 8 January 2022
Accepted: 24 January 2022
Published: 25 January 2022

1. Introduction

Over time, gastrointestinal tract diseases, or disorders, have increased in incidence and prevalence, becoming a huge problem for society, with millions of people affected. Gastrointestinal disorders also have a high economic impact and lead to decreased life quality [1–3]. Among gastrointestinal tract diseases, the burden of inflammatory bowel disease (IBD) is increasing worldwide. In 2015, in the United States of America, about 3.1 million people were diagnosed with IBD [4]. This disorder is characterized by the non-infectious chronic inflammation of the mucosa, and conditions such as fatigue and weight loss [5]. IBD includes various illnesses, namely, Crohn's disease and ulcerative colitis, which may later progress to small intestine cancer and colorectal cancer, respectively [6]. In the last few decades, we have witnessed significant progress in cancer research, namely, in its understanding, prevention, and treatment. According to the World Health Organization (WHO), in 2018, cancer was the second principal cause of death globally, with one in six deaths being caused by this pathology [7]. In fact, in a WHO statistical analysis, colorectal cancer was found to be one of the deadliest, corresponding, annually, to 940 thousand deaths, with a mortality rate above 50%.

The WHO estimates that about 30% to 50% of cancers can be prevented by changing or avoiding the main risk factors. Thus, cancer prevention represents an ideal strategy to reduce the burden of this disease, while also offering a more economical treatment strategy in the long term for cancer control [7]. Epidemiological studies established a positive correlation between the increased consumption of vegetables and a reduced risk of chronic degenerative diseases. This was attributed to the wide range of phytochemicals with important physiological properties [8,9]. Among the most promising compounds, phytochemicals from cruciferous vegetables stand out. As a clear example, a compound that is awakening a significant interest is phenylethyl or phenethyl isothiocyanate (PEITC), which, within what has been studied so far, exhibits interesting potential bioactivities (antioxidant, anti-inflammatory, anti-cancer). Indeed, this isothiocyanate is the target of this review article.

Here, we briefly review the current state of knowledge on the implications of PEITC for cancer prevention in the gastrointestinal tract. Simultaneously, we propose a framework to guide the research and development of sustainable nutraceutical solutions for prophylactic action on this global health system's priority problem. Figure 1 provides a general outline of the ideas put forward for this article.

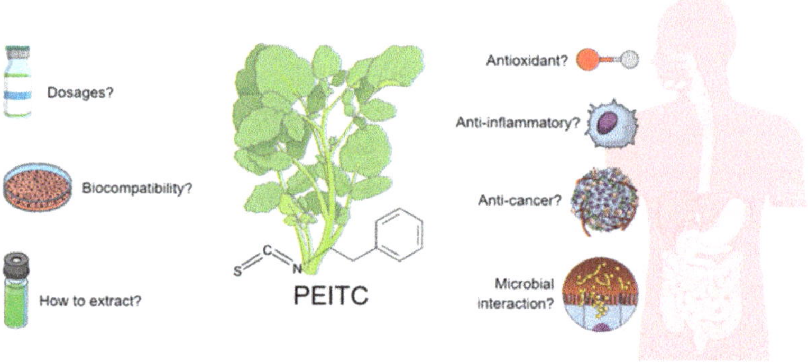

Figure 1. Outline of the general ideas for the literature review.

2. What Are Isothiocyanates and Their Natural Sources?

Isothiocyanates are the most abundant products of the natural enzymatic degradation of glucosinolates by the enzyme myrosinase [10,11]. These phytochemical molecules come from cruciferous vegetables (the *Brassicaceae* family), including plants such as watercress, cauliflower, broccoli, and brussels sprouts, among others [11]. Isothiocyanates lead to the characteristic spicy and bitter taste experienced when consuming these plants [12]. To date, a significant number of observational and intervention studies in humans have evaluated the benefits and safety of cruciferous vegetables and isothiocyanate intake [13]. Glucosinolates and their isothiocyanates are compounds proposed to be important contributors to the health benefits of these vegetables, with the anti-cancer effect being one of the main research focuses [13,14]. Phytochemicals from cruciferous plants protect against cancer by modulating the metabolism of carcinogens [13].

The metabolism of glucosinolates can occur with two different types of the stated enzyme: plant myrosinase, which coexists separately with glucosinolates in plants, being activated when the plant is damaged; and bacterial myrosinase, which acts mainly in the colon (comes from gut microbiota) [10]. This metabolism of glucosinolates by gut microbiota can occur when plant myrosinase is denatured. Indeed, the thermal inactivation of plant myrosinase can result in the preservation of some glucosinolates. This is particularly interesting in cooked cruciferous vegetables, as the glucosinolates, when ingested, can be partially absorbed in the stomach, and the remaining intact glucosinolates transit to the colon (due to their hydrophilic nature), where they can be extensively hydrolyzed by the intestinal microbiota and absorbed and/or excreted as isothiocyanates [10,13,14].

Glucosinolates are inert, anionic organic compounds, characterized by β-D-thioglucose, a sulfated oxime group ($C{=}NOSO_3^-$), and a variable side chain (-R), which will classify the glucosinolate as aliphatic, aromatic, or indole [15]. In response to mechanical or chemical stress, the glycosidic bond of the β-D-thioglucose present in glucosinolates is broken by the enzyme myrosinase, after which this compound converts to isothiocyanates, thiocyanates, and nitriles [16]. The amount of each product formed in this reaction can vary according to the specific proteins, pH, and/or temperature [12,16]. In Figure 2a, the described process is outlined.

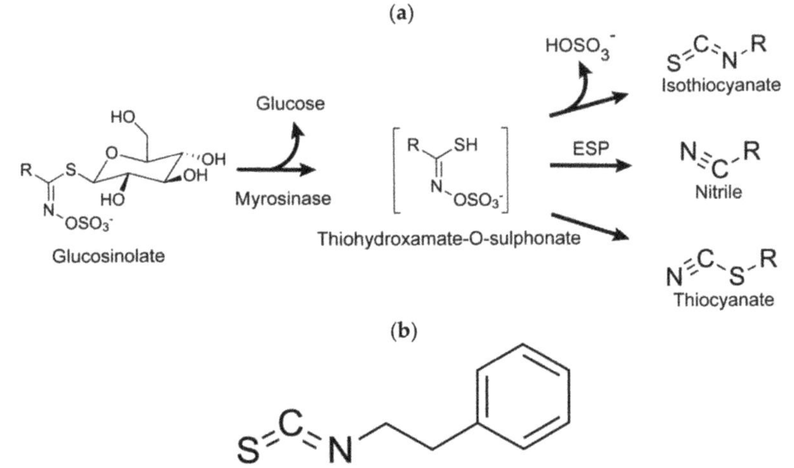

Figure 2. Glucosinolate hydrolysis and PEITC. Glucosinolate reaction catalyzed by the enzyme myrosinase (**a**). Phenylethyl isothiocyanate (**b**).

Regarding the chemical structure of isothiocyanates, a highly electrophilic carbon establishes two double bonds, one with sulfur and the other with nitrogen, and the radical is linked to the atom of nitrogen. This radical dictates the chemical properties of the active compound [11,17].

The production of isothiocyanates may vary depending on the conditions to which the plant is subjected, namely, the temperature and pH at which the reaction occurs, as well as the availability of ferrous ions and specifying proteins in the medium. Other factors will also determine the reaction's course and its final metabolites, such as plant species and age, place of cultivation, climatic conditions, storage, and processing [15].

3. Phenylethyl Isothiocyanate

Watercress (*Nasturtium officinale*) is a very accessible garden vegetable highly rich in glucosinolates [18]. The most characteristic glucosinolate is gluconasturtiin, an aromatic glucosinolate with an ethyl chain linked to benzene in its radical [19]. PEITC (Figure 2b) results from the hydrolysis of gluconasturtiin by the action of myrosinase. The final product is an isothiocyanate with a phenylethyl radical attached to a nitrogen atom [20].

PEITC is a bioactive compound involved in several biological mechanisms that are naturally related to protecting the plant against external factors. The plant produces PEITC in response to specific stress situations, since it presents biocidal activity against various pathogens, such as bacteria, fungi, insects, and other biotic stressors [11,21]. However, the physiological properties of PEITC are not limited to those exerted at the source of origin. In this way, PEITC can act in humans, combining a series of biological properties with antioxidant, anti-inflammatory, and anti-cancer action. PEITC's activity on the organism is justified by different bioactive mechanisms, namely, the generation of free radicals, reducing inflammation, and blocking the stages of carcinogenesis [22,23]. PEITC is also known to inhibit cell proliferation, stop the cell cycle, reduce the expression of carcinogenesis, or even tumor suppression via apoptosis and autophagy induction [15,24,25]. Since 2000, PEITC has been one of the main pure glucosinolate derivatives (9.1%) used in clinical trials, particularly to study its anti-cancer effects [10,26].

3.1. Antioxidant Action

Oxidative stress is an imbalance between antioxidant and oxidant species, in which the latter prevails over the others, causing an increase in the amount of reactive oxygen species (ROS) in the cell. When ROS concentration is high, damage may occur at the DNA level, increasing the carcinogenesis probability [15]. These ROS come from endogenous sources, such as mitochondrial reactions and cellular inflammatory mechanisms, and exogenous sources, such as exposure to UV radiation and electrophilic molecules [27]. ROS is related to the pathogenesis of diverse gastrointestinal diseases, including gastroesophageal reflux disease, gastritis, enteritis, colitis, associated cancers, pancreatitis, and liver cirrhosis [28].

As a chemo-preventive and antioxidant agent, PEITC may modulate the unregulated ROS concentration in cells, activating antioxidant defense mechanisms through the increased expression of detox enzymes, to lower ROS to basal levels [21,29]. Likewise, PEITC may act as a "selective" antagonist compound for tumor cells, since it acts simultaneously as an oxidizer, inducing ROS production and oxidative damage in tumor cells [17].

3.2. Anti-Inflammatory Action

In the gastrointestinal tract, cancer development is closely dependent on energy intake and nutrient availability, but it is also characterized by low-grade inflammation (a slight but chronic increase in the number of various inflammatory markers in the blood and organs) [30]. Inflammation is an immune response caused by several factors, such as infections or tissue damage, which can be subdivided into acute and chronic. Chronic inflammation, which results from an imbalance between pro- and anti-inflammatory cytokines, is closely associated with the pathogenesis of cancer diseases [31].

The excessive expression of a pro-inflammatory factor results in the damage of the epithelial barrier, initiating apoptosis of epithelial cells and the secretion of chemokine. Several researchers have demonstrated the anti-inflammatory properties of PEITC through the reduced expression of this protein by inhibiting NF-κB expression [17,32–34]. Pikarsky et al. (2004) and Greten et al. (2004) unequivocally demonstrated that NF-κB plays an essential role in developing liver and intestinal carcinogenesis, respectively [35,36]. In this sense, the prevention of NF-κB activation in hepatocytes was sufficient to inhibit the development of cancers in the livers of mice that were exposed to chronic liver inflammation for seven months [35]. On the other hand, classical colitis-induced carcinogenesis was abolished in mice when targeted to the NF-κB pathway [36]. Therefore, by preventing the activation of NF-κB, PEITC would inhibit cell proliferation and differentiation, and promote apoptosis, leading to cancer prevention.

3.3. Anti-Cancer Action

The consumption of the *Brassicaceae* family's cruciferous vegetables is associated with human health benefits, such as the reduced risk of chronic diseases and several types of cancer, including gastric and colon cancers [11,37]. Based on the literature available so far, PEITC exhibited its anti-cancer effects by inhibiting cell proliferation through cell cycle arrest and tumor cell apoptosis, as well as by resisting metastasis [15,23,24]. Moreover, cancer prevention comes from the effects described above (antioxidant and anti-inflammatory), since its action reduces the risk of developing these pathologies associated with cancer.

Carcinogens are subjected to metabolism and elimination, mainly by phase I and phase II biotransformation enzymes. In general, carcinogenesis occurs due to the bioactivation of carcinogens by phase I enzymes. Hence, phase I metabolism products are highly reactive intermediates that can be harmful by binding to critical macromolecules, such as DNA. In contrast, phase II enzymes play an essential role in the detoxification and excretion of carcinogens from the body. PEITC is involved in the inhibition of phase I enzymes and the induction of phase II enzymes, especially CYP enzymes and transferases, which may explain the chemo-preventive activity [38].

Moreover, it is known that cancer risk can be modified by the dietary intake of bioactive phytochemicals, such as PEITC, which can have epigenetic effects through the modulation of DNA/histone modification. However, currently, studies about the epigenetic mechanisms of these anti-tumor effects in human cells are so far very limited. Indeed, the future development of effective dietary regimes for cancer prevention will require a better understanding of the ingested phytochemicals' effects on DNA methylation, histone modification, and/or chromatin remodeling in developing tumors [23]. Concerning this topic, Park et al. (2017) reported that PEITC could exhibit chemo-preventive effects and inhibit colorectal cancer progression by inducing pro-apoptotic genes in tumor cells. In particular, PEITC induced stable changes in the tumor cell expression of epigenetic writers/erasers, the chromatin-binding of histone deacetylases, and the hypomethylation of the Polycomb group proteins, as well as other genes that are usually methylated in cancer, which contributed to restricting tumor development [23]. Furthermore, PEITC can be used as an adjunct to increase the potential for other cancer treatments. Giallourou et al. (2019) reported that the compounds extracted from watercress and PEITC improved the therapeutic results of radiotherapy, increasing the DNA damage caused by radiation in cancer cells and protecting non-tumorigenic cells from collateral damage [39].

Despite the encouragement of the foregoing, what is known so far is based on purely empirical studies conducted in controlled laboratory environments. According to the vague epidemiological studies carried out so far, the anticancer properties of PEITC are still under investigation. Up to date, just inverse proportionality correlations have been established between the consumption of PEITC-containing vegetables and the risk of chronic diseases, which makes clear the need to expand the study focus to more in-depth epidemiological studies to have a more complete understanding of the effect of PEITC consumption on cancer relations [15].

3.4. Microbial Interaction

Although the antimicrobial capacity of isothiocyanates to control the proliferation of plant and foodborne pathogens has been well documented, currently, there are still few studies that assess the action of these compounds to combat human infections. Furthermore, the information that exists mainly refers to the in vitro antimicrobial activity against bacterial pathogens, and little is known about the in vivo antimicrobial effects of isothiocyanates [14]. In this sense, there is little information about the interaction of PEITC with the microorganisms associated with human gastrointestinal infections and/or commensal flora.

PEITC may act as a potential bactericidal compound against some bacterial pathogens responsible for gastrointestinal infections through the disruption of the plasma membrane, the dysregulation of the enzymatic machinery, and cell death [40]. It is also able to reduce inflammation and inhibit urease activity from *Helicobacter pylori*, blocking its carcinogenic effects in the stomach [11]. These PEITC effects are very important because *H. pylori* infection is difficult to eradicate, and it is a major cause of gastritis and peptic ulcers, a condition that may be associated with the development of gastric cancer [14]. Concerning the antibacterial activities of PEITC against harmful intestinal bacteria, this phytochemical can strongly inhibit the growth of *Clostridium difficile* and *Clostridium perfringens*, two pathogenic agents of the genus *Clostridium* that can threaten human health [14,41]. Thus, it would be interesting to develop further studies to describe the potential clinical efficacy of PEITC as a therapeutic or preventive agent for the treatment of diseases caused by harmful gastrointestinal bacteria.

Furthermore, it is of particular interest to understand its interaction with the intestinal microbiota. Isothiocyanates, as products of myrosinase—based glucosinolate hydrolysis in the human gut, are important to health, particularly their anti-cancer properties and other beneficial roles in human health mentioned above [42]. Even though this is an emerging topic, the research studies on the impact of glucosinolates and their isothiocyanates on gut microbiota are still very scarce. However, a close relationship between the consumption of glucosinolates, their metabolism, and the intestinal microbiota composition has been suggested [10]. Indeed, in vitro studies have demonstrated the potential of *Bifidobacterium* sp., one of the common bacteria belonging to the human intestinal microflora, in the hydrolysis of glucosinolates [10,43]. Kellingray et al. (2017) examined whether a *Brassica*-rich diet in healthy adults was associated with changes in the gut microbiota composition. The study concluded that a diet rich in *Brassica* did not significantly alter the relative proportions of intestinal lactobacilli, but was associated with a reduction in the relative abundance of SRB [44]. Further studies about bacterial strains involved in the degradation of glucosinolates in the colon, the characterization of degradation products (particularly isothiocyanates), and their physiological effects on the intestinal microbiota are needed to understand the modulation of the gut microbiota by the metabolites of the cruciferous vegetables [43,44].

According to Kim and Lee (2009), PEITC showed not to have an antimicrobial action against commensal bacteria (*Bifidobacterium bifidum, Bifidobacterium breve, Bifidobacterium longum, Lactobacillus acidophilus,* and *Lactobacillus casei*), and thus did not negatively affect the intestinal microbiota [41]. This opens potential opportunities to explore whether these compounds can also contribute to health by interacting with the gut microbiota through several pathways, such as prebiotics.

In addition, PEITC exhibits strong antimicrobial potential against pathogens that compromise food safety, an essential public health issue that continues to be a significant concern for consumers, regulators, and food industries worldwide [45]. PEITC can interact with cell surface constitutes and consequently compromise the integrity of the cytoplasmatic membrane of bacteria, causing foodborne diseases such as *Escherichia coli, Listeria monocytogenes, Pseudomonas aeruginosa, Staphylococcus aureus,* and *Vibrio parahaemolyticus* [45–47].

4. Biocompatibility of PEITC

Currently, clinical trials have been carried out to prove the chemo-preventive properties of PEITC, for which it was necessary to analyze the toxicity in humans in order to establish limit values. In a study conducted in 2018, a dose of 40 to 80 mg of PEITC per day was administered to humans orally for 30 days, with no adverse effects [29]. However, when the dose consumed was increased to values between 120 and 160 mg of PEITC per day, during the same experimental period, some toxicity was observed, although it was not lethal [29]. Other authors concluded that the acceptable daily dose is only 40 mg in humans, which is consistent with the study previously presented [48]. Thus, there is still no definitive range of concentrations for levels of toxicity in humans. However, it is known that after specific doses, PEITC can be considered to be toxic and interact with other drugs that are being taken simultaneously. According to Abbaoui et al. (2018), for therapeutic effects, non-toxic doses of PEITC are sufficient, being safe for human consumption [16].

5. Dosages of PEITC

Knowing the dosages of a product is essential for its formulation, since it is essential to know which quantities should be ingested for a given purpose. In this article, we analyzed two aspects of the application in which the dose–effect is different, leading, on the one hand, to the prevention of pathologies that culminate in more severe problems, such as cancer, and, on the other hand, to the therapeutic effect of PEITC. To assess the best way forward, PEITC dosages for different cancer types were tested out on cell lines. Thus, the effects caused by tumor cells when exposed to different concentrations of PEITC were studied. A range of PEITC performance values was also established, from the more preventive to the most curative phase, ranging from 5 to 30 μM, corresponding to 0.82 to 4.90 mg L^{-1}. This information is gathered in Table 1.

Table 1. PEITC doses for different types of cancer and their mechanisms of action.

	PEITC [1]	PEITC [2]	Effect	References
Colon cancer	10.0	1.63	Attenuation of inflammation and cell proliferation	[17]
	10.0–40.0	1.63–6.53	Suppression of cell proliferation and loss of viability of tumor cells	[49]
			Apoptosis and anti-inflammatory action	[50]
	10.0	1.63	Tumor regression	[38]
	2.5–15.0	0.4–2.45	Inhibition of proliferation	
	1.0–5.0	0.16–0.82	Apoptosis	
	10.0	1.63	Anti-inflammatory action	
Gastric cancer	1.5	0.24	Apoptosis	
Cervical cancer	5.0–10.0	0.82–1.63	Cell proliferation inhibition and apoptosis induction	[24]
	15.0	2.45	Apoptosis	[51]
Breast cancer	20.0–30.0	3.26–4.90	Inhibition of cell proliferation and cell cycle arrest	[39]
Prostate cancer	5.0–7.5	0.82–1.22	Decreased expression of the NF-kB factor (anti-inflammatory action)	[50]
Lung cancer	12.5–20.0	2.04–3.26	Cell cycle arrest and apoptosis	
Laryngeal carcinoma	0.0–10.0	0.00–1.63	Inhibition of cell growth, cell cycle arrest, and apoptosis	
Leukemia	4.0	0.65	Beginning of apoptosis	[52]
	6.0–8.0	0.98–1.31	Significant increase in apoptosis	

[1] Values in μM. [2] Values in mg L^{-1}.

Thus, we can establish that an effective dose for prevention corresponds to a concentration of PEITC between 5 and 10 μM, equivalent to between 0.82 and 1.63 mg L^{-1}, since these dosages cause an anti-inflammatory action and the inhibition of cell proliferation. For PEITC to have a more therapeutic effect, the dosage must be higher. A concentration between 10 and 30 μM, which corresponds to between 1.63 and 4.90 mg L^{-1}, would be the most suitable, since there is an induction of apoptosis and a loss of viability for tumor cells.

As we have already established, in this article, the analysis of the state of the art on prophylactic and therapeutic properties of PEITC is focused on the gastrointestinal tract. In this sense, the most information on the chemo-active properties of PEITC comes from the study of colon cancer, being almost nil for gastric cancer. However, most of the studies are only at exploratory stages in vitro, which clarifies the need for progress to other phases to validate the findings.

6. PEITC Extraction

We have already discussed what is known so far about the potential properties of PEITC to prevent gastrointestinal disorders. PEITC is also a natural and sustainable compound, since it can be extracted from the by-products of watercress, a raw material that is not crucial to human needs [53]. This reasoning guarantees the preservation of ecosystems, allowing the label of "environmentally friendly" to be affixed. Additionally, it has a very high dose–effect relationship, since for the PEITC to have a preventive action, the necessary dose can be obtained through a tiny amount of watercress, which can be considered to be an economically viable process.

So far, the reported work on obtaining PEITC is scarce. Furthermore, few studies apply a sustainable approach [53–56]. The applied methodologies, in general, are analytical, and use polluting organic solvents that are not feasible at an industrial level, even more so when considering their use in the gastrointestinal tract. The few works with sustainable methodologies apply more complex and expensive techniques and are not easily scalable, e.g., the use of microwave-assisted ethanol extraction or supercritical fluids [37,54,57]. In Table 2, works with different methodologies and their advantages/disadvantages are reported. Through the analysis of Table 2, it appears that the extraction with aqueous micellar systems with non-ionic and biodegradable surfactants is the most advantageous. With this extraction method, the final product has no additives, does not contain toxic products, and is extracted sustainably. Likewise, it is a technique that needs to be further explored and optimized to increase its yield, with, for example, its incorporation in the process of external myrosinase to increase the conversion of glucosinolates, as was performed with the other methodologies. Simultaneously, by observing Table 2, it is possible to establish a range of values for PEITC in mg per 100 g of fresh watercress, between 10.5 and 68.8 mg.

Table 2. Advantages and disadvantages of the PEITC extraction methods.

Extraction Method.	Amount of Extracted PEITC [1]	Advantages	Disadvantages	References
Aqueous micellar systems with autolysis	10.5–14.0	It does not involve toxic solvents; reduced cost; sustainable; stabilized PEITC; "clean label" product	Depends on the amount of endogenous myrosinase present in the watercress	[53]
Organic solvent	23.3–68.8	Direct and ready-to-use technique; reduced costs	Toxic organic solvents; addition of external myrosinase; loss of active compound through filtration and evaporation	[57–59]
Pressurized fluid	33.5	A higher amount of extracted PEITC; does not involve toxic solvents; preserves the bioactivity of the compound	The use of high pressures; requires more sophisticated equipment	[37,57]

[1] Values in mg PEITC 100 g^{-1} fresh watercress.

7. Conclusions and Future Perspectives

In summary, PEITC, a product of glucosinolate hydrolysis found in cruciferous vegetables, has been extensively studied for its preventive and therapeutic effects in chronic diseases due to its antioxidant, anti-inflammatory, and anti-cancer properties. Moreover, PEITC can exhibit antibacterial activity against harmful bacteria in the gastrointestinal tract. Therefore, studying this compound as a potential natural antimicrobial agent against human infections might be interesting. Moreover, this topic is a promising area of study, especially considering the need to develop new antibacterial products, since drug-resistant infections are a significant threat to people's health [14]. Another emerging topic also includes the effects of PEITC on gut microbiota interactions, in line with the growing study of the gut–brain axis, which different bioactive compounds can modulate in different ways. This opens an opportunity for further investigation, as there is still a lot of lack of information on the subject.

The bioactive properties of PEITC are still under investigation, which creates a need for clinical studies to prove the safety and effectiveness of PEITC in humans. After validating its biological properties in vivo, PEITC could achieve promising integration in the pharmaceutical industry. Thus, the use of PEITC as an adjunct to existing medication for cancer treatment would reduce the amounts of drug administered, thus reducing its side effects [60]. Even so, for gastrointestinal disorders PEITC performance is underexplored. Its in-depth exploration would help the development of new products for the pharmaceutical and food industries with their nutraceutical lines.

Even though PEITC has great potential as a health-promoting compound, its industrial use has been limited because of its relative instability [56]. PEITC is a highly reactive electrophile, susceptible to attack by nucleophilic molecules [61]. Furthermore, PEITC is a compound with low molecular weight (MW = 163.2 g mol^{-1}) and considerable hydrophobicity (logP = 3.47). Its pharmacokinetic features include first-order linear absorption with a high protein binding nature [62]. Therefore, its stabilization becomes a technological challenge. An option to stabilize PEITC, and even increase its bioavailability in a food matrix, is micro/nanoencapsulation. However, PEITC and ITCs micro/nanoencapsulation has been poorly studied [63]. Till now, cyclodextrin and chitosan microparticles were reported to be plausible carriers for isothiocyanates [64,65]. Besides, PEITC was already stabilized with vegetable oils that protect non-polar isothiocyanates from decomposition or volatilization [56]. This opens a relevant research line aiming to identify the process conditions that could be used at an industrial level and explore or design different food and pharmaceutical matrices in which it can be incorporated. That is why the greatest opportunities for progress in the field are found in the search for alternatives, and the optimal stability of PEITC through different strategies for its use in the most diverse products remains possible. It is important to note that, in the pharmaceutical market, PEITC-based products are non-existent, while in the nutraceutical market, we hardly find low-purity watercress extracts. Therefore, after filling these gaps that still exist, an innovative strategy to respond to market needs would be the development of PEITC-based products associated with its biological properties as a preventive agent or as an adjuvant to existing treatments for cancer.

Author Contributions: Conceptualization, E.R.C.; methodology, E.R.C.; validation, E.R.C., C.A.R. and M.M.P.; formal analysis, E.R.C. and A.S.S.; investigation, E.R.C. and A.S.S.; resources, M.M.P.; data curation, E.R.C. and A.S.S.; writing—original draft preparation, E.R.C. and A.S.S.; writing—review and editing, E.R.C., C.A.R. and M.M.P.; supervision, M.M.P.; project administration, C.A.R. and M.M.P.; funding acquisition, C.A.R. and M.M.P. All authors have read and agreed to the published version of the manuscript.

Funding: This research was funded by the Foundation for Science and Technology (FCT) and the Competitiveness and Internationalization Operational Program, grant number 032094 "GastroCure–Bioactive Soybean and Cruciferous extracts towards application in gastrointestinal disorders: development, characterization and delivery".

Conflicts of Interest: The authors declare no conflict of interest.

References

1. O'Morain, N.; O'Morain, C. The burden of digestive disease across Europe: Facts and policies. *Dig. Liver Dis.* **2019**, *51*, 1–3. [CrossRef] [PubMed]
2. Ng, S.C.; Shi, H.Y.; Hamidi, N.; Underwood, F.E.; Tang, W.; Benchimol, E.I.; Panaccione, R.; Ghosh, S.; Wu, J.C.Y.; Chan, F.K.L.; et al. Worldwide incidence and prevalence of inflammatory bowel disease in the 21st century: A systematic review of population-based studies. *Lancet* **2017**, *390*, 2769–2778. [CrossRef]
3. Kudelka, M.R.; Stowell, S.R.; Cummings, R.D.; Neish, A.S. Intestinal epithelial glycosylation in homeostasis and gut microbiota interactions in IBD. *Nat. Rev. Gastroenterol. Hepatol.* **2020**, *17*, 597–617. [CrossRef] [PubMed]
4. Dahlhamer, J.M.; Zammitti, E.P.; Ward, B.W.; Wheaton, A.G.; Croft, J.B. Prevalence of Inflammatory Bowel Disease Among Adults Aged ≥18 Years–United States, 2015. *MMWR. Morb. Mortal. Wkly. Rep.* **2016**, *65*, 1166–1169. [CrossRef] [PubMed]
5. Fujii, T.; Watanabe, M. Definition and epidemiology of inflammatory bowel disease. *Nihon Rinsho* **2017**, *75*, 357–363.
6. Lucas López, R.; Grande Burgos, M.J.; Gálvez, A.; Pérez Pulido, R. The human gastrointestinal tract and oral microbiota in inflammatory bowel disease: A state of the science review. *APMIS* **2017**, *125*, 3–10. [CrossRef] [PubMed]
7. World Health Organization Cancer. Available online: https://www.who.int/health-topics/cancer#tab=tab_1 (accessed on 8 January 2022).
8. Boeing, H.; Bechthold, A.; Bub, A.; Ellinger, S.; Haller, D.; Kroke, A.; Leschik-Bonnet, E.; Müller, M.J.; Oberritter, H.; Schulze, M.; et al. Critical review: Vegetables and fruit in the prevention of chronic diseases. *Eur. J. Nutr.* **2012**, *51*, 637–663. [CrossRef]
9. Pinela, J.; Carvalho, A.M.; Ferreira, I.C.F.R. Watercress. In *Nutritional Composition and Antioxidant Properties of Fruits and Vegetables*; Elsevier: Amsterdam, The Netherlands, 2020; pp. 197–219; ISBN 9780128127803.
10. Marino, M.; Martini, D.; Venturi, S.; Tucci, M.; Porrini, M.; Riso, P.; Del Bo', C. An Overview of Registered Clinical Trials on Glucosinolates and Human Health: The Current Situation. *Front. Nutr.* **2021**, *8*, 730906. [CrossRef]
11. Palliyaguru, D.L.; Yuan, J.M.; Kensler, T.W.; Fahey, J.W. Isothiocyanates: Translating the Power of Plants to People. *Mol. Nutr. Food Res.* **2018**, *62*, 1700965. [CrossRef]
12. Bell, L.; Oloyede, O.O.; Lignou, S.; Wagstaff, C.; Methven, L. Taste and Flavor Perceptions of Glucosinolates, Isothiocyanates, and Related Compounds. *Mol. Nutr. Food Res.* **2018**, *62*, 1700990. [CrossRef]

13. Connolly, E.L.; Sim, M.; Travica, N.; Marx, W.; Beasy, G.; Lynch, G.S.; Bondonno, C.P.; Lewis, J.R.; Hodgson, J.M.; Blekkenhorst, L.C. Glucosinolates from Cruciferous Vegetables and Their Potential Role in Chronic Disease: Investigating the Preclinical and Clinical Evidence. *Front. Pharmacol.* **2021**, *12*, 767975. [CrossRef]
14. Romeo, L.; Iori, R.; Rollin, P.; Bramanti, P.; Mazzon, E. Isothiocyanates: An Overview of Their Antimicrobial Activity against Human Infections. *Molecules* **2018**, *23*, 624. [CrossRef]
15. Mitsiogianni, M.; Koutsidis, G.; Mavroudis, N.; Trafalis, D.T.; Botaitis, S.; Franco, R.; Zoumpourlis, V.; Amery, T.; Galanis, A.; Pappa, A.; et al. The Role of Isothiocyanates as Cancer Chemo-Preventive, Chemo-Therapeutic and Anti-Melanoma Agents. *Antioxidants* **2019**, *8*, 106. [CrossRef]
16. Abbaoui, B.; Lucas, C.R.; Riedl, K.M.; Clinton, S.K.; Mortazavi, A. Cruciferous Vegetables, Isothiocyanates and Bladder Cancer Prevention. *Mol. Nutr. Food Res.* **2018**, *62*, 1800079. [CrossRef]
17. Soundararajan, P.; Kim, J. Anti-Carcinogenic Glucosinolates in Cruciferous Vegetables and Their Antagonistic Effects on Prevention of Cancers. *Molecules* **2018**, *23*, 2983. [CrossRef]
18. Rubin, E.; Aziz, Z.A.; Surugau, N. Glucosinolates content in non-elicited plant culture, elicited plant culture and wild plant of watercress (Nasturtium officinale). *Trans. Sci. Technol.* **2018**, *5*, 40–45.
19. Farhana, N.; Aripin, B.; Surugau, N. Effects of Temperature and pH on Myrosinase Activity and Gluconasturtiin Hydrolysis Products in Watercress. *Trans. Sci. Technol.* **2016**, *3*, 449–454.
20. National Center for Biotechnology Information. PubChem Compound Summary for CID 16741, Phenethyl Isothiocyanate. 2020. Available online: https://pubchem.ncbi.nlm.nih.gov/compound/Phenethyl-isothiocyanate (accessed on 8 January 2022).
21. Yaqoob, M.; Aggarwal, P.; Kumar, M.; Purandare, N. Isothiocyanates; sources, physiological functions. *Plant Arch.* **2020**, *20*, 2758–2763.
22. Sundaram, M.K.; Preetha, R.; Haque, S.; Akhter, N.; Khan, S.; Ahmed, S.; Hussain, A. Dietary isothiocyanates inhibit cancer progression by modulation of epigenome. *Semin. Cancer Biol.* **2021**. [CrossRef]
23. Park, J.E.; Sun, Y.; Lim, S.K.; Tam, J.P.; Dekker, M.; Chen, H.; Sze, S.K. Dietary phytochemical PEITC restricts tumor development via modulation of epigenetic writers and erasers. *Sci. Rep.* **2017**, *7*, 1–13. [CrossRef]
24. Dai, M.; Wang, Y.; Chen, C.; Li, F.; Xiao, B.; Chen, S.; Tao, Z. Phenethyl isothiocyanate induces apoptosis and inhibits cell proliferation and invasion in Hep-2 laryngeal cancer cells. *Oncol. Rep.* **2016**, *35*, 2657–2664. [CrossRef] [PubMed]
25. Sharma, A.; Sharma, A.; Yadav, P.; Singh, D. Isothiocyanates in Brassica: Potential Anti Cancer Agents. *Asian Pac. J. Cancer Prev.* **2016**, *17*, 4507–4510.
26. Yuan, J.M.; Stepanov, I.; Murphy, S.E.; Wang, R.; Allen, S.; Jensen, J.; Strayer, L.; Adams-Haduch, J.; Upadhyaya, P.; Le, C.; et al. Clinical Trial of 2-Phenethyl Isothiocyanate as an Inhibitor of Metabolic Activation of a Tobacco-Specific Lung Carcinogen in Cigarette Smokers. *Cancer Prev. Res.* **2016**, *9*, 396–405. [CrossRef] [PubMed]
27. Dayalan Naidu, S.; Suzuki, T.; Yamamoto, M.; Fahey, J.W.; Dinkova-Kostova, A.T. Phenethyl Isothiocyanate, a Dual Activator of Transcription Factors NRF2 and HSF1. *Mol. Nutr. Food Res.* **2018**, *62*, 1700908. [CrossRef]
28. Kim, Y.J.; Kim, E.H.; Hahm, K.B. Oxidative stress in inflammation-based gastrointestinal tract diseases: Challenges and opportunities. *J. Gastroenterol. Hepatol.* **2012**, *27*, 1004–1010. [CrossRef]
29. Chikara, S.; Nagaprashantha, L.D.; Singhal, J.; Horne, D.; Awasthi, S.; Singhal, S.S. Oxidative stress and dietary phytochemicals: Role in cancer chemoprevention and treatment. *Cancer Lett.* **2018**, *413*, 122–134. [CrossRef] [PubMed]
30. Cani, P.D.; Jordan, B.F. Gut microbiota-mediated inflammation in obesity: A link with gastrointestinal cancer. *Nat. Rev. Gastroenterol. Hepatol.* **2018**, *15*, 671–682. [CrossRef] [PubMed]
31. Mori, N.; Shimazu, T.; Charvat, H.; Mutoh, M.; Sawada, N.; Iwasaki, M.; Yamaji, T.; Inoue, M.; Goto, A.; Takachi, R.; et al. Cruciferous vegetable intake and mortality in middle-aged adults: A prospective cohort study. *Clin. Nutr.* **2019**, *38*, 631–643. [CrossRef]
32. Debnath, T.; Kim, D.H.; Lim, B.O. Natural products as a source of anti-inflammatory agents associated with inflammatory bowel disease. *Molecules* **2013**, *18*, 7253–7270. [CrossRef]
33. Moon, P.D.; Kim, H.M. Anti-inflammatory effect of phenethyl isothiocyanate, an active ingredient of Raphanus sativus Linne. *Food Chem.* **2012**, *131*, 1332–1339. [CrossRef]
34. Taniguchi, K.; Karin, M. NF-B, inflammation, immunity and cancer: Coming of age. *Nat. Rev. Immunol.* **2018**, *18*, 309–324. [CrossRef]
35. Pikarsky, E.; Porat, R.M.; Stein, I.; Abramovitch, R.; Amit, S.; Kasem, S.; Gutkovich-Pyest, E.; Uriell-Shoval, S.; Galun, E.; Ben-Neriah, Y. NF-κB functions as a tumour promoter in inflammation-associated cancer. *Nature* **2004**, *431*, 461–466. [CrossRef]
36. Greten, F.R.; Eckmann, L.; Greten, T.F.; Park, J.M.; Li, Z.W.; Egan, L.J.; Kagnoff, M.F.; Karin, M. IKKβ links inflammation and tumorigenesis in a mouse model of colitis-associated cancer. *Cell* **2004**, *118*, 285–296. [CrossRef]
37. Pereira, L.; Silva, P.; Duarte, M.; Rodrigues, L.; Duarte, C.; Albuquerque, C.; Serra, A. Targeting Colorectal Cancer Proliferation, Stemness and Metastatic Potential Using Brassicaceae Extracts Enriched in Isothiocyanates: A 3D Cell Model-Based Study. *Nutrients* **2017**, *9*, 368. [CrossRef]
38. Gupta, P.; Wright, S.E.; Kim, S.H.; Srivastava, S.K. Phenethyl isothiocyanate: A comprehensive review of anti-cancer mechanisms. *Biochim. Biophys. Acta–Rev. Cancer* **2014**, *1846*, 405–424. [CrossRef]

39. Giallourou, N.S.; Rowland, I.R.; Rothwell, S.D.; Packham, G.; Commane, D.M.; Swann, J.R. Metabolic targets of watercress and PEITC in MCF-7 and MCF-10A cells explain differential sensitisation responses to ionising radiation. *Eur. J. Nutr.* **2019**, *58*, 2377–2391. [CrossRef]
40. Kaiser, S.J.; Mutters, N.T.; Blessing, B.; Günther, F. Natural isothiocyanates express antimicrobial activity against developing and mature biofilms of Pseudomonas aeruginosa. *Fitoterapia* **2017**, *119*, 57–63. [CrossRef]
41. Kim, M.G.; Lee, H.S. Growth-inhibiting activities of phenethyl isothiocyanate and its derivatives against intestinal bacteria. *J. Food Sci.* **2009**, *74*, M467–M471. [CrossRef]
42. Narbad, A.; Rossiter, J.T. Gut Glucosinolate Metabolism and Isothiocyanate Production. *Mol. Nutr. Food Res.* **2018**, *62*, e1700991. [CrossRef]
43. Cheng, D.L.; Hashimoto, K.; Uda, Y. In vitro digestion of sinigrin and glucotropaeolin by single strains of Bifidobacterium and identification of the digestive products. *Food Chem. Toxicol.* **2004**, *42*, 351–357. [CrossRef]
44. Kellingray, L.; Tapp, H.S.; Saha, S.; Doleman, J.F.; Narbad, A.; Mithen, R.F. Consumption of a diet rich in Brassica vegetables is associated with a reduced abundance of sulphate-reducing bacteria: A randomised crossover study. *Mol. Nutr. Food Res.* **2017**, *61*, 1–11. [CrossRef]
45. Borges, A.; Abreu, A.C.; Ferreira, C.; Saavedra, M.J.; Simões, L.C.; Simões, M. Antibacterial activity and mode of action of selected glucosinolate hydrolysis products against bacterial pathogens. *J. Food Sci. Technol.* **2015**, *52*, 4737–4748. [CrossRef] [PubMed]
46. Dias, C.; Aires, A.; Saavedra, M.J. Antimicrobial activity of isothiocyanates from cruciferous plants against methicillin-resistant Staphylococcus aureus (MRSA). *Int. J. Mol. Sci.* **2014**, *15*, 19552–19561. [CrossRef] [PubMed]
47. Hong, E.; Kim, G.H. Anticancer and antimicrobial activities of β-phenylethyl isothiocyanate in *Brassica rapa* L. *Food Sci. Technol. Res.* **2008**, *14*, 377–382. [CrossRef]
48. Lam-Ubol, A.; Fitzgerald, A.L.; Ritdej, A.; Phonyiam, T.; Zhang, H.; Myers, J.N.; Huang, P.; Trachootham, D. Sensory acceptable equivalent doses of β-phenylethyl isothiocyanate (PEITC) induce cell cycle arrest and retard the growth of p53 mutated oral cancer in vitro and in vivo. *Food Funct.* **2018**, *9*, 3640–3656. [CrossRef] [PubMed]
49. Chen, Y.; Li, Y.; Wang, X.Q.; Meng, Y.; Zhang, Q.; Zhu, J.Y.; Chen, J.Q.; Cao, W.S.; Wang, X.Q.; Xie, C.F.; et al. Phenethyl isothiocyanate inhibits colorectal cancer stem cells by suppressing Wnt/β-catenin pathway. *Phyther. Res.* **2018**, *32*, 2447–2455. [CrossRef]
50. Ramirez, C.N.; Li, W.; Zhang, C.; Wu, R.; Su, S.; Wang, C.; Gao, L.; Yin, R.; Kong, A.-N. In Vitro-In Vivo Dose Response of Ursolic Acid, Sulforaphane, PEITC, and Curcumin in Cancer Prevention. *AAPS J.* **2018**, *20*, 19. [CrossRef]
51. Wang, X.; Govind, S.; Sajankila, S.P.; Mi, L.; Roy, R.; Chung, F.L. Phenethyl isothiocyanate sensitizes human cervical cancer cells to apoptosis induced by cisplatin. *Mol. Nutr. Food Res.* **2011**, *55*, 1572–1581. [CrossRef]
52. Gao, N.; Budhraja, A.; Cheng, S.; Liu, E.H.; Chen, J.; Yang, Z.; Chen, D.; Zhang, Z.; Shi, X. Phenethyl isothiocyanate exhibits antileukemic activity in vitro and in vivo by inactivation of Akt and activation of JNK pathways. *Cell Death Dis.* **2011**, *2*, 1–9. [CrossRef]
53. Coscueta, E.R.; Reis, C.A.; Pintado, M. Phenylethyl Isothiocyanate Extracted from Watercress By-Products with Aqueous Micellar Systems: Development and Optimisation. *Antioxidants* **2020**, *9*, 698. [CrossRef]
54. Tanongkankit, Y.; Sablani, S.S.; Chiewchan, N.; Devahastin, S. Microwave-assisted extraction of sulforaphane from white cabbages: Effects of extraction condition, solvent and sample pretreatment. *J. Food Eng.* **2013**, *117*, 151–157. [CrossRef]
55. Fusari, C.M.; Ramirez, D.A.; Camargo, A.B. Simplified analytical methodology for glucosinolate hydrolysis products: A miniaturized extraction technique and multivariate optimization. *Anal. Methods* **2019**, *11*, 309–316. [CrossRef]
56. Pusateri, D.J.; Kizer, T.R.; Lowry, A.N. Extraction of Non-Polar Isothiocyanates from Plants. U.S. Patent 6,824,796, 30 November 2004.
57. Rodrigues, L.; Silva, I.; Poejo, J.; Serra, A.T.; Matias, A.A.; Simplício, A.L.; Bronze, M.R.; Duarte, C.M.M. Recovery of antioxidant and antiproliferative compounds from watercress using pressurized fluid extraction. *RSC Adv.* **2016**, *6*, 30905–30918. [CrossRef]
58. Palaniswamy, U.R.; McAvoy, R.J.; Bible, B.B.; Stuart, J.D. Ontogenic variations of ascorbic acid and phenethyl isothiocyanate concentrations in watercress (Nasturtium officinale R.Br.) leaves. *J. Agric. Food Chem.* **2003**, *51*, 5504–5509. [CrossRef]
59. Fahey, J.W. Method of Extraction of Isothiocyanates into Oil from Glucosinolate-Containing Plants and Method of Producing Products with Oil Containing Isothiocyanates Extracted from Glucosinolate-Containing Plants. U.S. Patent Application 11/302,118, 15 June 2006.
60. Wang, J.; Luo, B.; Li, X.; Lu, W.; Yang, J.; Hu, Y.; Huang, P.; Wen, S. Inhibition of cancer growth in vitro and in vivo by a novel ROS-modulating agent with ability to eliminate stem-like cancer cells. *Cell Death Dis.* **2017**, *8*, e2887. [CrossRef]
61. Kala, C.; Salman Ali, S.; Ahmad, N.; Jamal Gilani, S.; Ali Khan, N. Isothiocyanates: A Review. *Res. J. Pharmacogn.* **2018**, *5*, 71–89. [CrossRef]
62. Mohanty, S.; Sahoo, A.K.; Konkimalla, V.B.; Pal, A.; Si, S.C. Naringin in combination with isothiocyanates as liposomal formulations potentiates the anti-inflammatory activity in different acute and chronic animal models of rheumatoid arthritis. *ACS Omega* **2020**, *5*, 28319–28332. [CrossRef]
63. Zambrano, V.; Bustos, R.; Mahn, A. Insights about stabilization of sulforaphane through microencapsulation. *Heliyon* **2019**, *5*. [CrossRef]

64. Dagan, I.D.; Frisbee, A.R.; Newsome, P.W.; Baudet, M.P. Stabilized Sulforaphane. U.S. Patent 7879822B2, 1 February 2011.
65. Coscueta, E.R.; Sousa, A.S.; Reis, C.A.; Pintado, M. Chitosan-olive oil microparticles for phenylethyl isothiocyanate delivery: Optimal formulation. *PLoS ONE* **2021**, *16*, e0248257. [CrossRef]

molecules

MDPI

Article

Hydrolysate from Mussel *Mytilus galloprovincialis* Meat: Enzymatic Hydrolysis, Optimization and Bioactive Properties

Sara A. Cunha, Rita de Castro, Ezequiel R. Coscueta and Manuela Pintado *

CBQF—Centro de Biotecnologia e Química Fina—Laboratório Associado, Escola Superior de Biotecnologia, Universidade Católica Portuguesa Rua Diogo Botelho 1327, 4169-005 Porto, Portugal; scunha@ucp.pt (S.A.C.); rita_de_castro@live.com.pt (R.d.C.); ecoscueta@ucp.pt (E.R.C.)

* Correspondence: mpintado@ucp.pt

Abstract: Mussel production generates losses and waste since their commercialisation must be aligned with target market criteria. Since mussels are rich in proteins, their meat can be explored as a source of bioactive hydrolysates. Thus, the main objective of this study was to establish the optimal production conditions through two Box–Behnken designs to produce, by enzymatic hydrolysis (using subtilisin and corolase), hydrolysates rich in proteins and with bioactive properties. The factorial design allowed for the evaluation of the effects of three factors (hydrolysis temperature, enzyme ratio, and hydrolysis time) on protein/peptides release as well as antioxidant and anti-hypertensive properties of the hydrolysates. The hydrolysates produced using the optimised conditions using the subtilisin protease showed 45.0 ± 0.38% of protein, antioxidant activity via ORAC method of 485.63 ± 60.65 µmol TE/g of hydrolysate, and an IC_{50} for the inhibition of ACE of 1.0 ± 0.56 mg of protein/mL. The hydrolysates produced using corolase showed 46.35 ± 1.12% of protein, antioxidant activity of 389.48 ± 0.21 µmol TE/g of hydrolysate, and an IC_{50} for the inhibition of ACE of 3.7 ± 0.33 mg of protein/mL. Mussel meat losses and waste can be used as a source of hydrolysates rich in peptides with relevant bioactive properties, and showing potential for use as ingredients in different industries, such as food and cosmetics, contributing to a circular economy and reducing world waste.

Keywords: antioxidant; anti-hypertensive; proteins; sustainability; marine species; marine hydrolysates

Citation: Cunha, S.A.; de Castro, R.; Coscueta, E.R.; Pintado, M. Hydrolysate from Mussel *Mytilus galloprovincialis* Meat: Enzymatic Hydrolysis, Optimization and Bioactive Properties. *Molecules* **2021**, *26*, 5228. https://doi.org/10.3390/molecules26175228

Academic Editor: Jesus Simal-Gandara

Received: 2 August 2021
Accepted: 25 August 2021
Published: 28 August 2021

1. Introduction

Mussels are highly consumed in several countries. Asia and Europe are considered the leading producers, estimated to produce about 1.05 and 0.5 million tonnes of mussel per year, respectively [1,2]. Mussel consumption has several advantages, for both the environment and consumers. Environmentally, mussel farming can be done with minimal greenhouse gas emissions, and thus low carbon footprint and few environmental impacts [3]. Mussels farming produces about 0.6 kg of CO_2 emission/kg edible product, while beef produces about 19.0–36.7 kg of CO_2 emission/kg edible product [3]. For consumers, mussel meat has low fat and low calories. Still, more importantly, mussels are a rich source of sodium, selenium, vitamin B twelve, zinc [1], and an interesting source of proteins since they are composed of about 58.7% of protein on a dry weight basis [4]. Due to their protein-rich meat, mussels have been described as a source of bioactive peptides with relevant biological properties. Bioactive peptides are fragments that are inert when inside proteins but show different properties when broken from the original protein [5]. Thus, enzymatic hydrolysis with proteases seems an interesting approach for obtaining bioactive extracts since these enzymes may break mussel proteins into smaller peptides, which may be associated with other biological and functional properties [6]. Different enzymes have been used to produce bioactive peptides from mussels, such as pepsin [7], flavourzyme [8], papain [8], and trypsin [9]. Marine species have often been described as a source of bioactive peptides, and some bioactivities have been associated to mussel peptides, such as antioxidant [10], anti-hypertensive [11], antimicrobial [12], anticancer [7], anti-inflammatory [13], anticoagulant [14], antidiabetic [15], and antiviral [16]. As far as we know, *Mytilus galloprovincialis* bioactive extracts are not so extensively explored, with the main studies being developed with *Mytilus coruscus* and *Mytilus edulis*. The farming of *M. galloprovincialis*, also known as the Mediterranean mussel, has mainly been developing along the Spanish Atlantic coast and in the Mediterranean area [2].

Mussel commercialisation generates losses and waste since they are submitted to a pre-selection before being delivered for sale, resulting in the rejection of broken mussels or those which fail to meet established criteria in the target market [17–20]. It is estimated that about 27% of produced mussels are discarded [4,20]. Thus, mussel meat waste can be used to produce bioactive hydrolysates with interesting properties for food, cosmetic, pharmaceutical, and nutraceutical industrial applications.

In this work, *Mytilus galloprovincialis* meat was submitted to different conditions according to two factorial designs to produce hydrolysates with a maximum level of soluble rich proteins/peptides and bioactive properties, more specifically antioxidant and anti-hypertensive. The hydrolysates were produced by enzymatic hydrolysis using two different enzymes, subtilisin and corolase. Therefore, this study aims to valorise mussel meat by creating bioactive hydrolysates with potential for various industries.

2. Results

2.1. *Mytilus Galloprovincialis Characterisation*

The mussel's meat was received under refrigeration on the day of capture. It was characterised before being minced according to a few nutritional characteristics, showing a protein content of 70.50 ± 13.44%, 90.30 ± 4.24% moisture, and 5.00 ± 0.00% lipids.

2.2. *Optimisation of the Production of Hydrolysates Rich in Proteins and Bioactive Properties*

Enzymatic hydrolysis is one of the main used methods to produce bioactive extracts, and is described for the mussel species *M. coruscus*, *M. edulis*, and *M. galloprovincialis*. Several enzymes have been used in the mussel species, such as papain [8], flavourzyme [21], and the digestive enzymes pepsin [7] and trypsin [9]. In this work, two different proteases, subtilisin and corolase, were used to produce hydrolysates rich in proteins/peptides and with antioxidant and anti-hypertensive properties. To understand the conditions that allow to achieve the production of hydrolysates with a better protein% and higher bioactive properties, two experimental designs were made, one for each protease. Different combinations of factors in an enzymatic hydrolysis may lead to differing effectiveness. Thus, we have used a factorial design with 15 combinations of the enzyme (%), temperature (°C), and hydrolysis time (h) for each protease.

Mussel meat was initially minced until homogenised, thus creating uniformed biomass used for all the 60 hydrolysis reactions performed. Then, all the hydrolysis were performed using ultrapure water as the solvent, at 7.5 pH, with a ratio of 1:2 (*w:v*) (mussel biomass:water). The 60 hydrolysis reactions were performed using the factors combinations matrix generated by the experimental design, and protein/peptides and bioactivities were measured in the resulting supernatants.

In an enzymatic reaction, different factors combinations may lead to the production of extracts with different characteristics. Thus, the factorial designs allowed us to understand the best combination for optimising protein/peptide bioactive extract production. For both designs, the matrix and obtained results are presented as well as the Pareto charts obtained for each evaluated response, indicating the factors with the greatest influence for each variable studied. For each evaluated response, a multiple regression analysis of the experimental data allowed to obtain a model that can predict the responses and these are shown as Equations (1)–(6). An analysis of variance (ANOVA) was performed to evaluate the significance of each effect and to determine the factors that significantly affected protein % as well as antioxidant and anti-hypertensive properties.

2.2.1. Experimental Design with Subtilisin Protease

Table 1 shows the experimental design matrix and the results obtained for the Box–Behnken factorial design performed with the subtilisin protease. Figure 1 shows the Pareto charts obtained for the design performed with the subtilisin protease. Figure 2 illustrates the response surface graphics obtained for the independent variables tested, showing their interactions when studying each dependent variable.

Table 1. Box-Behnken factorial design matrix for three factors and three responses obtained for the subtilisin protease.

Run	Factors			Response [1]		
	% Enzyme	Hydrolysis Temperature (°C)	Hydrolysis Time (h)	Protein Content (%)	ORAC (µmol TE/mg)	ACE Inhibition (%)
1	0.5	40	2	45.56 ± 0.38	0.49 ± 0.08	49.89 ± 18.86
2	0.5	50	3	49.11 ± 0.41	0.60 ± 0.05	56.21 ± 2.39
3	1.5	40	2	47.17 ± 0.98	0.55 ± 0.06	44.51 ± 17.37
4	1.5	60	2	45.87 ± 2.79	0.72 ± 0.07	65.55 ± 1.77
5	0.5	50	1	48.83 ± 0.93	0.59 ± 0.02	63.86 ± 0.07
6	1	50	2	48.97 ± 1.86	0.60 ± 0.04	59.97 ± 3.64
7	1.5	50	3	48.03 ± 2.18	0.60 ± 0.05	72.74 ± 2.24
8	1	60	3	47.66 ± 2.21	0.66 ± 0.00	66.30 ± 0.41
9	1	40	1	45.04 ± 2.80	0.54 ± 0.02	61.31 ± 0.57
10	1	40	3	46.68 ± 2.71	0.56 ± 0.04	47.59 ± 3.73
11	1	60	1	45.66 ± 2.94	0.61 ± 0.07	69.48 ± 0.85
12	1.5	50	1	46.86 ± 1.25	0.59 ± 0.01	67.67 ± 0.48
13	0.5	60	2	48.80 ± 0.60	0.56 ± 0.04	67.62 ± 3.53
14	1	50	2	49.02 ± 0.02	0.64 ± 0.06	51.62 ± 2.04
15	1	50	2	49.12 ± 0.14	0.62 ± 0.04	60.41 ± 1.65

[1] Values expressed as mean ± SD of two replicates.

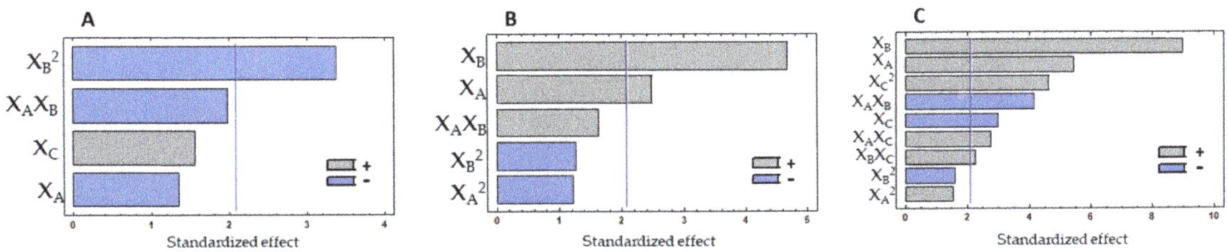

Figure 1. Pareto charts with the effect of three experimental factors, in decreasing order, obtained for protein % (**A**), antioxidant (**B**) and anti-hypertensive (**C**) properties in the experimental design with subtilisin, showing the most influent factors. The vertical lines in the pareto charts represent the level of significance (p = 0.05). X_A—Enzyme %; X_B—Hydrolysis temperature; X_C—Hydrolysis time.

Protein content did not show significant variations among all the combinations tested since the range of variation for results was 45.05–49.12% of protein in the final hydrolysate. The Pareto chart (Figure 1A) and the ANOVA analysis (Table 2) show that protein release was not influenced by the linear factors, but only by the quadratic effect of the temperature. The temperature quadratic coefficient showed a negative effect, indicating an increase in protein % at intermediate values. Since none of the factors had a significant effect, the model was adjusted to best fit. Thus, only the linear effect of enzyme% and time, as well as the linear interaction between enzyme% and temperature and the quadratic effect of temperature were considered (Table 2). By multiple regression, the predicted response for the protein % could be expressed by the model in Equation (1).

$$\text{Protein } \% = -13.1089 + 10.2225 \times X_A + 0.635 \times X_C - 0.22625 \times X_A X_B - 0.0200714 \times X_B{}^2 \quad (1)$$

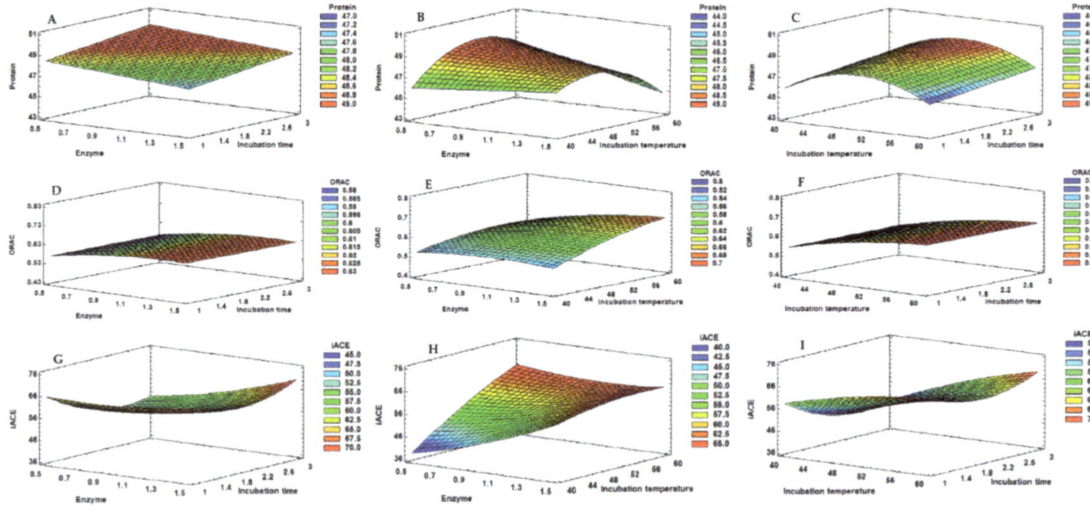

Figure 2. Response surface graphs corresponding to the combined effect of enzyme % (X_A), hydrolysis temperature (X_B) and hydrolysis time (X_C), on the protein (**A–C**), ORAC (**D–F**) and iACE (**G–I**) responses obtained for the mussel extracts produced with subtilisin.

Table 2. Analysis of variance (ANOVA) for protein % obtained for the subtilisin Box-Behken design.

Model	Sum of Squares	DF	Mean Square	F Value	*p*-Value
X_A (Enzyme %)	4.7524	1	4.7524	1.80	0.1955
X_C (Time)	6.4516	1	6.4516	2.44	0.1345
X_AX_B	10.2378	1	10.2378	3.88	0.0637
$X_B{}^2$	30.0804	1	30.0804	11.39	0.0032
Lack of fit	7.94221	6	1.3237	0.50	0.7995
Pure error	50.16	19	2.64		
Total	109.624	29			
$R^2 = 47.00$, Adj $- R^2 = 38.52$, CV = 1.62					

The final adjusted model showed a significant fit ($p > 0.050$). However, the $R^2 = 38.52\%$, indicating that only 38.52% of the variability observed may be explained by the model. This indicates that, for obtaining hydrolysates rich in proteins/peptides, the minimum studied variables could be used (0.5% enzyme, 40 °C, 1 h) to achieve protein release superior to 40%. Although, since this work aims to produce hydrolysates with bioactive properties, the antioxidant and anti-hypertensive properties must be analysed.

The antioxidant property, measured in the soluble hydrolysates by ORAC, was only significantly influenced by the linear effect of the hydrolysis temperature (X_B) and enzyme % (X_A) ($p < 0.050$), with hydrolysis time (X_C) not showing a significant effect on this property (Figure 1B). Temperature and enzyme % positively affected antioxidant property, meaning that the response is directly proportional to the tested values. For ORAC, the interactions between variables did not show a significant effect, the same being verified for the quadratic effect of the three studied variables. By multiple regression, the predicted response for the ORAC could be obtained by the model in Equation (2).

$$\text{ORAC} = -0.0333894 - 0.0436106 \times X_A + 0.0206284 \times X_B \quad (2)$$

Analysing the ANOVA results (Table 3), the final adjusted model showed a significant fit ($p > 0.050$). However, the $R^2 = 45.65\%$, indicating that only about 45.65% of the variability in the antioxidant activity may be explained by the model.

Table 3. Analysis of variance (ANOVA) for ORAC results obtained for the subtilisin Box-Behken design.

Model	Sum of Squares	DF	Mean Square	F Value	*p*-Value
X_A (Enzyme %)	0.0120506	1	0.0120506	6.20	0.0212
X_B (Temperature)	0.0422508	1	0.0422508	21.75	0.0001
Lack of fit	0.0124529	3	0.00415097	2.14	0.1259
Pure error	0.0407955	21	0.00194264		
Total	0.118375	29			
$R^2 = 55.02$, Adj $- R^2 = 45.65$, CV = 0.044					

The ACE inhibition % was evaluated using a concentration of 10 mg hydrolysate/mL. The linear effects of the three studied factors (X_A, X_B and X_C) were significant for the variations observed on the ACE inhibition %, as well as the linear interaction between all the variables (X_AX_B, X_AX_C, and X_BX_C) and the quadratic effect of the hydrolysis time (X_C^2) (Figure 1C). Thus, seven effects showed a $p < 0.050$, while the other two effects (X_A^2 and X_B^2) were not significant and consequently removed from the model. Temperature and enzyme % showed a high impact on the response. The linear effect of hydrolysis time (X_C) and the linear interaction between temperature and enzyme % showed a negative contribution, which means that there is an increase of the ACE inhibition at intermediate values. Thus, the longer the reaction, the lower the inhibition of ACE will be, which can mean that the higher extension of the hydrolysis leads to the formation of peptides with less activity against ACE. The quadratic effect of hydrolysis time (X_C^2) showed a positive effect on the response, which means that, considering the negative effect verified for X_C, higher values of iACE were achieved near the minimum values studied. The other linear factors with a significant contribution positively affected the results, meaning that the anti-hypertensive potential is enhanced by the increase in enzyme % and by the higher temperatures.

By multiple regression, the predicted response for the iACE could be obtained by the model in Equation (3).

$$\text{iACE} = -24.3485 + 38.3047 \times X_A + 3.41606 \times X_B - 45.3303 \times X_C - 1.13715 \times X_A \times X_B + 6.3675 \times X_A \times X_C - 0.0200471 \times X_B{}^2 + 0.2635 \times X_B \times X_C + 5.83804 \times X_C{}^2 \quad (3)$$

Analysing the ANOVA results (Table 4), the final adjusted model showed a significant fit ($p > 0.050$) and an $R^2 = 81.30\%$, indicating the variability observed in terms of ACE inhibition is highly explained by the model.

Table 4. Analysis of variance (ANOVA) for ACE inhibition (%) obtained for the subtilisin Box-Behken design.

Model	Sum of Squares	DF	Mean Square	F Value	*p*-Value
X_A (Enzyme %)	315.868	1	315.868	29.62	0.0001
X_B (Temperature)	855.895	1	855.895	80.27	0.0000
X_C (Time)	94.9163	1	94.9163	8.90	0.0093
$X_A X_B$	184.73	1	184.73	17.32	0.0008
$X_A X_C$	81.0901	1	81.0901	7.60	0.0147
$X_B X_C$	55.5458	1	55.5458	5.21	0.0375
X_C^2	230.418	1	230.418	21.61	0.0003
Lack of fit	85.5525	3	28.5175	2.67	0.0848
Pure error	159.948	15	10.6632		
Total	1969.71	27			
$R^2 = 87.54$, Adj $- R^2 = 81.30$, CV = 3.26					

After analysing the results for each variable, we intended to maximise the hydrolysis in order to achieve higher protein content and antioxidant and anti-hypertensive properties. For that, a Derringer's desirability analysis was performed [22] (Table 5). The optimum conditions predicted were 52.5 °C, 1.5% of subtilisin, and an hydrolysis time of 3 h (Table 5).

Table 5. Optimal conditions predicted by the experimental design to maximise protein/peptide release and antioxidant and anti-hypertensive activities of the hydrolysates, and Derringer desirability to predict the optimal conditions for a multiple response.

Factors	Response			
	Protein	ORAC	iACE	Erringer Desirability
Temperature (°C)	52.8	60.0	59.9	52.5
Protease (%)	0.5	1.5	1.5	1.5
Hydrolysis time (h)	3.0	2.0	3.0	3.0

The experiments were validated in triplicate, using the same biomass quantities and solvent volumes but adapting temperature to 52 °C, to work practically. The obtained protein content, ORAC, and ACE inhibition values were 46.70 ± 0.36%, 0.54 ± 0.029 µmol TE/mg hydrolysate and 70.21 ± 2.9%, respectively (Table 6). When comparing the results predicted by the factorial design, the obtained results were similar to those predicted by the design.

Table 6. Results predicted by the model, and results obtained in a validation and a scaled-up enzymatic hydrolysis, performed with the optimal conditions described in Table 5.

Evaluated Characteristics	Predicted Results	Obtained Results	
		Validation	Scaled-Up
Protein (%)	48.22	46.70 ± 0.36	45.0 ± 0.38
Antioxidant activity (ORAC) (µmol TE/mg hydrolysate)	0.64	0.54 ± 0.029	0.49 ± 0.061
Anti-hypertensive activity (% inhibition at 2.5 mg/mL)	71.87	70.21 ± 2.9	——
Anti-hypertensive activity (IC_{50} mg of protein/mL)	——	——	1.0 ± 0.56

After optimising the enzymatic hydrolysis reaction, a scale-up was performed, in triplicate, increasing the amount of mussel biomass and solvent by 15 times and maintaining the tested ratio. The temperature was adjusted to 50 °C, to be easily adapted to an industrial scale. The obtained protein content, ORAC, and IC_{50} for ACE inhibition values were 45.0 ± 0.38, 0.49 ± 0.061 µmol TE/mg hydrolysate, and 1.0 ± 0.56 mg protein/mL, respectively (Table 6). The obtained results indicated that the increase in the proportions seems to influence the evaluated responses negatively. Nevertheless, the obtained hydrolysates showed interesting protein/peptide values and antioxidant potential, making these extracts an interesting protein source and a potential ingredient for functional food or cosmetic formulations focused on anti-ageing properties.

2.2.2. Experimental Design with the Corolase Protease

The experimental design matrix and the responses obtained for the Box–Behnken factorial design performed with the corolase protease are shown in Table 7. Figure 3 shows the Pareto charts, and Figure 4 shows the response surface graphics obtained for the independent variables tested, showing their interactions when studying each dependent variable.

Table 7. Box-Behnken factorial design matrix for three factors and three responses obtained for the corolase protease.

Run	Factors			Response [1]		
	% Enzyme	Hydrolysis Temperature (°C)	Hydrolysis Time (h)	Protein Content (%)	ORAC (μmol TE/mg)	ACE Inhibition (%)
1	3	40	2	47.27 ± 0.21	0.76 ± 0.04	61.10 ± 5.43
2	2	50	3	47.38 ± 1.37	0.46 ± 0.06	46.79 ± 7.94
3	2	40	2	45.84 ± 0.28	0.57 ± 0.02	46.13 ± 6.34
4	2	60	2	48.33 ± 0.65	0.59 ± 0.01	48.44 ± 2.09
5	3	50	1	47.39 ± 0.02	0.55 ± 0.00	59.40 ± 1.36
6	2	50	2	45.56 ± 2.40	0.57 ± 0.01	43.43 ± 9.55
7	3	50	3	49.77 ± 0.49	0.60 ± 0.17	49.34 ± 17.73
8	1	60	3	47.26 ± 2.29	0.52 ± 0.02	38.80 ± 4.13
9	1	40	1	46.42 ± 1.28	0.42 ± 0.06	40.40 ± 8.85
10	1	40	3	44.64 ± 0.99	0.63 ± 0.11	32.19 ± 3.76
11	2	60	1	47.38 ± 1.09	0.67 ± 0.16	49.66 ± 7.55
12	2	50	1	45.85 ± 0.74	0.67 ± 0.15	50.21 ± 4.48
13	2	60	2	48.23 ± 0.91	0.72 ± 0.00	52.33 ± 4.90
14	1	50	2	44.30 ± 0.41	0.46 ± 0.04	45.72 ± 4.09
15	3	50	2	47.85 ± 0.85	0.59 ± 0.02	54.23 ± 3.00

[1] Values expressed as mean ± SD of two replicates.

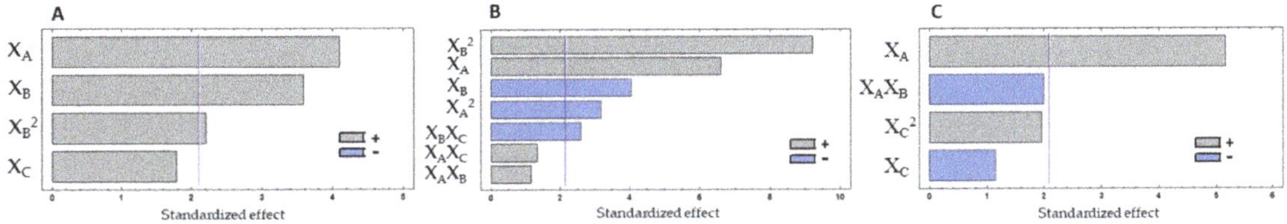

Figure 3. Pareto charts with the effect of three experimental factors, in decreasing order, obtained for protein % (**A**), antioxidant (**B**) and anti-hypertensive (**C**) properties in the experimental design with corolase, showing the most influent factors. The vertical lines in the pareto charts represent the level of significance (p = 0.05). X_A—Enzyme %; X_B—Hydrolysis temperature; X_C—Hydrolysis time.

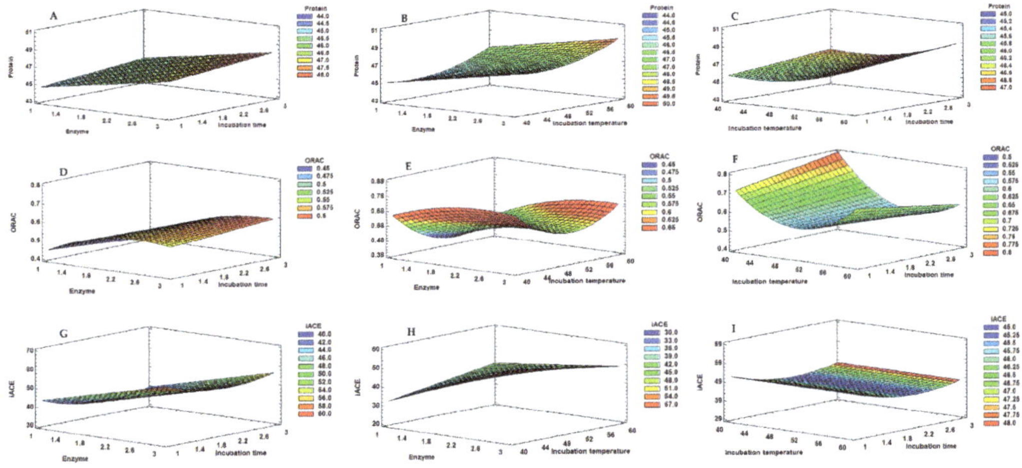

Figure 4. Response surface graphs corresponding to the combined effect of enzyme % (XA), hydrolysis temperature (XB) and hydrolysis time (XC), on the protein (**A–C**), ORAC (**D–F**) and iACE (**G–I**) responses obtained for the mussel extracts produced with corolase.

The protein content of the hydrolysates showed a variation in the range of 44.30–49.77%. By analysing the Pareto chart (Figure 3A), protein% was positively influenced by the linear effect of enzyme% (X_A) and temperature (X_B), meaning that the increase in enzyme concentration and temperature leads to an increase of protein release from the mussel biomass. The time of the hydrolysis did not significantly affect the protein % ($p > 0.050$). The quadratic effect of the temperature was also positively significant ($p < 0.050$). By multiple regression, the predicted response for the protein % could be obtained by the model in Equation (4).

$$\text{Protein }\% = 61.4391 + 1.20687 \times X_A - 0.846161 \times X_B + 0.00951786 \times {X_B}^2 \quad (4)$$

The final adjusted model showed a significant fit ($p > 0.050$) (Table 8). Still, the R^2 indicates that only about 60% of the variability observed in relation to the protein content is explained by the model, indicating that the increase of corolase %, temperature, and hydrolysis time is not beneficial to produce protein-rich hydrolysates.

Table 8. Analysis of variance (ANOVA) for protein % obtained for the corolase Box-Behken design.

Model	Sum of Squares	DF	Mean Square	F Value	*p*-Value
X_A (enzyme %)	23.3048	1	23.3048	16.79	0.0008
X_B (Temperature)	17.8506	1	17.8506	12.86	0.0023
${X_B}^2$	6.76402	1	6.76402	4.87	0.0413
Lack of fit	3.91492	8	0.489365	0.35	0.9315
Pure error	23.5968	17	1.38805		
Total	79.8516	29			
$R^2 = 65.54$, Adj $- R^2 = 60.03$, CV = 1.17					

The antioxidant activity was significantly influenced by the linear effect of enzyme% (X_A) and temperature (X_B) ($p < 0.050$), while the hydrolysis time (X_C) did not show a significant effect ($p > 0.050$). Enzyme% showed a positive effect, meaning that the increase in enzyme concentration results in higher ORAC values. On the other hand, temperature had a negative effect. Thus, the higher the temperature the lower the ORAC values will be. Furthermore, the quadratic effect of both factors also showed a significant effect. The linear interaction between temperature and hydrolysis time had a significant negative effect (Figure 3B).

By multiple regression, the predicted response for the ORAC could be obtained by the model in Equation (5).

$$\text{ORAC} = 4.11383 + 0{,}158065 \times X_A - 0.153981 \times X_B - 0.0548188 \times {X_A}^2 + 0.00154318 \times {X_B}^2 - 0.0045 \times X_B \times X_C \quad (5)$$

The final adjusted model highly explains the antioxidant activity, showing a significant fit ($p > 0.050$) and $R^2 = 87.09\%$ (Table 9).

Table 9. Analysis of variance (ANOVA) for ORAC obtained for the corolase Box-Behken design.

Model	Sum of Squares	DF	Mean Square	F Value	*p*-Value
X_A (enzyme%)	0.0781744	1	0.0781744	43.67	0.0000
X_B (Temperature)	0.0293709	1	0.0293709	16.41	0.0014
${X_A}^2$	0.0181707	1	0.0181707	10.15	0.0072
${X_B}^2$	0.152151	1	0.152151	85.00	0.0000
$X_B X_C$	0.01215	1	0.01215	6.79	0.0218
Lack of fit	0.00655055	5	0.00131011	0.73	0.6122
Pure error	0.02327	13	0.00179		
Total	0.320938	25			
$R^2 = 90.71$, Adj $- R^2 = 87.09$, CV = 0.04					

The ACE inhibition (iACE) % was evaluated in a concentration of 10 mg hydrolysate/mL. The only variable that showed a significant effect on iACE was the enzyme% (X_A), with a positive effect (Figure 3C). Thus, the increase in enzyme concentration increases the anti-hypertensive potential of the hydrolysates, which may be explained by the formation of more peptides with the ability to inhibit the ACE.

By multiple regression, the predicted response for the iACE could be obtained by the model in Equation (6).

$$iACE = 4.11383 + 0.158065 \times X_A \quad (6)$$

The ANOVA results for the adjusted model was verified to have a significant fit ($p > 0.050$) (Table 10). However, the model only explained 57.90% of the variability in the anti-hypertensive results.

Table 10. Analysis of variance (ANOVA) for ACE inhibition (%) obtained for the corolase Box-Behken design.

Model	Sum of Squares	DF	Mean Square	F Value	*p*-Value
X_A (enzyme %)	1120.74	1	1120.74	26.62	0.0001
Lack of fit	58.3703	6	9.72839	0.23	0.9612
Pure error	799.952	19	42.1027		
Total	2364.9	29			
R^2 = 63.71, Adj − R^2 = 57.90, CV = 6.49					

The Box–Behnken design allowed to optimise the conditions that would enable higher results for the individual responses (Table 11). However, a Derringer's desirability analysis was performed to optimise multiple responses of the design (Table 11). Thus, the hydrolysis of the minced mussel meat with 3% of the enzyme, at 40 °C for 3 h, seems to represent the best conditions to obtain the higher results in terms of hydrolysate proteins/peptides content as well as antioxidant and anti-hypertensive properties.

Table 11. Optimal conditions predicted by the experimental design to maximise protein/peptide release and antioxidant and anti-hypertensive activities of the hydrolysates, and Derringer desirability to predict the optimal conditions for a multiple response.

Factors	Response			
	Protein	ORAC	iACE	Derringer Desirability
Temperature (°C)	60.0	40.0	40.0	40.1
Protease (%)	3.0	2.7	3.0	3.0
Hydrolysis time (h)	3.0	3.0	1.0	3.0

An enzymatic hydrolysis was performed, in triplicate, using the optimised conditions according to the design for the purpose of validation. The experiment was performed using the exact quantities used in the design experiments. The temperature was adjusted to 40 °C to work practically. The obtained hydrolysates were freeze-dried and then evaluated regarding their protein content and antioxidant and anti-hypertensive potential. The hydrolysates showed a mean of 47.36 ± 1.02% of protein content, antioxidant activity of 0.65 ± 0.062 µmol TE/mg hydrolysate, and ability to inhibit the activity of ACE in 55.36 ± 2.12% (at 10 mg hydrolysate/mL). The obtained results, although slightly lower, were not so different from the predicted ones. Furthermore, a scale-up hydrolysis was performed, in triplicate, with an increase of 15 times the amount of mussel biomass and solvent, maintaining the ratio used in the experimental design. The final hydrolysates showed a mean of 46.35 ± 1.12% of protein content, antioxidant activity of 0.389 ± 0.021, and IC_{50} for ACE inhibition of 3.7 ± 0.33 mg protein/mL (Table 12). The scaled-up results were verified to be slightly lower than the predicted ones and the validation hydrolysates regarding protein content. However, the antioxidant activity showed a pronounced decrease.

Table 12. Results predicted by the model, and results obtained in a validation and scaled-up enzymatic hydrolysis, performed with the optimal conditions described in Table 11.

Evaluated Characteristics	Predicted Results	Obtained Results	
		Validation	Scaled-Up
Protein (%)	48.01	47.36 ± 1.06	46.35 ± 1.12
Antioxidant activity (ORAC) (μmol TE/mg hydrolysate)	0.82	0.65 ± 0.062	0.389 ± 0.021
Anti-hypertensive activity (% inhibition at 5 mg/mL)	61.10	55.36 ± 2.12	——
Anti-hypertensive activity (IC_{50} mg of protein/mL)	——	——	3.7 ± 0.33

3. Discussion

Mussel meat has a high protein content, making it interesting to produce bioactive hydrolysates rich in proteins and bioactive peptides. However, the mussel *Mytilus galloprovincialis* is less exploited regarding its bioactive potential when compared to other mussel species, such as *M. coruscus* and *M. edulis*. Since we wanted to create a food-grade method, we chose two food-grade proteases to carry out enzymatic hydrolysis. Thus, to explore this mussel potential, we have performed two Box–Behken experimental designs, with two different proteases, aiming to obtain hydrolysates with interesting potential for industrial applications. Furthermore, we have not found studies with mussels from the genus *Mytilus* performing enzymatic hydrolysis with subtilisin or corolase. The most frequent enzymes found were mainly gastric enzymes, such as pepsin and trypsin, and non-gastric enzymes, such as papain and flavourzyme.

The mussel meat biomass used showed 70.50 ± 13.44% of protein on a dry weight (DW) basis and a moisture content of 90.30 ± 4.24%. These results show higher values of protein when compared to other studies with *Mytilus* sp. from Portugal and Spain that showed protein content variation from 39.17–42.94 (DW) and moisture % of 81.71–87.59% [23]. However, these results are in line with the possible variations in protein content that can occur in different months, as shown by Çelik [24] in a study with *Mytilus galloprovincialis* indicating higher protein levels (74.64%) in February.

The protein % of the hydrolysates does not seem to be highly influenced by the determined models, indicating that enzymatic hydrolysis with both enzymes can produce hydrolysates with protein contents in the range of 40–48% (DW). So, to obtain mussel hydrolysates with a content of above 40%, the most economical and fastest conditions can probably be used.

The subtilisin protease optimised method was an enzymatic hydrolysis with 1.5% of enzyme with a duration of 3 h at 52 °C. In a scale-up test with these conditions, the final hydrolysates showed protein content, ORAC, and IC50 for ACE inhibition values of 45.0 ± 0.38, 0.49 ± 0.061 245 μmol TE/mg hydrolysate, and 1.0 ± 0.56 mg protein/mL, respectively. With the corolase, the optimised method was an enzymatic hydrolysis with 3.0% of enzyme with a duration of 3 h at 40 °C, obtaining scale-up hydrolysates with protein content, ORAC, and IC_{50} for ACE inhibition values of 46.35 ± 1.12, 0.389 ± 0.021 μmol TE/mg hydrolysate, and 3.7 ± 0.33 mg protein/mL, respectively. The experimental design responses were not highly explained by the models, indicating that the system is highly variable, as necessary to enhance the process, or a plateau may have been quickly reached, which challenges the explanation of the variability in the models. However, the hydrolysates showed potential as proteins/peptides sources with antioxidant properties, bringing interest to the results. In both experiments, interesting protein values were obtained with a few hours of hydrolysis, which is in line with other studies showing that enzymatic hydrolysis with papain for 2 h was enough for achieving the maximum protein extraction [25]. The obtained protein content (450 and 463 mg protein/g hydrolysate) was close to those obtained for *Mytilus edulis* by Vareltzis and Undeland (430 and 580 mg protein/g with acid and alkaline process, respectively) [26], but lower than those obtained by Neves et al. (735.45 ± 11.45 mg protein/g hydrolysate) [15]. The subtilisin method needs a lower enzyme% to obtain higher bioactive properties than the corolase, with the main difference being observed for the anti-hypertensive potential. Even though the corolase hydrolysate shows a higher protein %, this does not bring much potential for this hydrolysate due to the small difference compared to the subtilisin hydrolysate. So, mussel meat hydrolysate produced with the subtilisin protease appears to have more potential for further studies as an active ingredient, at least regarding the antioxidant and anti-hypertensive potential. However, it is important to highlight that the obtained values for the anti-hypertensive property are not very significant since $IC_{50} \geq 1000$ µg protein/mL [27]. The hydrolysate produced with corolase shows the lowest potential with an $IC_{50} = 3700$ µg protein/mL. The subtilisin hydrolysate seems to be more promising, with an $IC_{50} = 1000$ µg protein/mL. Bioactive peptides usually have a molecular weight (MW) less than 6 KDa [28], and the most efficient anti-hypertensive peptides are usually associated with MW lower than 3 KDa [29]. Several marine derived peptides with MW lower than 3 KDa have been described, such as the microalgae *Chlorella vulgaris* VECYGPNRPQF peptide (1.3 KDa; IC_{50} of 29.6 µM) [30] and *C. ellipsoidea* VEGY peptide (467 Da; IC_{50} of 128.4 µM) [31]; the macroalgae *Gracilariopsis lemaneiformis* TGAPCR peptide (604 Da; IC_{50} of 23.94 µM) [32] and *Nannochloropsis oculata* LEQ peptide (369 Da; IC_{50} of 173 µM) [33]. Thus, to increase the anti-hypertensive potential of the produced hydrolysates, a future approach may be to submit them to a ultrafiltration system using 3-KDa cut-offs, to concentrate peptides with lower MW [34]. Furthermore, the production of low MW peptides may also increase the antimicrobial potential of hydrolysates, thus presenting new possible pplications.

All the hydrolysis performed used the same mussel batch, initially minced and stored at −20 °C. The main goal was to assure that all the hydrolysis were performed with minimum mussel internal variations, since we wanted to compare a large number of extracts to optimize the hydrolysate production. The validation and scaled-up hydrolysis were also performed with the same batch, allowing us to precisely compare these extracts with those obtained using the experimental design, excluding possible mussel internal chemical variations. However, it is important to point out that mussel meat biochemical composition varies with the harvesting season, due to their reproductive cycle, environmental conditions, growth, and food availability [24]. Çelik et al. [24] showed that mussel protein content is highly related with the spawning seasons, with decreased protein levels being observed during this season, which increases after spawning time. So, different harvesting seasons lead to variations in the biochemical composition, which may be reflected in differences in mussel protein and amino acids, not only in terms of quantity, but also quality. Consequently, the enzyme action will produce different peptides over the seasons. Therefore, it would be expected that the ORAC and iACE results obtained for hydrolysates produced with the presented methods may differ between different mussel batches, depending on their harvesting season and other external factors. Furthermore, the mussel's digestive gland produces proteases, which also seems to be influenced by their diet [35], and mussels seem able to modulate their digestive enzyme activities in response to limited feeding and thermal stress [36]. Since endogenous proteases may also have either a proteolytic effect or serve as an enzymatic substrate in the hydrolysis, the amount of endogenous proteases may also contribute to the variability of results. Thus, in the future, it would be of great interest to perform the same hydrolysis in different mussel batches, harvested in different months, and perhaps from different locations, to examine the variability of the produced hydrolysates when influenced by the expected biochemical composition differences.

The production of multifunctional extracts from mussels may be an interesting approach for food applications since they are not only a source of proteins, but also present bioactivities that can enhance consumer health, useful for the creation of functional food. Moreover, they may also be used as nutraceuticals or as cosmetic ingredients. Antioxidant food and nutraceuticals may help reduce levels of radical oxygen species that are constantly produced by the human organism, especially during high exposure to external factors, such as alcohol, tobacco smoke, and stress [37]. Hypertension has been associated as one of the main causes of cardiovascular diseases [38], with the angiotensin-converting enzyme (ACE) being one of the major enzymes involved in the process of blood pressure regulation [39]. Thus, multifunctional extracts may be incorporated in food matrices with health claims, to facilitate sale as functional food. However, for claiming health benefits, it is important to study the bioavailability of food matrices incorporating these hydrolysates [40], by analysing their resistance to the gastrointestinal (GI) tract enzymes and conditions, to verify if their properties are maintained throughout the GI tract passage [41]. In cosmetics, antioxidants are especially important for anti-ageing purposes since free radicals are highly associated with skin ageing. Thus, natural antioxidant hydrolysates used as active cosmetic ingredients may help decrease free radical damage and work as an alternative for synthetic antioxidant ingredients.

Furthermore, mussel protein and peptide hydrolysates are frequently associated with other properties, especially antimicrobial properties [42,43], but also anticancer [44], anti-inflammatory [45], anticoagulant [14], antidiabetic [15], and antiviral [46]. Thus, in the future, it would be interesting to study these hydrolysates for other bioactivities. Additionally, the water-soluble nature of these extracts makes it easy to incorporate them in several matrices. Although the freeze-drying process may lead to a loss of bioactivity, it is important for a better preservation of the hydrolysates, facilitating their incorporation in both solid and liquid matrices.

4. Materials and Methods

4.1. Materials

The enzymes used were subtilisin kindly supplied by Aquitex, and the commercial digestive-enzyme complex Corolase PP purchased from AB Enzymes GmbH (Darmstadt, Germany). The mussels were kindly supplied by Testa & Cunhas (Gafanha da Nazaré, Portugal).

4.2. Mytilus Galloprovincialis Meat Characterisation

The *Mytilus galloprovincialis* meat used was characterised, in triplicate, before being minced, according to a few nutritional characteristics. Total fat, protein, and moisture content were measured in accordance with the established standards PE.Q.AC.04 Ed.06, PE.Q.AC.03 Ed.07 (ISO 1871:2009), and PE.Q.AC.01 Ed.06 (NP 2282:2009), respectively.

4.3. Enzymatic Hydrolysis Procedures

When received, mussels were clean, and the meat was separated from the shell. Mussel meat was then minced until homogenised and stored at −20 °C for further analysis. A preliminary study was performed with different conditions, with variations on the enzymes concentration (0.5–4%), hydrolysis time (from 30 min to 4 h), and mussel/water ratio (*w:v*) (1:1, 1:2, 1:3), to understand the limits to be established for the experimental design. Concerning the experimental design, all the hydrolysis reactions for both enzymes were prepared using the previously stored mussel meat minced biomass. Briefly, mussel biomass was mixed with ultrapure water in a ratio of mussel:water of 1:2 (*w:v*) and pH was adjusted to 7.5. Then, the enzyme was added in the intended test concentration and the mixtures were incubated at the test temperature in an orbital shaker (Thermo Scientific™ MaxQ™ 6000) (conditions tested at Tables 1 and 7). The pH was verified and adjusted to 7.5 every 15 min. To stop the hydrolysis reaction, the samples were incubated at 90 °C for 10 min to inactivate the enzymes. Samples were centrifuged at 5000× *g* for 30 min, and the supernatant was collected and freeze-dried for further analysis.

4.4. Experimental Design

Two experimental designs, one with corolase and the other with the subtilisin protease, were implemented to establish the most influential factors that could produce a hydrolysate rich in proteins and bioactive properties. For that, a Box–Behnken design was selected. The factors evaluated were enzyme %, hydrolysis temperature (°C), and hydrolysis time (hours). Enzyme % and hydrolysis time were chosen according to single-factor experiments (data not shown). The ORAC assay was performed for each hydrolysate. The temperature and the pH tested were selected according to the functioning range of the enzymes. The levels of the factors coded as −1, 0, and 1 were established and are shown in Table 13. The selected response variables were protein content as well as antioxidant and anti-hypertensive potential. Each design resulted in an arrangement of 15 treatments, executed in duplicate (a total of 30 runs). Each hydrolysis was performed as described before.

Table 13. Levels for 3 experimental factors for the two experimental designs.

Factors	Subtilisin			Corolase		
	−1	0	1	−1	0	1
% Enzyme (X_A)	0.5	1	1.5	1	2	3
Hydrolysis temperature (°C) (X_B)	40	50	60	40	50	60
Hydrolysis time (h) (X_C)	1	2	3	1	2	3

4.5. Statistical Analysis and Statistical Model

The optimisation analysis was performed using Statgraphic Centurion software. All data were expressed as means ± standard deviation (S.D.). Means were considered statistically significant using a significance level of 0.05. Responses were adjusted to the second-order polynomial model (Equation (7)):

$$Y = \beta_0 + \beta_A X_A + \beta_B X_B + \beta_C X_C + \beta_{A,B} X_A X_B + \beta_{A,C} X_A X_C + \beta_{B,C} X_B X_C + \beta_{A,A} X_A{}^2 + \beta_{B,B} X_B{}^2 + \beta_{C,C} X_C{}^2 + \varepsilon \quad (7)$$

where Y is the measured response; β0 is the constant; βA–βC are the coefficients associated with linear, quadratic, and interaction effects of the variables X_A (enzyme %), X_B (Temperature), and X_C (Time), respectively, and ε is the residual error. In the final models for each variable, only the significant effects appear ($p < 0.05$). To optimise the multiple responses obtained, a Derringer's desirability function was applied to the results of each design [22].

4.6. Protein Quantification

Total nitrogen content was determined by the micro-Kjeldahl method. Briefly, 0.2 g of freeze-dried hydrolysate were digested with 1 g of Kjeldahl catalyst and 4 mL of H_2SO_4 (ρ20 = 1.84 g/mL) at 400 °C for 2 h. The reaction was stopped with 20 mL of deionised water. The samples were distilled using 30 mL of NaOH 10 M. A boric acid solution with bromocresol and methyl red was used as indicator. The resulting solution was titrated with HCl 0.1 M. The total nitrogen and protein percentage were determined using the Equations (8) and (9), respectively, where f (HCl 0.1 m) = 0.0014 and Kjeldahl factor = 6.25.

$$\text{Total nitrogen (\%)} = f \times (V_{sample} - V_{blank}) \times (100/\text{sample weight}) \quad (8)$$

$$\text{Protein content (\%)} = \text{Total nitrogen (\%)} \times 6.25 \quad (9)$$

4.7. Antioxidant Activity

The antioxidant activity was measured by the Oxygen radical absorbance capacity (ORAC) assay, performed in a black 96-well microplate (Nunc, Denmark) according to the method described by Coscueta et al. (2020) [47]. Briefly, the reaction was carried out in 75 mM phosphate buffer (pH 7.4) at 40 °C. The final assay mixture was 200 μL, containing fluorescein (70 nM, final concentration in well), 2′-Azobis (2-methylpropionamidine) dihydrochloride (AAPH) (12 mM, final concentration in well), and either Trolox (1–8 μM, final concentration in well), for the calibration curve, or sample. A control with PBS instead of the antioxidant solution was used. Before adding AAPH, the mixture was pre-incubated for 10 min at 37 °C. AAPH solution was added rapidly. The fluorescence was recorded at intervals of 1min for 80 min in a multidetection plate reader (Synergy H1; BioTek, Winooski VT, USA) with excitation and emission wavelengths of 485 nm and 528 nm, respectively. The equipment was controlled by the Gen5 BioTek software version 3.04. Antioxidant curves (fluorescence versus time) were first normalised to the curve of the blank corresponding to the same assay by multiplying original data by the factor fluorescence blank, t = 0/fluorescence control, t = 0. The area under the fluorescence decay curve (AUC) was calculated according to the trapezoidal method from the normalised curves. The final AUC values were calculated by subtracting the AUC of the blank from all the results. Regression equations between net AUC and antioxidant concentration were calculated.

4.8. Anti-Hypertensive Activity

The ACE-inhibitory activity was performed in a black 96-well microplate (Nunc, Denmark) according to the method described by Sentandreu & Toldrá (2006) [48] with some modifications [27]. This method is based on the ability of the angiotensin-I converting enzyme (ACE) to hydrolyse a specific substrate (o-aminobenzoylglycyl-p-nitrophenylalanyl-proline (Abz–Gly–Phe(NO2)–Pro)), generating the fluorescent product o-aminobenzoylglycine (Abz–Gly). A commercial Angiotensin-I converting enzyme (EC 3.4.15.1, 5.1 U/mg), purchased from Sigma Chemical (St. Louis, MO, USA), was previously diluted in 5 mL of a glycerol solution in 50% ultra-pure water. Then, ACE was diluted 1:24 with a 150 mM Tris buffer solution (pH 8.3), containing 1 μM of ZnCL2, reaching a final concentration of 42 mU/mL). A total of 40 μL of ultrapure water or ACE working solution was added to each microtiter-plate well, then adjusted to 80 μL by adding ultrapure water to blank, control, or samples. The reaction was initiated with the addition of 160 μL of the substrate solution (0.45 mM solution of ABz-Gly-Phe(NO2)-Pro (Bachem Feinchemikalien, Bubendorf, Switzerland) dissolved in 150 mM Tris buffer (pH 8.3) containing 1.125 M NaCl). The mixture was incubated at 37 °C for 30 min, and the fluorescence generated was measured using a multidetection plate reader (Synergy H1; BioTek, Winooski VT Vermont, USA) with excitation and emission wavelengths of 350 nm and 420 nm, respectively. To obtain the IC_{50} of the inhibitory activity, which is the concentration of the sample that is required to inhibit the original ACE activity by 50%, serial dilutions of each sample were performed (1/1 to 1/32). A non-linear modelling of the obtained data was used to calculate the IC_{50} values, using the 5 Parameter curve fit method and the Interpolate function from Gen5 software (BioTek Instruments).

5. Conclusions

Although marine species have often been described as a source of bioactive hydrolysates and bioactive peptides, the mussel *Mytilus galloprovincialis* has been less exploited. Due to its high protein level, this marine specie seems to be an interesting potential source of bioactive peptides. Thus, in this work, the factorial designs allowed to confirm the combination of experimental factors that leads to the production of the most efficient hydrolysate from the mussel *Mytilus galloprovincialis,* with the highest levels of proteins/peptides as well as antioxidant and anti-hypertensive activity. The use of enzymatic hydrolysis with food-grade enzymes presents the opportunity to create active ingredients that can be further explored to produce functional food, nutraceuticals, and cosmetics. Furthermore, the use of discarded mussels to produce functional ingredients for food, cosmetic, and pharmaceutic industries may contribute to the valorisation of world waste in a circular economy context.

Author Contributions: Conceptualisation, S.A.C. and M.P.; methodology, S.A.C., R.d.C. and M.P.; software, E.R.C.; validation, S.A.C. and R.d.C.; formal analysis, S.A.C., E.R.C. and M.P.; investigation, S.A.C. and M.P.; resources, M.P.; data curation, S.A.C. and E.R.C.; writing—original draft preparation, S.A.C.; writing—review and editing, S.A.C., E.R.C. and M.P.; visualisation, S.A.C. and E.R.C.; supervision, M.P.; project administration, M.P.; funding acquisition, S.A.C. and M.P. All authors have read and agreed to the published version of the manuscript.

Funding: This research and the APC was funded by Fundo Europeu de Desenvolvimento Regional (FEDER), through the Programa Operacional Competitividade e Internacionalização (POCI) under the project VALORMAR: Valorização Integral dos recursos Marinhos: Potencial, Inovação Tecnológica e Novas Aplicações (POCI-01-0247-FEDER-024517); CBQF under the FCT project UIDB/50016/2020; and the individual FCT PhD research grant (ref. SFRH/BD/144155/2019) for the author Sara A. Cunha.

Institutional Review Board Statement: Not applicable.

Informed Consent Statement: Not applicable.

Data Availability Statement: Not applicable.

Acknowledgments: The authors acknowledge Testa & Cunhas (Gafanha da Nazaré, Portugal) for the *Mytilus galloprovincialis* supply; and Aquitex S.A. for the subtilisin protease supply.

Conflicts of Interest: The authors declare no conflict of interest.

Sample Availability: Samples of the compounds are not available from the authors.

References

1. *Fao Globefish Research Programme Food and Agriculture Organization of the United Nations Products The European Market for Mussels*; FAO: Rome, Italy, 2014; Volume 115.
2. Tamburini, E.; Turolla, E.; Fano, E.A.; Castaldelli, G. Sustainability of Mussel (Mytilus galloprovincialis) farming in the Po River delta, northern Italy, based on a life cycle assessment approach. *Sustainability* **2020**, *12*, 3814. [CrossRef]
3. Yaghubi, E.; Carboni, S.; Snipe, R.M.J.; Shaw, C.S.; Fyfe, J.J.; Smith, C.M.; Kaur, G.; Tan, S.-Y.; Hamilton, D.L. Farmed Mussels: A Nutritive Protein Source, Rich in Omega-3 Fatty Acids, with a Low Environmental Footprint. *Nutrients* **2021**, *13*, 1124. [CrossRef]
4. Naik, A.S.; Hayes, M. Bioprocessing of mussel by-products for value added ingredients. *Trends Food Sci. Technol.* **2019**, *92*, 111–121. [CrossRef]
5. Lemes, A.C.; Sala, L.; Ores, J.D.C.; Braga, A.R.C.; Egea, M.B.; Fernandes, K.F. A review of the latest advances in encrypted bioactive peptides from protein-richwaste. *Int. J. Mol. Sci.* **2016**, *17*, 950. [CrossRef]
6. Harnedy, P.A.; FitzGerald, R.J. Bioactive peptides from marine processing waste and shellfish: A review. *J. Funct. Foods* **2012**, *4*, 6–24. [CrossRef]
7. Kim, E.K.; Joung, H.J.; Kim, Y.S.; Hwang, J.W.; Ahn, C.B.; Jeon, Y.J.; Moon, S.H.; Song, B.C.; Park, P.J. Purification of a novel anticancer peptide from enzymatic hydrolysate of Mytilus coruscus. *J. Microbiol. Biotechnol.* **2012**, *22*, 1381–1387. [CrossRef] [PubMed]
8. Kim, E.K.; Oh, H.J.; Kim, Y.S.; Hwang, J.W.; Ahn, C.B.; Lee, J.S.; Jeon, Y.J.; Moon, S.H.; Sung, S.H.; Jeon, B.T.; et al. Purification of a novel peptide derived from Mytilus coruscus and in vitro/in vivo evaluation of its bioactive properties. *Fish Shellfish Immunol.* **2013**, *34*, 1078–1084. [CrossRef]
9. Oh, R.; Lee, M.J.; Kim, Y.O.; Nam, B.H.; Kong, H.J.; Kim, J.W.; Park, J.Y.; Seo, J.K.; Kim, D.G. Myticusin-beta, antimicrobial peptide from the marine bivalve, Mytilus coruscus. *Fish Shellfish Immunol.* **2020**, *99*, 342–352. [CrossRef] [PubMed]
10. Jung, W.K.; Qian, Z.J.; Lee, S.H.; Sun, Y.C.; Nak, J.S.; Byun, H.G.; Kim, S.K. Free radical scavenging activity of a novel antioxidative peptide isolated from in vitro gastrointestinal digests of Mytilus coruscus. *J. Med. Food* **2007**, *10*, 197–202. [CrossRef] [PubMed]
11. Je, J.Y.; Park, P.J.; Byun, H.G.; Jung, W.K.; Kim, S.K. Angiotensin I converting enzyme (ACE) inhibitory peptide derived from the sauce of fermented blue mussel, Mytilus edulis. *Bioresour. Technol.* **2005**, *96*, 1624–1629. [CrossRef]
12. Qin, C.L.; Huang, W.; Zhou, S.Q.; Wang, X.C.; Liu, H.H.; Fan, M.H.; Wang, R.X.; Gao, P.; Liao, Z. Characterization of a novel antimicrobial peptide with chiting-biding domain from Mytilus coruscus. *Fish Shellfish Immunol.* **2014**, *41*, 362–370. [CrossRef]
13. Kim, Y.S.; Ahn, C.B.; Je, J.Y. Anti-inflammatory action of high molecular weight Mytilus edulis hydrolysates fraction in LPS-induced RAW264.7 macrophage via NF-κB and MAPK pathways. *Food Chem.* **2016**, *202*, 9–14. [CrossRef]
14. Jung, W.K.; Kim, S.K. Isolation and characterisation of an anticoagulant oligopeptide from blue mussel, Mytilus edulis. *Food Chem.* **2009**, *117*, 687–692. [CrossRef]
15. Neves, A.C.; Harnedy, P.A.; FitzGerald, R.J. Angiotensin Converting Enzyme and Dipeptidyl Peptidase-IV Inhibitory, and Antioxidant Activities of a Blue Mussel (Mytilus edulis) Meat Protein Extract and Its Hydrolysates. *J. Aquat. Food Prod. Technol.* **2016**, *25*, 1221–1233. [CrossRef]

16. Balseiro, P.; Falcó, A.; Romero, A.; Dios, S.; Martínez-López, A.; Figueras, A.; Estepa, A.; Novoa, B. Mytilus galloprovincialis Myticin C: A Chemotactic Molecule with Antiviral Activity and Immunoregulatory Properties. *PLoS ONE* **2011**, *6*, e23140. [CrossRef]
17. Medina Uzcátegui, L.U.; Vergara, K.; Martínez Bordes, G. Sustainable alternatives for by-products derived from industrial mussel processing: A critical review. *Waste Manag. Res.* **2021**, 734242X21996808. [CrossRef]
18. Murphy, J.N.; Hawboldt, K.; Kerton, F.M. Enzymatic processing of mussel shells to produce biorenewable calcium carbonate in seawater. *Green Chem.* **2018**, *20*, 2913–2920. [CrossRef]
19. Barnaby, C. An investigation into the Reuse of Organic Waste Produced by the New Zealand Mussel Industry. Doctoral Dissertation, Auckland University of Technology, Auckland, New Zealand, 2004.
20. Naik, A.S.; Mora, L.; Hayes, M. Characterisation of Seasonal Mytilus edulis By-Products and Generation of Bioactive Hydrolysates. *Appl. Sci.* **2020**, *10*, 6892. [CrossRef]
21. Kim, E.K.; Kim, Y.S.; Hwang, J.W.; Kang, S.H.; Choi, D.K.; Lee, K.H.; Lee, J.S.; Moon, S.H.; Jeon, B.T.; Park, P.J. Purification of a novel nitric oxide inhibitory peptide derived from enzymatic hydrolysates of Mytilus coruscus. *Fish Shellfish Immunol.* **2013**, *34*, 1416–1420. [CrossRef] [PubMed]
22. Suich, R.; Derringer, G.C. Is the regression equation adequate?—A further note. *Technometrics* **1980**, *22*, 125–126. [CrossRef]
23. Oliveira, A.R.; Sykes, A.V.; Hachero-Cruzado, I.; Azeiteiro, U.M.; Esteves, E. A sensory and nutritional comparison of mussels (Mytilus sp.) produced in NW Iberia and in the Armona offshore production area (Algarve, Portugal). *Food Chem.* **2015**, *168*, 520–528. [CrossRef]
24. Çelik, M.Y.; Karayücel, S.; Karayücel, I.; Öztürk, R.; Eyüboğlu, B. Meat yield, condition index, and biochemical composition of mussels (Mytilus galloprovincialis Lamarck, 1819) in sinop, south of the Black Sea. *J. Aquat. Food Prod. Technol.* **2012**, *21*, 198–205. [CrossRef]
25. Zhou, Y.; He, Q.; Zhou, D. Optimization Extraction of Protein from Mussel by High-Intensity Pulsed Electric Fields. *J. Food Process. Preserv.* **2017**, *41*, e12962. [CrossRef]
26. Vareltzis, P.K.; Undeland, I. Protein isolation from blue mussels (Mytilus edulis) using an acid and alkaline solubilisation technique–process characteristics and functionality of the isolates. *J. Sci. Food Agric.* **2012**, *92*, 3055–3064. [CrossRef] [PubMed]
27. Coscueta, E.R.; Amorim, M.M.; Voss, G.B.; Nerli, B.B.; Picó, G.A.; Pintado, M.E. Bioactive properties of peptides obtained from Argentinian defatted soy flour protein by Corolase PP hydrolysis. *Food Chem.* **2016**, *198*, 36–44. [CrossRef] [PubMed]
28. Karami, Z.; Akbari-Adergani, B. Bioactive food derived peptides: A review on correlation between structure of bioactive peptides and their functional properties. *J. Food Sci. Technol.* **2019**, *56*, 535–547. [CrossRef]
29. Alcaide-Hidalgo, J.M.; Romero, M.; Duarte, J.; López-Huertas, E. Antihypertensive Effects of Virgin Olive Oil (Unfiltered) Low Molecular Weight Peptides with ACE Inhibitory Activity in Spontaneously Hypertensive Rats. *Nutrients* **2020**, *12*, 271. [CrossRef] [PubMed]
30. Sheih, I.C.; Fang, T.J.; Wu, T.K. Isolation and characterisation of a novel angiotensin I-converting enzyme (ACE) inhibitory peptide from the algae protein waste. *Food Chem.* **2009**, *115*, 279–284. [CrossRef]
31. Ko, S.C.; Kang, N.; Kim, E.A.; Kang, M.C.; Lee, S.H.; Kang, S.M.; Lee, J.B.; Jeon, B.T.; Kim, S.K.; Park, S.J.; et al. A novel angiotensin I-converting enzyme (ACE) inhibitory peptide from a marine Chlorella ellipsoidea and its antihypertensive effect in spontaneously hypertensive rats. *Process Biochem.* **2012**, *47*, 2005–2011. [CrossRef]
32. Deng, Z.; Liu, Y.; Wang, J.; Wu, S.; Geng, L.; Sui, Z.; Zhang, Q. Antihypertensive effects of two novel angiotensin i-converting enzyme (ace) inhibitory peptides from gracilariopsis lemaneiformis (Rhodophyta) in spontaneously hypertensive rats (SHRs). *Mar. Drugs* **2018**, *16*, 299. [CrossRef]
33. Samarakoon, K.; Jeon, Y.J. Bio-functionalities of proteins derived from marine algae—A review. *Food Res. Int.* **2012**, *48*, 948–960. [CrossRef]
34. Raghavan, S.; Kristinsson, H.G. ACE-inhibitory activity of tilapia protein hydrolysates. *Food Chem.* **2009**, *117*, 582–588. [CrossRef]
35. Van Weel, P.B. The comparative physiology of digestion in molluscs. *Integr. Comp. Biol.* **1961**, *1*, 245–252. [CrossRef]
36. Connor, K.M.; Sung, A.; Garcia, N.S.; Gracey, A.Y.; German, D.P. Modulation of digestive physiology and biochemistry in Mytilus californianus in response to feeding level acclimation and microhabitat. *Biol. Open* **2016**, *5*, 1200. [CrossRef] [PubMed]
37. Lobo, V.; Patil, A.; Phatak, A.; Chandra, N. Free radicals, antioxidants and functional foods: Impact on human health. *Pharmacogn. Rev.* **2010**, *4*, 118–126. [CrossRef]
38. Nova, P.; Pimenta-Martins, A.; Laranjeira Silva, J.; Silva, A.M.; Gomes, A.M.; Freitas, A.C. Health benefits and bioavailability of marine resources components that contribute to health—what's new? *Crit. Rev. Food Sci. Nutr.* **2020**, *60*, 3680–3692. [CrossRef]
39. Cao, D.; Lv, X.; Xu, X.; Yu, H.; Sun, X.; Xu, N. Purification and identification of a novel ACE inhibitory peptide from marine alga Gracilariopsis lemaneiformis protein hydrolysate. *Eur. Food Res. Technol.* **2017**, *243*, 1829–1837. [CrossRef]
40. Nova, P.; Martins, A.P.; Teixeira, C.; Abreu, H.; Silva, J.G.; Silva, A.M.; Freitas, A.C.; Gomes, A.M. Foods with microalgae and seaweeds fostering consumers health: A review on scientific and market innovations. *J. Appl. Phycol.* **2020**, *32*, 1789–1802. [CrossRef]
41. Bhandari, D.; Rafiq, S.; Gat, Y.; Gat, P.; Waghmare, R.; Kumar, V. A Review on Bioactive Peptides: Physiological Functions, Bioavailability and Safety. *Int. J. Pept. Res. Ther.* **2020**, *26*, 139–150. [CrossRef]
42. Semreen, M.H.; El-Gamal, M.I.; Abdin, S.; Alkhazraji, H.; Kamal, L.; Hammad, S.; El-Awady, F.; Waleed, D.; Kourbaj, L. Recent updates of marine antimicrobial peptides. *Saudi Pharm. J.* **2018**, *26*, 396–409. [CrossRef]

43. Tsankova, G.; Todorova, T.; Ermenlieva, N.; Merdzhanova, A.; Panayotova, V.; Dobreva, D.; Peytcheva, K. Antibacterial activity of different extracts tracts of black mussel (Mytilus galloprovincialis) from the black sea, Bulgaria. *J. IMAB* **2021**, *27*, 3506–3509. [CrossRef]
44. Kim, E.K.; Hwang, J.W.; Kim, Y.S.; Ahn, C.B.; Jeon, Y.J.; Kweon, H.J.; Bahk, Y.Y.; Moon, S.H.; Jeon, B.T.; Park, P.J. A novel bioactive peptide derived from enzymatic hydrolysis of Ruditapes philippinarum: Purification and investigation of its free-radical quenching potential. *Process Biochem.* **2013**, *48*, 325–330. [CrossRef]
45. Park, S.Y.; Kim, Y.S.; Ahn, C.B.; Je, J.Y. Partial purification and identification of three antioxidant peptides with hepatoprotective effects from blue mussel (Mytilus edulis) hydrolysate by peptic hydrolysis. *J. Funct. Foods* **2016**, *20*, 88–95. [CrossRef]
46. Novoa, B.; Romero, A.; Álvarez, Á.L.; Moreira, R.; Pereiro, P.; Costa, M.M.; Dios, S.; Estepa, A.; Parra, F.; Figueras, A. Antiviral Activity of Myticin C Peptide from Mussel: An Ancient Defense against Herpesviruses. *J. Virol.* **2016**, *90*, 7692–7702. [CrossRef] [PubMed]
47. Coscueta, E.R.; Reis, C.A.; Pintado, M. Phenylethyl Isothiocyanate Extracted from Watercress By-Products with Aqueous Micellar Systems: Development and Optimisation. *Antioxidants* **2020**, *9*, 698. [CrossRef] [PubMed]
48. Sentandreu, M.Á.; Toldrá, F. A rapid, simple and sensitive fluorescence method for the assay of angiotensin-I converting enzyme. *Food Chem.* **2006**, *97*, 546–554. [CrossRef]

molecules

Review

Phages and Enzybiotics in Food Biopreservation

José Ramos-Vivas [1,2], María Elexpuru-Zabaleta [1], María Luisa Samano [1,2], Alina Pascual Barrera [2], Tamara Y. Forbes-Hernández [3], Francesca Giampieri [4,5,*] and Maurizio Battino [4,6,*]

1 Research Group on Foods, Nutritional Biochemistry and Health, Universidad Europea del Atlántico, 39011 Santander, Spain; jose.ramos@uneatlantico.es (J.R.-V.); maria.elexpuru@uneatlantico.es (M.E.-Z.); marialuisa.samano@unini.edu.mx (M.L.S.)
2 Department of Project Management, Universidad Internacional Iberoamericana, Campeche 24560, Mexico; alina.pascual@unini.edu.mx
3 Department of Analytical and Food Chemistry, CITACA, CACTI, University of Vigo, 36310 Vigo, Spain; tforbes@uvigo.es
4 Department of Clinical Sciences, Polytechnic University of Marche, 60131 Ancona, Italy
5 Department of Biochemistry, Faculty of Sciences, King Abdulaziz University, Jeddah 21589, Saudi Arabia
6 International Research Center for Food Nutrition and Safety, Jiangsu University, Zhenjiang 212013, China
* Correspondence: f.giampieri@univpm.it (F.G.); m.a.battino@univpm.it (M.B.); Tel.: +339-071-220-4136 (F.G.); +339-071-220-4646 (M.B.)

Abstract: Presently, biopreservation through protective bacterial cultures and their antimicrobial products or using antibacterial compounds derived from plants are proposed as feasible strategies to maintain the long shelf-life of products. Another emerging category of food biopreservatives are bacteriophages or their antibacterial enzymes called "phage lysins" or "enzybiotics", which can be used directly as antibacterial agents due to their ability to act on the membranes of bacteria and destroy them. Bacteriophages are an alternative to antimicrobials in the fight against bacteria, mainly because they have a practically unique host range that gives them great specificity. In addition to their potential ability to specifically control strains of pathogenic bacteria, their use does not generate a negative environmental impact as in the case of antibiotics. Both phages and their enzymes can favor a reduction in antibiotic use, which is desirable given the alarming increase in resistance to antibiotics used not only in human medicine but also in veterinary medicine, agriculture, and in general all processes of manufacturing, preservation, and distribution of food. We present here an overview of the scientific background of phages and enzybiotics in the food industry, as well as food applications of these biopreservatives.

Keywords: bacteriophage; endolysin; enzybiotics; biopreservation

Citation: Ramos-Vivas, J.; Elexpuru-Zabaleta, M.; Samano, M.L.; Barrera, A.P.; Forbes-Hernández, T.Y.; Giampieri, F.; Battino, M. Phages and Enzybiotics in Food Biopreservation. *Molecules* **2021**, *26*, 5138. https://doi.org/10.3390/molecules26175138

Academic Editors: Manuela Pintado, Ezequiel Coscueta, María Emilia Brassesco and Ricardo Calhelha

Received: 13 July 2021
Accepted: 20 August 2021
Published: 25 August 2021

Publisher's Note: MDPI stays neutral with regard to jurisdictional claims in published maps and institutional affiliations.

1. Introduction

Food preservation by suitable means is key in food safety and quality. There are several traditional and well-known physical preservation techniques such as refrigeration and pasteurization, but the modern industry is always looking for new procedures for food preservation to increase the product's shelf-life by minimizing the loss of nutritional quality and organoleptic properties. Presently, some modern biopreservation techniques rely on naturally occurring microorganisms (i.e., lactic acid bacteria) and their metabolites. These food preservatives are mainly used to produce safer food for the consumer, preventing the action of pernicious microbes which can cause food deterioration or even toxicity and therefore be dangerous to human health.

Moreover, bacteria -including multidrug-resistant bacteria- can reach food at different points in the food supply chain, from farm to postharvest, and processing such as slaughtering, fermentation, packaging and storage [1–5].

As most natural foods are highly perishable, by extending their half-life we can also control their native microbiota for proper preservation, maintaining their safety and quality.

As microorganisms produce a long list of molecules ranging from classic antibiotics to antibacterial enzymes, the control of indigenous populations in food can be achieved by adding these products directly. The paradigm of bacterial molecules used in the food industry as biopreservatives is Nisin, a bacteriocin produced by the Gram-positive bacterium *Lactococcus lactis*, one of the lactic acid bacteria most extensively used for the manufacture of dairy products [6]. Other well-known bacteriocins, such as Pediocin, Natamycin, Enterocin, and Leucocin [7], also have inhibitory properties against other microorganisms which makes them very interesting for use in the food industry. Some bacteria that produce these compounds have been used as probiotics. Current research on probiotics is quite promising and modern fashion trends push probiotics and bacteriocins from modulation of the gut microbiota toward a wide range of other health-promoting activities away from food, such as cancer treatment, skin health care, periodontal health, or allergies [8–11].

In addition, the use of bacteriocin producing strains or those that can compete against pathogens in the context of the food industry needs new approaches, mainly due to the increase in foodborne infections, the appearance of new production processes, the massive demand for food, and the changing consumer trends. Moreover, the extensive use of antibiotics against animal and human pathogens has also led to an increase in foodborne pathogens resistant to antibiotics, which makes the picture not reassuring at all [12–14].

Goodridge and Abedon published an article in 2003 where they proposed to use the terms "phage biocontrol" and "phage bioprocessing" to differentiate the application of bacteriophages in the farm or crops from their use in the food industry [15]. Several years later, Greer published a review of the control of foodborne bacteria using phages, including the effects of these microorganisms on food storage and preservation [16].

At that time, the excellent properties of endolysins to kill bacteria were already known, but their use to protect food from foodborne pathogens had not yet been effectively tested. One of the first murein hydrolases to be studied concerning food-related bacteria was that of the *Lactobacillus helveticus* bacteriophage 0303 [17]. This endolysin exhibited a broad spectrum of activity, killing different bacterial species such *Pediococcus acidilactici*, *Lactobacillus delbrueckii* subsp. *bulgaricus*, *Lactobacillus delbrueckii* subsp. *lactis*, *Lactobacillus acidophilus*, *Bacillus subtilis*, *Enterococcus faecium*, and several strains of *Lactobacillus helveticus*.

Problems of deterioration of the organoleptic properties have been described after physical treatments; also, consumers are increasingly demanding low-processed foods. One of the advantages of phages over the usual physical treatments is that phages do not modify any organoleptic properties of foods. Moreover, even with common treatments such as heat, team and UV light, a relatively high percentage of food products are lost due to subsequent microbial spoilage or microbial contamination; when food becomes contaminated, it will lead to food spoilage, and such food will no longer be fit for consumption.

Thanks to their ability to control or to inactivate spoilage and/or foodborne bacteria selectively, bacteriophages have great potential as food biopreservatives. Additionally, in terms of food biopreservation, enzybiotics are beginning to be increasingly studied in the field of food microbiology, taking advantage of the pull that in vitro successes have displayed against very important multidrug-resistant human and animal pathogens [18–20].

In this review, we discuss the use of phages and their lytic enzymes as a tool to eliminate or reduce spoilage bacteria and common foodborne bacterial pathogens.

2. Why Bacteriophages?

Bacteriophages are an alternative to antimicrobials in the fight against bacteria, mainly because they have a practically unique host range, which gives them great specificity. Apart from their selective activity, bacteriophages have been successfully tested to eliminate or weaken biofilms formed by different classes of both Gram-negative and Gram-positive pathogens in the food industry [21–24]. Biofilms are consortia of bacteria that persist on different surfaces or pipelines within the food industries that contaminate food at some point in the processing or packaging chain.

In addition to their potential ability to specifically control strains and biofilms of pathogenic bacteria, their use does not generate a negative environmental impact like in the case of antibiotics or disinfectants [25]. Other advantages of these viruses are: (i) safety—as they are not toxic to eukaryotic cells, (ii) the preservation of the organoleptic properties of food, and (iii) the control of multi-resistant bacteria since the tolerance of some strains to phages can often be overcome with the use of phage cocktails [14]. In addition, phages can be used in combination with antibiotics, bacteriocins, or even with probiotics.

The main limitations of bacteriophages as biopreservative tools in foods derive from the scarce knowledge of their genetics since the use of strains that may contain virulence factors, lysogeny, or antibiotic resistance genes is inadvisable. As an example, studies prior to this decade did not have the modern and inexpensive sequencing techniques that almost all laboratories can afford today. Furthermore, in some cases, it is necessary to use phage cocktails that are more difficult to characterize than individual strains. Additionally, we need to learn much more about their behavior within solid and liquid food matrices to optimize the amount of phage to be used in each case. The method of releasing phages on food is also important, since the phages must reach the largest number of bacteria possible so that they can effectively control them and reduce their number to safe values. In other words, phages and bacteria must be in contact with liquid but also with solid foods; moreover, as much bacterial contamination occurs initially at low numbers (a minimum bacterial density is a prerequisite) sometimes we must apply a large number of phages to those foods. Knowing the optimal number of viral particles (multiplicity of infection, MOI) to use for each food, as well as their infection kinetics in each food matrix, it is essential to understand how these phages are acting on their target pathogens [26–32]. Minimum host threshold requirement has been demonstrated for phages of different food pathogens [33,34]. As successful biopreservative agents, it is also important to consider phages' stability in food matrices under different environmental conditions such as water activity, salinity, temperature, pH, osmotic shock, and light (visible and UV). According to several authors, phages have a remarkable stability in foods [35–37]. Phage propagation on a susceptible host, purification, and phage or cocktail formulation are very relevant parameters too.

In some studies, in which a high number of phages are used, the bacterial lysis 'from without' can occur because many viral particles bind to the bacterial surface, leading to the production of numerous holes in the cell wall [38,39]. All these concepts must be better studied and understood in order to apply phages to food pathogens.

Although the application of phages will continue, there is a phenomenon that must always be kept in mind, the emergence of phage-resistant strains. When infecting bacterial cells, phages already face a range of antiviral mechanisms (i.e., restriction modification systems/enzymes), and they have evolved multiple tactics to avoid these mechanisms. In this co-evolution between bacteria and phages, most authors agree that phages can effectively raise a counter-resistance. Therefore, finding a new phage that can infect a bacterium will always be easier than finding an entirely novel family of antibiotics.

We do not know much about how often these resistant variants of phages used in the food industry appear, as few publications include assays to study this phenomenon. It is likely that researchers prioritize the study of efficacy over safety. Moreover, multidrug resistance, where a bacterium has obtained resistance mechanisms against several different families of antibiotics, is increasingly common, but this phenomenon does not occur when phages are used. Additionally, many studies suggest that phage combinations can be optimized to limit the emergence and persistence of resistance, therefore promoting the long-term usefulness of phage therapy. With regards to this issue, enzybiotics offer the advantage that they do not generate resistance because they act on essential targets for the bacteria's viability, so, it is difficult for bacteria to modify them.

The other most important issue in addition to the development of phage-resistant strains is phage spread. As bacteriophages applied to food can be easily transferred between facilities in the food industry, we must pay particular attention to the number of phages used, and above all, to how they are applied to food. An undesirable effect would be the inactivation of starter cultures that initiate the fermentation processes. Despite the narrow spectrum of a specific phage, the problem of the phages spread within the food industries is real because it is not convenient; for example, to collaterally eliminate some species of lactic acid bacteria that confer characteristic properties to the products in which they are present [40].

As with isolated phages, phage cocktails can be used directly on food or surfaces and food handling tools in chain processing plants. Another advantage of phage cocktails is that they can be modified quickly and conveniently to deal with specific strains that may appear in a particular food manufacturing facility [41]. No articles were reviewed here where more than three bacteriophages or cocktails containing undefined strains were used because in the last few years there have been excellent reviews on that scope [26,41–43]. Moreover, Theuretzbacher's recent article in the currently available weaponry against superbugs indicates that more than 20 different bacteriophage-based products have been approved for the control of pathogenic bacteria related to the food industries or direct food contamination [44].

Our review of approximately 100 bacteriophages indicates that three families (*Myoviridae*, *Siphoviridae*, and *Podoviridae*) account for the majority of virulent phages for the most common food-borne pathogen species. Much work has focused on the biocontrol or biopreservation of foods with six of the most important food-borne pathogens: *E. coli* (mainly serotype 015:H7), *Listeria monocytogenes*, *Staphylococcus aureus*, *Clostridium* spp., *Campilobacter jejuni*, and *Salmonella* spp., (Table 1). In addition to those six important food pathogens, phages against many other bacteria capable of causing foodborne infections should begin to be studied. This would allow us to identify not only new phages but also interesting enzybiotics.

Table 1. Phages tested against food-borne pathogens and their proposed use as food biopreservatives.

Target Bacteria	Phage/s	Source	Characterization Method	Genome Length	[2] Family	Food Application	Reference
Aeromonas hydrophila	AH-1 AH-4 AH-5	Sewage samples	TEM	ND	*Myoviridae*	Depuration of artificially contaminated cockles	[45]
Bacillus cereus	PBC1	Sewage sample	TEM, sequencing	41,164 bp	*Siphoviridae*	Inhibition of *B. cereus* growth in boiled rice	[46,47]
Brochothrix thermosphacta	A3	Spoiled retail rib steaks	TEM	ND	ND	Control of bacterial strains during refrigerated storage	[16,48]
Campylobacter jejuni	Cj6	Unknown	-	ND	ND	Control of pathogens in liquid foods	[36,49]
C. jejuni	2	Unknown	[1] dsDNA	~140 kb	*Myoviridae*	Reduction of *C. jejuni* contamination of retail poultry products	[50,51]
C. jejuni	CP8 CP30	Poultry excreta	TEM, dsDNA	~140 kb	*Myoviridae*	Reduction of food-borne bacteria and biofilms	[52,53]
C. jejuni	12673	NCTC (UK)	TEM, DNA sequencing	~135 kb	*Myoviridae*	Reduction of bacterial contamination on chicken carcass surfaces	[54,55]
Clostridium tyrobutyricum and *C. sporogenes*	CTP1	Landfill	TEM, DNA sequencing	59,199 bp	*Siphoviridae*	Cheese manufacturing	[56]
Crcnobacter (*Enterobacter*) *sakazakii*	ESP 1–3 ESP 732–1	Sewage treatment plant	TEM, dsDNA	ND	Siphoviridae *Myoviridae*	Control of *E. sakazakii* in reconstituted infant formula	[57]
Escherichia coli	PE37	Bovine intestine samples	TEM, DNA sequencing	166,423 bp	*Myoviridae*	Biocontrol of *E. coli* STEC O157:H7 and ESBLEC.	[58]
E. coli	EC6 EC9 EC11	Sewage	TEM, dsDNA	ND	Siphoviridae *Myoviridae* *Podoviridae*	Biocontrol against *E. coli* in UHT and raw bovine milk	[59]
E. coli (STEC) O145	Ro145clw	Non-fecal compost samples	TEM, DNA sequencing	42,031 bp	*Siphoviridae*	Control of foodborne STEC O145	[60]
E. coli O157:H7	vB_EcoS_FFH_1 _vB_EcoS_FFH_1	Wastewater treatment plants	TEM, sequencing	108,483 bp 139,020 bp	Siphoviridae *Myoviridae*	Reduction of contamination in ground beef	[61,62]
E. coli O157:H7	e11/2 e4/1c PP01	Bovine farmyard Slurry samples Swine stool samples	TEM, dsDNA	ND ND ~140 kb	Myoviridae *Siphoviridae* *Myoviridae*	Elimination or reduction of *E. coli* O157:H7 bacteria from meat carcasses	[63–65]
E. coli O157:H7	FAHEc1	Raw screened sewage	TEM, dsDNA	~90 kb	*Myoviridae*	Inactivation of *E. coli* O157:H7 on beef	[34,66]

Table 1. *Cont.*

Target Bacteria	Phage/s	Source	Characterization Method	Genome Length	[2] Family	Food Application	Reference
E. coli O157:H7	KH1 KH4 KH5	Cattle and sheep fecal samples	-	ND	ND	Elimination of O157:H7 from foods under refrigerated conditions. Reduction of *E. coli* on surfaces.	[67,68]
E. coli O157:H7	ECML-4 ECML-117 ECML-134	Fresh and salt water environments	TEM, DNA sequencing	157,308 bp 66,854 bp 166,783 bp	*Myoviridae*	Reduction of contamination of hard surfaces and foods contaminated by *E. coli* O157:H7	[69,70]
E. coli strains, *Salmonella* and *Shigella* spp.	C203 P206	Cottage cheese and from poultry liver	TEM, DNA sequencing	138,073 bp	*Myoviridae*	Biocontrol agent against *E. coli* EHEC O157	[37]
Shigatoxigenic *E. coli* Enteropathogenic *E. coli*	DT1 DT5 DT6	Stool samples of patients with diarrhea	TEM	ND	*Myoviridae*	Control of pathogenic *E.coli* in meat products and during milk fermentation	[71,72]
E. coli strains including serotype O157:H7	OSY-SP	Manure from cattle, sheep, and horse farms	Pulsed-field gel electrophoresis (PFGE)	~150 Kb	*Myoviridae*	Reduction of *E. coli* in fresh produce type (cut green pepper or spinach leaves)	[73]
Lactobacillus brevis	SA-C12	fresh silage	TEM	ND	*Myoviridae*	Control of *L. brevis* beer-spoilage	[74]
Leuconostoc gelidum	ggg	vacuum-packaged pork	TEM	ND	*Siphoviridae*	Inhibition of *Leuconostoc* in raw pork	[75]
Listeria monocytogenes	A500 ATCC® 23074-B1™	*L. monocytogenes* isolated from Guinea pig	TEM	38,867 bp	*Siphoviridae*	Control of L-forms of *L. monocytogenes* on surfaces	[76,77]
L. monocytogenes	H387 H387-A 2671	-	TEM	ND	*Siphoviridae*	Disinfection of working surfaces of food processing plants	[78,79]
L. monocytogenes	LiMN4L LiMN4p LiMN17	Seafood waste water treatment unit	ND	ND	ND	Control of *L. monocytogenes* on stainless steel in seafood processing environments	[22]
L. monocytogenes	A511	Sewage from a sewage treatment plant	Phage typing, TEM, sequencing	134,494 bp	*Myoviridae*	Ready-to-eat foods from plant and animal origin including cheeses	[80–84]
L. monocytogenes	FWLLm1	Sheep feces	TEM,	ND	ND	Reduction of *L. monocytogenes* growth in ready-to-eat poultry products	[85]
L. monocytogenes	IZSAM-1	Floor drain-water from an Italian blue cheese dairy factory	TEM, sequencing	~50 kb	*Siphoviridae*	Biocontrol of *L. monocytogenes* within cheese industrial facilities	[86,87]

Table 1. *Cont.*

Target Bacteria	Phage/s	Source	Characterization Method	Genome Length	[2] Family	Food Application	Reference
Listeria spp.	P100	Sewage effluent from a dairy plant	TEM, sequencing	131,384 bp	*Myoviridae*	Biocontrol of contaminated surfaces, the surface of soft cheeses, ready-to-eat foods, fresh-cut fruit, and fruit juices, raw fish fillets,	[88–92]
Pseudomonas fragi	Wy	Ground Beef	TEM, dsDNA	ND	-	Reduction of *P. fragi* in refrigerated raw milk	[93–95]
Pseudomonas sp.	C35	spoiled retail beef	-	ND	-	Biological control of beef spoilage	[96,97]
Pseudomonas lactis	HU1	sludge obtained after passing raw cow's milk through a centrifugal clarifier	TEM, dsDNA	~48 Kb	*Podoviridae*	Control of *P. lactis* in Raw Cow's Milk	[98]
Pseudomonas fluorescens *E. cloacae* strains	PspYZU5415 EcpYZU01	Sewage samples	TEM, sequencing	39,636 bp 39,767 bp	*Siphoviridae* *Corticoviridae*	Growth inhibition of *E. cloacae* and *P. fluorescens* in cucumber juice with different salt concentrations	[43]
P. fluorescens	IBB-PF7A	Sewage treatment plant	TEM, dsDNA	~42 kbp	*Podoviridae*	Biocontrol of *P. fluorescens* in dairy and other food industries	[99,100]
Salmonella Enteritidis, *S.* Typhimurium	wksl3	Chicken by-product samples	TEM, sequencing	42,766 bp	*Siphoviridae*	Control *Salmonella* on chicken skin. from broiler carcasses	[101]
Salmonella serovars	LPSEYT	Water samples	TEM, sequencing	53,387 bp	*Myoviridae*	Biocontrol of *Salmonella* in food matrices including milk and lettuce	[42]
Salmonella Enteritidis	CAU-SEP-1 CAU-SEP-2 CAU-SEP-3 CAU-SEP-4	River water samples	TEM	ND	*Myoviridae* and *Siphoviridae*	Control of *S. Enteritidis* in chicken breast meat	[102]
Salmonella Enteritidis	CNPSA 1 CNPSA3 CNPSA4	free-range chickens	TEM, dsDNA	ND	tailed dsDNA phages	Reduction of *Salmonella* Enteritidis in Contaminated Chicken Cuts	[103–105]
Salmonella Enteritidis	P29C	Raw human sewage	-	ND	*Siphoviridae*	Reduction of bacterial contamination on chicken carcass surfaces	[54,106]
Salmonella spp.	PSE5	Poultry slaughterhouse wastewater	plaque morphology and RAPD analysis	ND	ND	Reduction of contamination in raw chicken eggs	[107]

Table 1. *Cont.*

Target Bacteria	Phage/s	Source	Characterization Method	Genome Length	[2] Family	Food Application	Reference
Salmonella spp.	LPSTLL LPST94 LPST153	Environmentally water samples	TEM	ND	Siphoviridae *Ackermannviridae* *Podoviridae*	Reduction of *Salmonella* counts in milk and chicken breast and on stainless still surfaces	[108,109]
Salmonella strains	LPSE1	Environmental samples	TEM, dsDNA, sequencing	41,854 bp	*Siphoviridae*	Control of *Salmonella* in ready-to-eat foods	[110]
Salmonella strains	Felix O1/Felix O1-E2	Feces of paratyphoid B patients	TEM, Sequencing	86,155 bp/~84 kb	*Myoviridae*	Suppression of *Salmonella* growth on chicken frankfurters, poultry products, and ready-to-eat foods	[111–114]
Salmonella strains	PHL4	Wastewater treatment plant	-	ND	ND	Reduction of *Salmonella* growth poultry products	[115]
Salmonella strains	vB_SalS_SJ_3	Wastewater	TEM, DNA sequencing	162,910 bp	*Siphoviridae*	Biocontrol of *Salmonella* in contaminated Eggs and Pork	[116–118]
Salmonella strains	Pu20	Sewage samples	TEM, sequencing	59,435 bp	*Podoviridae*	Growth inhibition of *Salmonella* strains in liquid egg white and yolk	[119]
Salmonella strains	D1-2	Environmental samples	TEM, sequencing	86,878 bp	*Myoviridae*	Growth inhibition of *Salmonella* strains in liquid egg white and yolk	[120]
Salmonella Typhimurium	P22 [Argo4]	*Salmonella enterica* subsp. *enterica* serovar Typhimurium	TEM, sequencing. Reference strain ATCC® 97540™	41,724 bp	*Podoviridae*	Prevention of attachment to food surfaces and food matrices	[121–124]
Salmonella Typhimurium	P7	Unknown	-	ND	ND	Control of pathogens in liquid foods	[36]
Salmonella serovars	LPST153	Water samples	TEM, sequencing	39,176 bp	*Autographivirinae*	Control of *Salmonella* in raw milk and raw beef sausages	[125]
S. enterica serovar Typhimurium *S. enterica* serovar Enteritidis	UAB_Phi 20 UAB_Phi78 UAB_Phi87	Chicken Chicken pig	TEM, dsDNA, sequencing	41,809 bp 44,110 bp 87,669 bp	*Podoviridae* *Podoviridae* *Myoviridae*	Reduction of *Salmonella* on foods and reduction of *Salmonella* Colonization of poultry	[126–128]
Salmonella Enteritidis	SP-1 SP-3	Intestinal content of broiler chickens	TEM, dsDNA, PCR amplification	~86 kb ~88 kb	Podoviridae *Siphoviridae*	Biocontrol of *Salmonella* in cooked chicken meat	[35,129,130]
Salmonella Enteritidis	SJ2	Chicken egg	ND	ND	ND	Reduction of *Salmonella* counts in Cheddar cheese made from both raw and pasteurized milk, and in contaminated eggs and pork	[131]

Table 1. *Cont.*

Target Bacteria	Phage/s	Source	Characterization Method	Genome Length	[2] Family	Food Application	Reference
Salmonella Enteritidis	vBSenM-PA13076 (PA13076) vBSenM-PC2184 (PC2184)	Chicken sewage	TEM	52,474 bp ND	*Myoviridae*	Biocontrol of *Salmonella* in foods (chicken breast, pasteurized whole milk, Chinese cabbage)	[132,133]
Salmonella and *E. coli* O157:H7	PS5	Raw chicken products	TEM, sequencing	158,400 bp	*Myoviridae*	Reduction of viable counts on solid and liquid foods	[134]
Salmonella Oranienburg	SSP5 SSP6	Sewage samples	TEM	ND	*Myoviridae* *Siphoviridae*	Control of *Salmonella* Oranienburg on alfalfa seeds	[135]
S. Typhimurium *S. Enteritidis* *S. Montevideo*	A B	sewage treatment plant	TEM	ND	*Myoviridae* *Siphoviridae*	Control of *Salmonella* in mustard and broccoli seeds	[136]
Salmonella strains, including MDR *Salmonella*	T156	Waste water	TEM, dsDNA, sequencing	123,849 bp	*Siphoviridae*	Microencapsulated bacteriophage applied in skim milk and lettuce for biocontrol of *Salmonella*	[137]
Staphylococcus aureus	K	Deposited by EA Asheshov	ATCC® 19685-B1™	139,831 bp	*Myoviridae*	Removing *S. aureus* biofilms	[138,139]
S. aureus	H5 (phiPLA88) A72 (phiPLA35)	Raw milk	TEM, dsDNA, sequencing	42,526 bp 45,344 bp	*Siphoviridae*	Curd manufacturing, fresh and hard-type cheeses	[140–142]
S. aureus	SA46-CTH2	Food samples	TEM	17,505 bp	*Podoviridae*	Inactivation of *S. aureus* planktonic cells in pasteurized milk and biofilms on stainless steel surfaces	[143]
S. aureus	SA13m	Temperate phage SA13 isolated from a goat fecal sample	TEM, sequencing	42,652 bp	*Siphoviridae*	Biocontrol of *S. aureus* in pasteurized whole milk at refrigeration and ambient temperatures	[144]
Shewanella baltica and *S. putrefaciens*	SppYZU01 to SppYZU10	Wastewater from freshwater and marine product marketplaces	TEM, sequencing	SppYZU01 (43,567 bp) SppYZU05 (54,319 bp)	*Myoviridae* *Siphoviridae*	Biopreservation of chilled channel catfish	[145]
Shigella spp.	SF-A2 SD-11 SS-92	Spiced chicken Pig farm effluent Pig farm effluent	TEM	ND	*Myoviridae*	inactivation of foodborne *Shigella* on ready-to-eat chicken	[146]
Vibrio parahaemolyticus	vB_VpaS_OMN (designated as phage OMN)	Sea water	TEM, sequencing	42,202 bp	*Podoviridae*	Inactivation of *V. parahaemolyticus* in oyster meat	[147]

[1] Nuclease digestion tests and/or Random Amplified Polymorphic DNA Analyses (RAPD), [2] Family designated by the authors, ND: not determined, TEM: Transmission Electron Microscopy.

According to the articles analyzed, the phages of the family *Myoviridae* were preferentially used to control *E. coli*. Other important food pathogens such as *C. jejuni*, *Salmonella*, *L. monocytogenes*, and *S. aureus* were controlled by *Siphoviridae* and *Myoviridae*. The analyzed studies showed that the *Podoviridae* family can infect all these species, but fewer phage strains of this family have been found to control bacteria in the different foods tested. Comparative genomics and morphological observation by transmission electron microscopy revealed that the phage LPSEYT, able to infect *Salmonella*, represents a new genus within the *Myoviridae* family [42]. This last example shows that if we go a little deeper into the genomic characterization of the isolated strains, we will be able to advance in the knowledge of the taxonomy of phages. Most of the phages used to control these pathogen species in food were isolated from wastewater, sewage, or other environmental samples; but many have also been isolated from different foods. One phage strain (EcpYZU01) of the *Corticoviridae* family was isolated from sewage samples and tested against *Enterobacter cloacae* in cucumber juice [43]. Finally, a phage (LPST94) from the *Ackermannviridae* family isolated from water was effective against *Salmonella* in foods [108,109]. This newly assigned family was recently added to the list of the International Committee on Taxonomy of Viruses ICTV catalog. The isolation of phages from sewage and water samples is common due to their abundance in these ecosystems. However, Scattolini et al., pointed out that the search and characterization of phages isolated in the same foods in which the pathogens can hide could be a good way "to integrate this control measure in an innovative, cost-effective, safe and environmentally friendly way" [86]. Therefore, it seems like a good idea to use phages in food safety which in turn come from food, especially for the consumer, who can identify fewer drawbacks than when consuming phages or their genetically manipulated enzybiotics.

Bacteriophages can also be used to prevent or to reduce colonization of domesticated livestock with bacterial pathogens before they enter the production chain [148]. After that, phages can be used to decontaminate inanimate surfaces made, for example, of stainless steel or to fight bacterial biofilms. Finally, phages can be used directly on food, both in unprocessed or ready-to-eat foods as well as processed foods, even stored at temperatures ranging from 4 °C to 20 °C.

Several cofactors tested with phages used in the control of *L. monocytogenes* in the food industry have been recently reviewed by Kawacka and coworkers [26,149]. Among those factors, we can find other bacterial cultures such as *Lactobacillus* spp., *Gluconocbacter assii*, the bacteriocins Nisin, Enterocin and Pediocin, and several compounds such as lauric arginate, potassium lactate, sodium diacetate, sucrose monolaurate.

3. Spatial Distribution of Phages

Bacteriophages' ubiquity is another advantage. It is estimated that there are 10 bacteriophages for every bacterium present on our planet, representing a virtually unlimited source, not only of virions but also of lytic enzymes. Phages are especially abundant in seawater and soil and have also been found in large quantities in wastewater. The potential use of bacteriophages as indicators of environmental contamination has also been investigated in the last few decades [150–155]. Perhaps the most impressive figures are that phages kill bacteria at rates of up to 40% of the total population of marine bacteria per day and that carbon flux through phage biomass is estimated at 145 gigatonnes per year, playing a crucial role in our planet's global carbon cycle [156,157]. They are also easily found on any animal or plant surfaces as they are part of the microbiota of most living things. Phages have also been isolated from a variety of foods, including ready-to-eat foods, fish and shellfish, milk products, meat, and vegetables [33,158–162]. Because of this, consumers are already in contact with food bacteriophages every day. Therefore, if researchers could offer an adequate explanation, it would help consumers to increase their acceptance of the use of food bacteriophages. In other words, they should accept their use as biopreservatives if we can explain well what this class of virus really is and how exactly they are used to fight "bad" bacteria in food.

4. Morphology and Classification

Initially, phages were characterized by transmission electron microscopy (TEM), followed by pulse-field gel electrophoresis and restriction endonuclease analysis. However, although TEM continues to be essential in publications on bacteriophage viruses, the quality of the images in many of the articles is questionable [163]. Most studies use the work of Ackermann or the criteria of the International Committee on Taxonomy of Viruses (ICTV) [164] to identify their phage isolates [165–167]. For further taxonomic classification and phage characterization, more detailed information, such as genomic data, has begun to be included in scientific publications [168–171].

Most phages belong to the order Caudovirales. Based on the tail morphology, Caudovirales are divided into three families: *Myoviridae*, *Siphoviridae*, and *Podoviridae*. *Myoviridae* phages are characterized by long straight contractile tails, *Siphoviridae* phages possess long flexible non-contractile tails, and *Podoviridae* phages have short, non-contractile tails [172].

Alternatively, we can also use the PCR technique and subsequent sequencing to partially characterize the isolated phages. For example, some authors used specific primers to detect the Major Capsid Protein (MCP) of reported *Salmonella* phages [158,159].

Augustine et al., also used PCR or multiplex PCR to perform a screening of virulence factors in DNA obtained from phages [35]. Tomat et al. used PCR to detect virulence factor genes (from diarrheagenic *E. coli* toxins) in two phages (DT1 and DT6) isolated from stool samples of patients with diarrhea [72].

Presently, full genome sequencing and analysis provide the key tool for taxonomic classification and for alerting the presence of "dangerous" genes that phage genomes may contain. We believe that it is necessary to sequence phage genomes to obtain information on the presence of antibiotic-resistant genes or virulence factors before determining their suitability for food applications. An outline with the steps followed for the isolation and characterization of phages for food biopreservation is shown in Figure 1.

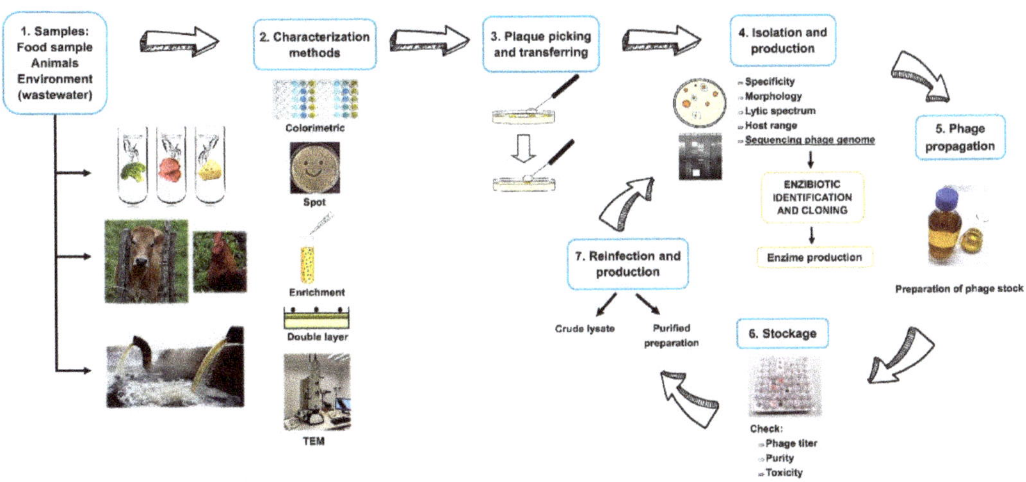

Figure 1. Steps followed for the isolation and characterization of phages.

DNA genomes of Caudovirales range in size from about 15 up to 500 kbp [173]. The study of the genome of phages is crucial today, but most investigations analyzed before to the last 10 years do not include the sequencing or annotation of these genomes. The complete genomes of phages are already included as a technique of characterization and phylogeny, but the in-depth analysis of these genomes has only been carried out very recently; this even allows us to discover new subfamilies and new genera of phages infecting food pathogens [43,125].

5. Phage's Life Cycle

To perpetuate themselves, phages must infect their host bacteria by binding to specific receptors on them. After injecting their nucleic acid into the bacterium's cytoplasm, phages can hijack the bacterium's cellular machinery to induce their own replication, through a process called the "lytic cycle", giving rise to hundreds or thousands of complete viral particles that will leave the cell after killing it (Figure 2). Alternatively, if the phage nucleic acid is inserted into the chromosome or within a plasmid of the bacterium, it can remain in a kind of dormant state known as the "lysogenic cycle," which will not produce new virus particles until conditions are favorable, or their genes are activated by some external stimulus. Lytic bacteriophages are the first choice to selectively kill bacteria in foods because lysogenic phages remain in the bacterial chromosome and will not multiply until the environment in which the bacterium is found allows for it, making lysogenic phages difficult to control.

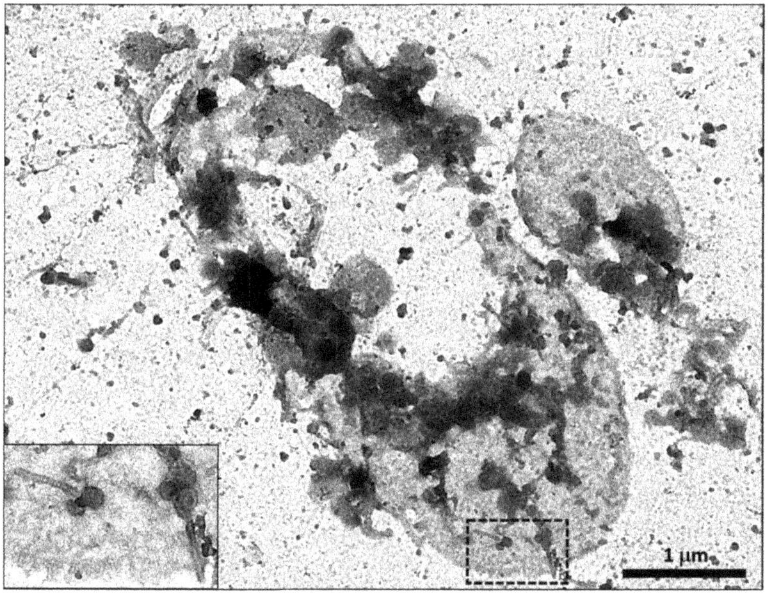

Figure 2. Gram-negative bacterium after lysis by phages. Numerous complete or incomplete phage heads and tails can be seen in the image. Inset: Detail of the boxed area showing two phages of the *Siphoviridae* family. Original magnification ×25,000.

6. Enzybiotics

There are three classes of bacterial cell wall hydrolases: animal lysozymes, bacterial autolysins, and phage lysins. All animal lysozymes share the ability to hydrolyze the β-(1,4)-glycosidic bond between the alternating N-acetylmuramic acid and N-acetylglucosamine residues of the bacterial cell wall polymer called peptidoglycan. Their biological role is mainly antibacterial defense, but some lysozymes also work as food digestive enzymes in animal guts [174]. Bacterial cell wall hydrolases are involved in carefully remodeling the cell wall to maintain cell integrity but also participate actively in processes such as cell division, bacterial surface appendages' assembly, and the facilitation of bacterial secretion systems' stabilization [175,176]. Most of these autolysins are peptidoglycan hydrolases (PGHs) that can provoke bacterial autolysis, so their expression and activity need to be tightly regulated.

The third class of cell wall hydrolases are phage endolysins, enzymes that directly target bonds in the peptidoglycan of the bacterial cell wall. These so-called enzybiotics (for ENZYme antiBIOTICS) are synthesized at the end of the bacteriophage lytic cycle to lyse the bacterium they parasitize, producing a lysis "from within" in Gram-negative bacteria [177]. Most endolysins contain one or two enzymatically active domains (EAD) in the N-terminus (which cleave one of the bonds in the bacterial peptidoglycan) and one cell wall-binding domain (CBD) in the C-terminal region (which is involved in host bacterial recognition). Based on their EAD, enzybiotics can be broadly divided into three types: endopeptidases, amidases, and glycosidases.

On the other hand, in Gram-positive bacteria, endolysins are also able to lyse bacteria "from outside" during the phage adsorption at the bacterial surface [178,179].

Endolysins have an extensive structural variation and a diverse cleavage predilection for the molecules with glycosidic, amide, or peptide bonds present in the bacterial peptidoglycan [180,181]. The structure of endolysins can be either globular or modular. Globular endolysins are unique for phages infecting Gram-negative bacteria, whereas modular endolysins are found in phages with a Gram-positive host. Another class of phage enzymes is virion-associated peptidoglycan hydrolases which share a similar mode of action on the bacterial peptidoglycan [182–185]. A good example of these newly studied antibacterial molecules is the virion-associated peptidoglycan hydrolase HydH5 of *Staphylococcus aureus* bacteriophage vB_SauS-phiIPLA88 [186]. Additionally, some phages can produce depolymerases to overcome bacterial protective layers such as proteinaceous S-layers [187] or polysaccharide capsules [188].

Among the advantages of enzybiotics, we include the possibility of totally or partially breaking the structure of bacterial biofilms. A biofilm can be defined as a structured community of bacterial cells enclosed in a self-produced polymeric matrix and adherent to an inert or living surface. Growth in biofilms enhances the survival of bacterial populations in the food industry environments, increasing the probability of causing food-borne infections. Due to the presence of extracellular material that protects biofilms, many phages have limited access to bacteria inside these structures. This can be solved using phages expressing exopolysaccharide depolymerases and endolysins. Endolysins can act effectively irrespective of the metabolic status of the cells (exponential and stationary phase cells) and are capable of killing planktonic cells as well as sessile cells. In this way, phage endolysins have been shown to be effective in eliminating biofilms formed by tenacious pathogens on different surfaces commonly used in the food industry [189–192]. Moreover, endolysins can be evaluated in combination with depolymerases or even with antibiotics to kill the underlying pathogen that formed the biofilm. On the other hand, as many pathogens build their biofilms based on different substances that form the biofilm matrix, it would be advisable to evaluate the activity of endolysins against biofilms that present a different proportion of proteins, nucleic acids, sugars or lipids.

Additionally, endolysins can kill "persister" bacteria that escape conventional antibiotics and even can kill the dreaded multi-resistant strains. Although there are not many studies in this regard, endolysins also offer the possibility of being used in combination with other molecules or with other solutions for the food industry, such as bacteriocins or probiotics. Furthermore, as gene-encoded proteins, enzybiotics are amenable to bioengineering strategies, both to optimize specificity and to increase yields [193,194]. An example is the construction of hybrid proteins consisting of LysSA11 -an endolysin of the *S. aureus* phage SA11 and the enzymatically active domain of LysB4- and endolysin from the *Bacillus cereus* phage B4 [195].

The search, characterization, and practical use of these phage-derived lysins have received less attention than phages, basically because they are more difficult for many laboratories to study. However, there is a growing body of work on these enzymes, particularly in the field of human and animal pathogens, which has encouraged researchers in other fields, including food safety, to begin promising work with enzybiotics. Not surprisingly, many enzybiotics have been successfully tested as biopreservatives or have been proposed by their discoverers as good candidates to be used in food against Gram-negative and Gram-positive bacteria (Table 2). The study of these enzymes in phages that do not belong to the "selective group" of food pathogens could provide a wide range of new proteins with different properties and varied spectra.

Table 2. Enzybiotics tested against food-borne pathogens and their proposed use in foods.

Target Bacteria	Enzybiotic	Source	Food Application	Reference
Bacillus cereus, *B. subtilis* and *L. monocytogenes*	LysB4	*B. cereus* phage B4	antibacterial agent to control foodborne pathogens.	[196]
Clostridium tyrobutyricum and *C. sporogenes*	Ctp1L	Bacteriophage CTP1 isolated from landfill	Cheese manufacture, reduction of clostridial activity in cheese	[56,197]
C. tyrobutyricum *C. acetobutylicum*	CS74L	Lytic bacteriophage (ATCC® 8074-B1™) of *C. sporogenes*	Biocontrol of clostridia strains in foods	[198]
C. perfringens	Ply3626	*C. perfringens* ATCC 3626	Control of anaerobic spore-formers	[199]
C. perfringens	LysCPAS15	*C. perfringens* phage CPAS-15	Inhibition of *C. perfringens* in sterilized milk	[200]
Bacillus subtilis *B. megaterium* *L. monocytogenes*	PLY118, PLY500 PLY 511	Phages from *Listeria monocytogenes*	Production of airy starter cultures with biopreservation properties	[201–203]
E. coli O157:H7	PlyEc2	Phage from *E. coli*	Reduction of *E. coli* O157:H7 on contaminated lettuce	[204]
Lactococcus lactis, *Pediococcus acidilactici* and *P. pentosaceus*	LysA2	*L. casei* bacteriophage A2	Ripening of fermented products	[205]
Lactobacilli, lactococci, pediococci, *B. Subtilis* *Brevibacterium linens* *Enterococcus faecium*	Mur-LH	Phage 0303 from *Lactobacillus helveticus* CNRZ 303	Preventing the growth of spoilage microbes	[17]
L. monocytogenes *B. subtilis*	PlyP100	Phage from *L. monocytogenes*	Antimicrobial biopreservative in fresh cheese.	[206]
L. monocytogenes	LysZ5	Phage from *L. monocytogenes*	Control pathogens in soya milk	[207]
L. monocytogenes	PlyLM	Phage from *L. monocytogenes strain 4b*	Proposed control of *L. monocytogenes* in food matrices and processing facilities	[208]
L. monocytogenes	HPL118 HPL500 HPL511 HPLP35	Recombinant endolysins from *L. monocytogenes* phages	Reduction of *L. monocytogenes* viable counts in iceberg lettuce. Promising perspectives in production and packaging environments	[201,209, 210]
Methicillin-resistant *Staphylococcus aureus*	LysGH15	Phage isolated from Sewage samples	Biopreservative in whole and skim milk	[211,212]
methicillin-resistant *S. aureus*	LysSA11	*Staphylococcus aureus* phage SA11	Biocontrol of *S. aureus* on strain in pasteurized milk or ham and utensils	[213]
S. aureus *Bacillus cereus*	Hybrid LysB4EAD-LysSA11	Phage SA11 from *S. aureus* phage B4 S from *B. cereus*	Biocontrol of *S. aureus* and *B. cereus* in boiled rice	[195]
S. aureus	LysH5	Staphylococcal bacteriophage phi-SauS-IPLA88	Disinfection process of industrial food facilities. Elimination of *S. aureus* in pasteurized milk	[190,214]
S. aureus	CHAPSH3b	Chimeric protein (CHAP domain from peptidoglycan hydrolase HydH5 and the SH3b cell wall-binding domain from lysostaphin)	*S. aureus* biofilm elimination	[215]
S. aureus	$CHAP_K$	Truncated derivative of the phage lysin LysK from the staphylococcal bacteriophage K	Reduction of biofilm formation in processing systems	[189]
S. aureus	HydH5 HydH5Lyso HydH5SH3b CHAPSH3b and lysostaphin	*S. aureus* bacteriophage vB_SauS-phiIPLA88	Biocontrol of *S. aureus* in dairy products	[216]
Streptococcus spp.	λSA2	*Streptococcus agalactiae* (serotype III GBS strain 3330) bacteriophage B30	Inactivation of *Strepcococcus* spp. in cow milk	[217,218]
S. Typhimurium	LysSTG2	*Salmonella*-lytic bacteriophage STG2	Combating *S.* Typhimurium biofilms in food industries	[219]
Salmonella strains	LysSE24	*Salmonella* phage LPSE1	Food Control of *Salmonella* strains	[220]
Several Gram-negative pathogens, particularly against *Salmonella* Typhimurium	Lys68	*Salmonella* phage phi68 isolated from feces from a poultry farm	Combat Gram-negative pathogens in the food industry	[221]

Furthermore, enzybiotics can improve the narrow host spectrum of phages against both Gram-positive and Gram-negative bacteria. Therefore, the narrow host range of phages should be used to control specific spoilage or pathogenic bacteria, while the broadest spectrum of enzybiotics can be used to control different strains or species. Some of the newly isolated and characterized endolysins have a broad spectrum so they could be candidates for use in the food industry. An example is endolysin M4Lys, which has a peculiar mosaic structure [222].

The main limitation in the use of phage enzybiotics in food is their complicated production and purification, since relatively large amounts of proteins are needed even to be studied in in vitro assays. Another problem is their low resistance to high temperatures used in different processes in the food industry, such as disinfection. However, the search for new enzymes with new properties will make it possible to find thermostable and easy-to-produce forms in heterologous hosts such as *E. coli* and *Lactococcus lactis* [21,221,223–225].

7. Concluding Remarks

Many natural and eco-friendly methodologies for food preservation have been proposed in the last few years, but only limited data are available about the usefulness of most of them under industrial scale conditions, which needs proper attention to satisfy the requirements of the industry as well as the demand of the consumers [226–230]. Consequently, studies about the ability of the reported biopreservative agents to control the development of undesirable microorganisms when applied at the industrial scale are greatly required.

Studies on the biocontrol of food-borne pathogens in foods have generally produced very good results. However, not all are lights in the use of phages against pathogenic bacteria in food, there are also shadows. There are assays in which it was not possible to reduce the number of pathogenic bacteria in food using bacteriophages [136,231,232].

The use of phages in human and veterinary medicine has received much more attention than their use in the food industry; but the increasing appearance of antibiotic-resistant strains in the food industry has begun to make these viruses be seriously taken into account when seeking their (application for food safety), also in this context. Similarly, their lytic enzymes have not been sufficiently exploited in the food industry to date. However, this is beginning to change; indeed, after the successful use of lysozyme (animal) or Nisin (bacteria), enzymes are beginning to be seriously valued in the food industry. Phages offer new and interesting possibilities when planning the control of annoying microorganisms in food manufacturing, food biopreservation, or food processing. Additionally, their lytic enzymes, easily modifiable through molecular biology processes, offer a very wide range of possibilities both for direct application against bacteria, as well as for inclusion in food matrices or the preparation of antibacterial surfaces generated by biotechnology [233].

Virulent bacteriophages are naturally present in foods, therefore both phages and their enzybiotics would be exploited in different ways for food safety as the consumer demand for the use of ecofriendly biopreservatives is increasing. Contamination of ready-to-eat products with pathogenic bacteria is a more serious problem than the contamination of food that will then be cooked before being consumed since many of the cooking methods reduce the number of these bacteria. In this context, both phage and enzybiotics have been tested in ready-to-eat meals. However, not only is the use of phages and their enzymes in food is not only an area of incipient research, but the whole biology of phages is experiencing a new boom in all domains of research, mainly in human and veterinary health, where spectacular achievements have already been reached in some patients and farm animals.

Along with this increasing amount of isolation and characterization of phage strains capable of controlling important food-borne pathogens—it is always desirable to increase our armament against superbugs—we must make a parallel effort to understand more in-depth their interaction with target pathogens, as well as their biology and ecology in food if we want to apply them in the different stages of the production chain, increasing their biopreservation capacity. At the molecular level, we must better characterize enzybiotics, study the possibility of applying them in different processes, and optimize their production so that their application is profitable for food producers and does not raise the price too much for consumers.

Furthermore, the safety and ubiquity of phages must be well explained to both food producers and consumers to avoid rejection of "the unknown" [234,235]. Bacteriophages are the most abundant microorganisms on the planet and even in our guts, with approximately 10^{14} phage particles in our body [236]. As we have seen in this review, phages and their enzybiotics can be found in the environment, in animals, and in food we eat every day. Finally, some phage-based products for the control of pathogens in food are already being used in different countries after being approved by competent authorities, even in ready-to-eat products. Those products mainly include a cocktail of phages, for example against *E. coli* (EcoShield™), *L. monocytogenes* (ListShield™ and PhageGuard Listex™), and *Salmonella* spp. (SalmoFresh™) [237].

Author Contributions: Conceptualization, J.R.-V.; writing—original draft preparation, J.R.-V., M.E.-Z., M.L.S.; writing—review and editing, J.R.-V., M.E.-Z., M.L.S., A.P.B.; Visualization: T.Y.F.-H.; supervision F.G., M.B. All authors have read and agreed to the published version of the manuscript.

Funding: Research in our group was supported by SODERCAN (Project RH20-XX-032, FAGOFOOD).

Acknowledgments: Tamara Y. Forbes-Hernández is supported by a "Juan de la Cierva-Formación" post-doctoral contract.

Conflicts of Interest: The authors declare no conflict of interest. The funders had no role in the design of the study, or in the writing of the manuscript.

References

1. Lundén, J.; Björkroth, J.; Korkeala, H. Contamination Routes and Analysis in Food Processing Environments. In *Handbook of Hygiene Control in the Food Industry*; Lelieved, H.L.M., Holah, M.A., Eds.; Woodhead Publishing Series in Food Science, Technology and Nutrition; Woodhead Publishing: Cambridge, UK, 2005; pp. 539–555.
2. Alegbeleye, O.O.; Singleton, I.; Sant'Ana, A.S. Sources and Contamination Routes of Microbial Pathogens to Fresh Produce during Field Cultivation: A Review. *Food Microbiol.* **2018**, *73*, 177–208. [CrossRef]
3. Olaimat, A.N.; Holley, R.A. Factors Influencing the Microbial Safety of Fresh Produce: A Review. *Food Microbiol.* **2012**, *32*, 1–19. [CrossRef]
4. Kim, N.H.; Cho, T.J.; Rhee, M.S. Current Interventions for Controlling Pathogenic *Escherichia coli*. *Adv. Appl. Microbiol.* **2017**, *100*, 1–47. [CrossRef] [PubMed]
5. Rajan, K.; Shi, Z.; Ricke, S.C. Current Aspects of *Salmonella* Contamination in the US Poultry Production Chain and the Potential Application of Risk Strategies in Understanding Emerging Hazards. *Crit. Rev. Microbiol.* **2017**, *43*, 370–392. [CrossRef] [PubMed]
6. Laroute, V.; Tormo, H.; Couderc, C.; Mercier-Bonin, M.; Le Bourgeois, P.; Cocaign-Bousquet, M.; Daveran-Mingot, M.-L. From Genome to Phenotype: An Integrative Approach to Evaluate the Biodiversity of *Lactococcus Lactis*. *Microorganisms* **2017**, *5*, 27. [CrossRef] [PubMed]
7. Klaenhammer, T.R. Genetics of Bacteriocins Produced by Lactic Acid Bacteria. *FEMS Microbiol. Rev.* **1993**, *12*, 39–85. [CrossRef]
8. Kang, M.-S.; Lee, D.-S.; Lee, S.-A.; Kim, M.-S.; Nam, S.-H. Effects of Probiotic Bacterium *Weissella cibaria* CMU on Periodontal Health and Microbiota: A Randomised, Double-Blind, Placebo-Controlled Trial. *BMC Oral Health* **2020**, *20*, 243. [CrossRef] [PubMed]
9. Jeong, J.H.; Lee, C.Y.; Chung, D.K. Probiotic Lactic Acid Bacteria and Skin Health. *Crit. Rev. Food Sci. Nutr.* **2016**, *56*, 2331–2337. [CrossRef] [PubMed]
10. Esber, N.; Mauras, A.; Delannoy, J.; Labellie, C.; Mayeur, C.; Caillaud, M.-A.; Kashima, T.; Souchaud, L.; Nicolis, I.; Kapel, N.; et al. Three Candidate Probiotic Strains Impact Gut Microbiota and Induce Anergy in Mice with Cow's Milk Allergy. *Appl. Environ. Microbiol.* **2020**, *86*, e01203-20. [CrossRef] [PubMed]
11. Paparo, L.; Nocerino, R.; Di Scala, C.; Della Gatta, G.; Di Costanzo, M.; Buono, A.; Bruno, C.; Berni Canani, R. Targeting Food Allergy with Probiotics. *Adv. Exp. Med. Biol.* **2019**, *1125*, 57–68. [CrossRef]
12. de Dapkevicius, M.L.E.; Sgardioli, B.; Câmara, S.P.A.; Poeta, P.; Malcata, F.X. Current Trends of Enterococci in Dairy Products: A Comprehensive Review of Their Multiple Roles. *Foods* **2021**, *10*, 821. [CrossRef]
13. De Silva, L.a.D.S.; Wickramanayake, M.V.K.S.; Heo, G.-J. Virulence and Antimicrobial Resistance Potential of *Aeromonas* Spp. Associated with Shellfish. *Lett. Appl. Microbiol.* **2021**, *73*, 176–186. [CrossRef]
14. Luque-Sastre, L.; Arroyo, C.; Fox, E.M.; McMahon, B.J.; Bai, L.; Li, F.; Fanning, S. Antimicrobial Resistance in *Listeria* Species. *Microbiol. Spectr.* **2018**, *6*. [CrossRef]

15. Goodridge, L.; Abedon, S.T. Bacteriophage Biocontrol and Bioprocessing: Application of Phage Therapy to Industry. *Soc. Ind. Microbiol. News* **2003**, *53*, 254–262.
16. Greer, G.G. Bacteriophage Control of Foodborne Bacteriat. *J. Food Prot.* **2005**, *68*, 1102–1111. [CrossRef] [PubMed]
17. Deutsch, S.-M.; Guezenec, S.; Piot, M.; Foster, S.; Lortal, S. Mur-LH, the Broad-Spectrum Endolysin of *Lactobacillus helveticus* Temperate Bacteriophage Phi-0303. *Appl. Environ. Microbiol.* **2004**, *70*, 96–103. [CrossRef]
18. Röhrig, C.; Huemer, M.; Lorgé, D.; Luterbacher, S.; Phothaworn, P.; Schefer, C.; Sobieraj, A.M.; Zinsli, L.V.; Mairpady Shambat, S.; Leimer, N.; et al. Targeting Hidden Pathogens: Cell-Penetrating Enzybiotics Eradicate Intracellular Drug-Resistant *Staphylococcus aureus*. *mBio* **2020**, *11*, e00209-20. [CrossRef]
19. Dams, D.; Briers, Y. Enzybiotics: Enzyme-Based Antibacterials as Therapeutics. *Adv. Exp. Med. Biol.* **2019**, *1148*, 233–253. [CrossRef]
20. Gerstmans, H.; Rodríguez-Rubio, L.; Lavigne, R.; Briers, Y. From Endolysins to Artilysin®s: Novel Enzyme-Based Approaches to Kill Drug-Resistant Bacteria. *Biochem. Soc. Trans.* **2016**, *44*, 123–128. [CrossRef]
21. Gutiérrez, D.; Rodríguez-Rubio, L.; Martínez, B.; Rodríguez, A.; García, P. Bacteriophages as Weapons Against Bacterial Biofilms in the Food Industry. *Front. Microbiol.* **2016**, *7*, 825. [CrossRef]
22. Ganegama Arachchi, G.J.; Cridge, A.G.; Dias-Wanigasekera, B.M.; Cruz, C.D.; McIntyre, L.; Liu, R.; Flint, S.H.; Mutukumira, A.N. Effectiveness of Phages in the Decontamination of *Listeria monocytogenes* Adhered to Clean Stainless Steel, Stainless Steel Coated with Fish Protein, and as a Biofilm. *J. Ind. Microbiol. Biotechnol.* **2013**, *40*, 1105–1116. [CrossRef]
23. Soni, K.A.; Nannapaneni, R. Removal of *Listeria monocytogenes* Biofilms with Bacteriophage P100. *J. Food Prot.* **2010**, *73*, 1519–1524. [CrossRef]
24. Sillankorva, S.; Neubauer, P.; Azeredo, J. Phage Control of Dual Species Biofilms of *Pseudomonas fluorescens* and *Staphylococcus lentus*. *Biofouling* **2010**, *26*, 567–575. [CrossRef]
25. Ghosh, C.; Sarkar, P.; Issa, R.; Haldar, J. Alternatives to Conventional Antibiotics in the Era of Antimicrobial Resistance. *Trends Microbiol.* **2019**, *27*, 323–338. [CrossRef]
26. Kawacka, I.; Olejnik-Schmidt, A.; Schmidt, M.; Sip, A. Effectiveness of Phage-Based Inhibition of *Listeria monocytogenes* in Food Products and Food Processing Environments. *Microorganisms* **2020**, *8*, 1764. [CrossRef]
27. Zaburlin, D.; Quiberoni, A.; Mercanti, D. Changes in Environmental Conditions Modify Infection Kinetics of Dairy Phages. *Food Environ. Virol.* **2017**, *9*, 270–276. [CrossRef] [PubMed]
28. Payne, R.J.; Phil, D.; Jansen, V.A. Phage Therapy: The Peculiar Kinetics of Self-Replicating Pharmaceuticals. *Clin. Pharmacol. Ther.* **2000**, *68*, 225–230. [CrossRef] [PubMed]
29. Shao, Y.; Wang, I.-N. Bacteriophage Adsorption Rate and Optimal Lysis Time. *Genetics* **2008**, *180*, 471–482. [CrossRef] [PubMed]
30. Gáspár, S.; Rontó, G.; Müller, G. Determination of the Biological Parameters of Bacterium-Phage Complexes. *Z. Allg. Mikrobiol.* **1979**, *19*, 163–169. [CrossRef] [PubMed]
31. Abedon, S.T. Kinetics of Phage-Mediated Biocontrol of Bacteria. *Foodborne Pathog. Dis.* **2009**, *6*, 807–815. [CrossRef]
32. Abedon, S.T.; Katsaounis, T.I. Basic Phage Mathematics. *Methods Mol. Biol.* **2018**, *1681*, 3–30. [CrossRef] [PubMed]
33. Hudson, J.A.; Billington, C.; Carey-Smith, G.; Greening, G. Bacteriophages as Biocontrol Agents in Food. *J. Food Prot.* **2005**, *68*, 426–437. [CrossRef]
34. Hudson, J.A.; Billington, C.; Wilson, T.; On, S.L.W. Effect of Phage and Host Concentration on the Inactivation of *Escherichia coli* O157:H7 on Cooked and Raw Beef. *Food Sci. Technol. Int.* **2015**, *21*, 104–109. [CrossRef] [PubMed]
35. Augustine, J.; Louis, L.; Varghese, S.M.; Bhat, S.G.; Kishore, A. Isolation and Partial Characterization of ΦSP-1, a *Salmonella* Specific Lytic Phage from Intestinal Content of Broiler Chicken. *J. Basic Microbiol.* **2013**, *53*, 111–120. [CrossRef] [PubMed]
36. Bigwood, T.; Hudson, J.A.; Billington, C. Influence of Host and Bacteriophage Concentrations on the Inactivation of Food-Borne Pathogenic Bacteria by Two Phages. *FEMS Microbiol. Lett.* **2009**, *291*, 59–64. [CrossRef]
37. Sváb, D.; Falgenhauer, L.; Rohde, M.; Chakraborty, T.; Tóth, I. Identification and Characterization of New Broad Host-Range RV5-like Coliphages C203 and P206 Directed against Enterobacteria. *Infect. Genet. Evol.* **2018**, *64*, 254–261. [CrossRef] [PubMed]
38. Tarahovsky, Y.S.; Ivanitsky, G.R.; Khusainov, A.A. Lysis of *Escherichia coli* Cells Induced by Bacteriophage T4. *FEMS Microbiol. Lett.* **1994**, *122*, 195–199. [CrossRef]
39. Delbrück, M. The growth of bacteriophage and lysis of the host. *J. Gen. Physiol.* **1940**, *23*, 643–660. [CrossRef]
40. Sommer, J.; Trautner, C.; Witte, A.K.; Fister, S.; Schoder, D.; Rossmanith, P.; Mester, P.-J. Don't Shut the Stable Door after the Phage Has Bolted-The Importance of Bacteriophage Inactivation in Food Environments. *Viruses* **2019**, *11*, 468. [CrossRef]
41. Molina, F.; Simancas, A.; Ramírez, M.; Tabla, R.; Roa, I.; Rebollo, J.E. A New Pipeline for Designing Phage Cocktails Based on Phage-Bacteria Infection Networks. *Front. Microbiol.* **2021**, *12*, 564532. [CrossRef] [PubMed]
42. Yan, T.; Liang, L.; Yin, P.; Zhou, Y.; Sharoba, A.M.; Lu, Q.; Dong, X.; Liu, K.; Connerton, I.F.; Li, J. Application of a Novel Phage LPSEYT for Biological Control of *Salmonella* in Foods. *Microorganisms* **2020**, *8*, 400. [CrossRef]
43. Zheng, X.-F.; Yang, Z.-Q.; Zhang, H.; Jin, W.-X.; Xu, C.-W.; Gao, L.; Rao, S.-Q.; Jiao, X.-A. Isolation of Virulent Phages Infecting Dominant Mesophilic Aerobic Bacteria in Cucumber Pickle Fermentation. *Food Microbiol.* **2020**, *86*, 103330. [CrossRef] [PubMed]
44. Theuretzbacher, U.; Outterson, K.; Engel, A.; Karlén, A. The Global Preclinical Antibacterial Pipeline. *Nat. Rev. Microbiol.* **2020**, *18*, 275–285. [CrossRef]
45. Duarte, J.; Pereira, C.; Costa, P.; Almeida, A. Bacteriophages with Potential to Inactivate *Aeromonas hydrophila* in Cockles: In Vitro and In Vivo Preliminary Studies. *Antibiotics* **2021**, *10*, 710. [CrossRef] [PubMed]

46. Kong, M.; Kim, M.; Ryu, S. Complete Genome Sequence of *Bacillus cereus* Bacteriophage PBC1. *J. Virol.* **2012**, *86*, 6379–6380. [CrossRef] [PubMed]
47. Kong, M.; Ryu, S. Bacteriophage PBC1 and Its Endolysin as an Antimicrobial Agent against *Bacillus cereus*. *Appl. Environ. Microbiol.* **2015**, *81*, 2274–2283. [CrossRef]
48. Greer, G.G. Psychrotrophic *Brocothrix thermosphacta* Bacteriophages Isolated from Beef. *Appl. Environ. Microbiol.* **1983**, *46*, 245–251. [CrossRef] [PubMed]
49. Carey-Smith, G.V. The Use of Bacteriophages as a Biocontrol Mechanism for *Campylobacter* and *Salmonella* Contaminants of Food. Master's Thesis, School of Biological Sciences, University of Canterbury, Christchurch, New Zealand, 2004.
50. Atterbury, R.J.; Connerton, P.L.; Dodd, C.E.R.; Rees, C.E.D.; Connerton, I.F. Application of Host-Specific Bacteriophages to the Surface of Chicken Skin Leads to a Reduction in Recovery of *Campylobacter jejuni*. *Appl. Environ. Microbiol.* **2003**, *69*, 6302–6306. [CrossRef] [PubMed]
51. Sails, A.D.; Wareing, D.R.; Bolton, F.J.; Fox, A.J.; Curry, A. Characterisation of 16 *Campylobacter jejuni* and *C. coli* Typing Bacteriophages. *J. Med. Microbiol.* **1998**, *47*, 123–128. [CrossRef]
52. Siringan, P.; Connerton, P.L.; Payne, R.J.H.; Connerton, I.F. Bacteriophage-Mediated Dispersal of *Campylobacter jejuni* Biofilms. *Appl. Environ. Microbiol.* **2011**, *77*, 3320–3326. [CrossRef]
53. Loc Carrillo, C.; Atterbury, R.J.; el-Shibiny, A.; Connerton, P.L.; Dillon, E.; Scott, A.; Connerton, I.F. Bacteriophage Therapy to Reduce *Campylobacter jejuni* Colonization of Broiler Chickens. *Appl. Environ. Microbiol.* **2005**, *71*, 6554–6563. [CrossRef] [PubMed]
54. Goode, D.; Allen, V.M.; Barrow, P.A. Reduction of Experimental *Salmonella* and *Campylobacter* Contamination of Chicken Skin by Application of Lytic Bacteriophages. *Appl. Environ. Microbiol.* **2003**, *69*, 5032–5036. [CrossRef]
55. Kropinski, A.M.; Arutyunov, D.; Foss, M.; Cunningham, A.; Ding, W.; Singh, A.; Pavlov, A.R.; Henry, M.; Evoy, S.; Kelly, J.; et al. Genome and Proteome of *Campylobacter jejuni* Bacteriophage NCTC 12673. *Appl. Environ. Microbiol.* **2011**, *77*, 8265–8271. [CrossRef]
56. Mayer, M.J.; Payne, J.; Gasson, M.J.; Narbad, A. Genomic Sequence and Characterization of the Virulent Bacteriophage PhiCTP1 from *Clostridium tyrobutyricum* and Heterologous Expression of Its Endolysin. *Appl. Environ. Microbiol.* **2010**, *76*, 5415–5422. [CrossRef]
57. Kim, K.-P.; Klumpp, J.; Loessner, M.J. *Enterobacter sakazakii* Bacteriophages Can Prevent Bacterial Growth in Reconstituted Infant Formula. *Int. J. Food Microbiol.* **2007**, *115*, 195–203. [CrossRef]
58. Son, H.M.; Duc, H.M.; Masuda, Y.; Honjoh, K.-I.; Miyamoto, T. Application of Bacteriophages in Simultaneously Controlling *Escherichia coli* O157:H7 and Extended-Spectrum Beta-Lactamase Producing *Escherichia coli*. *Appl. Microbiol. Biotechnol.* **2018**, *102*, 10259–10271. [CrossRef] [PubMed]
59. McLean, S.K.; Dunn, L.A.; Palombo, E.A. Phage Inhibition of *Escherichia coli* in Ultrahigh-Temperature-Treated and Raw Milk. *Foodborne Pathog. Dis.* **2013**, *10*, 956–962. [CrossRef]
60. Liao, Y.-T.; Salvador, A.; Harden, L.A.; Liu, F.; Lavenburg, V.M.; Li, R.W.; Wu, V.C.H. Characterization of a Lytic Bacteriophage as an Antimicrobial Agent for Biocontrol of Shiga Toxin-Producing *Escherichia coli* O145 Strains. *Antibiotics* **2019**, *8*, 74. [CrossRef] [PubMed]
61. Hong, Y.; Pan, Y.; Ebner, P.D. Meat Science and Muscle Biology Symposium: Development of Bacteriophage Treatments to Reduce *Escherichia coli* O157:H7 Contamination of Beef Products and Produce. *J. Anim. Sci.* **2014**, *92*, 1366–1377. [CrossRef]
62. Hong, Y.; Pan, Y.; Harman, N.J.; Ebner, P.D. Complete Genome Sequences of Two *Escherichia coli* O157:H7 Phages Effective in Limiting Contamination of Food Products. *Genome Announc.* **2014**, *2*, e00519-14. [CrossRef]
63. O'Flynn, G.; Ross, R.P.; Fitzgerald, G.F.; Coffey, A. Evaluation of a Cocktail of Three Bacteriophages for Biocontrol of *Escherichia coli* O157:H7. *Appl. Environ. Microbiol.* **2004**, *70*, 3417–3424. [CrossRef]
64. Akusobi, C.; Chan, B.K.; Williams, E.S.C.P.; Wertz, J.E.; Turner, P.E. Parallel Evolution of Host-Attachment Proteins in Phage PP01 Populations Adapting to *Escherichia coli* O157:H7. *Pharmaceuticals* **2018**, *11*, 60. [CrossRef] [PubMed]
65. Morita, M.; Tanji, Y.; Mizoguchi, K.; Akitsu, T.; Kijima, N.; Unno, H. Characterization of a Virulent Bacteriophage Specific for *Escherichia coli* O157:H7 and Analysis of Its Cellular Receptor and Two Tail Fiber Genes. *FEMS Microbiol. Lett.* **2002**, *211*, 77–83. [CrossRef] [PubMed]
66. Hudson, J.A.; Billington, C.; Cornelius, A.J.; Wilson, T.; On, S.L.W.; Premaratne, A.; King, N.J. Use of a Bacteriophage to Inactivate *Escherichia coli* O157:H7 on Beef. *Food Microbiol.* **2013**, *36*, 14–21. [CrossRef] [PubMed]
67. Kudva, I.T.; Jelacic, S.; Tarr, P.I.; Youderian, P.; Hovde, C.J. Biocontrol of *Escherichia coli* O157 with O157-Specific Bacteriophages. *Appl. Environ. Microbiol.* **1999**, *65*, 3767–3773. [CrossRef] [PubMed]
68. Sharma, M.; Ryu, J.-H.; Beuchat, L.R. Inactivation of *Escherichia coli* O157:H7 in Biofilm on Stainless Steel by Treatment with an Alkaline Cleaner and a Bacteriophage. *J. Appl. Microbiol.* **2005**, *99*, 449–459. [CrossRef]
69. Abuladze, T.; Li, M.; Menetrez, M.Y.; Dean, T.; Senecal, A.; Sulakvelidze, A. Bacteriophages Reduce Experimental Contamination of Hard Surfaces, Tomato, Spinach, Broccoli, and Ground Beef by *Escherichia coli* O157:H7. *Appl. Environ. Microbiol.* **2008**, *74*, 6230–6238. [CrossRef]
70. Ferguson, S.; Roberts, C.; Handy, E.; Sharma, M. Lytic Bacteriophages Reduce *Escherichia coli* O157: H7 on Fresh Cut Lettuce Introduced through Cross-Contamination. *Bacteriophage* **2013**, *3*, e24323. [CrossRef]
71. Tomat, D.; Mercanti, D.; Balagué, C.; Quiberoni, A. Phage Biocontrol of Enteropathogenic and Shiga Toxin-Producing *Escherichia coli* during Milk Fermentation. *Lett. Appl. Microbiol.* **2013**, *57*, 3–10. [CrossRef]

72. Tomat, D.; Migliore, L.; Aquili, V.; Quiberoni, A.; Balagué, C. Phage Biocontrol of Enteropathogenic and Shiga Toxin-Producing *Escherichia coli* in Meat Products. *Front. Cell Infect. Microbiol.* **2013**, *3*, 20. [CrossRef]
73. Snyder, A.B.; Perry, J.J.; Yousef, A.E. Developing and Optimizing Bacteriophage Treatment to Control Enterohemorrhagic *Escherichia coli* on Fresh Produce. *Int. J. Food Microbiol.* **2016**, *236*, 90–97. [CrossRef]
74. Deasy, T.; Mahony, J.; Neve, H.; Heller, K.J.; van Sinderen, D. Isolation of a Virulent *Lactobacillus brevis* Phage and Its Application in the Control of Beer Spoilage. *J. Food Prot.* **2011**, *74*, 2157–2161. [CrossRef] [PubMed]
75. Greer, G.G.; Dilts, B.D.; Ackermann, H.-W. Characterization of a *Leuconostoc gelidum* Bacteriophage from Pork. *Int. J. Food Microbiol.* **2007**, *114*, 370–375. [CrossRef]
76. Hibma, A.M.; Jassim, S.A.; Griffiths, M.W. Infection and Removal of L-Forms of *Listeria monocytogenes* with Bred Bacteriophage. *Int. J. Food Microbiol.* **1997**, *34*, 197–207. [CrossRef]
77. Klumpp, J.; Loessner, M.J. Listeria Phages: Genomes, Evolution, and Application. *Bacteriophage* **2013**, *3*, e26861. [CrossRef] [PubMed]
78. Roy, B.; Ackermann, H.W.; Pandian, S.; Picard, G.; Goulet, J. Biological Inactivation of Adhering *Listeria monocytogenes* by Listeriaphages and a Quaternary Ammonium Compound. *Appl. Environ. Microbiol.* **1993**, *59*, 2914–2917. [CrossRef] [PubMed]
79. Ackermann, H.W.; DuBow, M.S. *Viruses of Prokaryotes: General Properties of Bacteriophages*; CRC Press Inc.: Boca Raton, FL, USA, 1987; pp. 49–85.
80. Zink, R.; Loessner, M.J. Classification of Virulent and Temperate Bacteriophages of *Listeria* Spp. on the Basis of Morphology and Protein Analysis. *Appl. Environ. Microbiol.* **1992**, *58*, 296–302. [CrossRef] [PubMed]
81. Loessner, M.J.; Goeppl, S.; Busse, M. Comparative Inducibility of Bacteriophage in Naturally Lysogenic and Lysogenized Strains of *Listeria* Spp. by u.v. Light and Mitomycin C. *Lett. Appl. Microbiol.* **1991**, *12*, 196–199. [CrossRef]
82. Guenther, S.; Huwyler, D.; Richard, S.; Loessner, M.J. Virulent Bacteriophage for Efficient Biocontrol of *Listeria monocytogenes* in Ready-to-Eat Foods. *Appl. Environ. Microbiol.* **2009**, *75*, 93–100. [CrossRef]
83. Klumpp, J.; Dorscht, J.; Lurz, R.; Bielmann, R.; Wieland, M.; Zimmer, M.; Calendar, R.; Loessner, M.J. The Terminally Redundant, Nonpermuted Genome of *Listeria* Bacteriophage A511: A Model for the SPO1-like Myoviruses of Gram-Positive Bacteria. *J. Bacteriol.* **2008**, *190*, 5753–5765. [CrossRef]
84. Guenther, S.; Loessner, M.J. Bacteriophage Biocontrol of *Listeria monocytogenes* on Soft Ripened White Mold and Red-Smear Cheeses. *Bacteriophage* **2011**, *1*, 94–100. [CrossRef]
85. Bigot, B.; Lee, W.-J.; McIntyre, L.; Wilson, T.; Hudson, J.A.; Billington, C.; Heinemann, J.A. Control of *Listeria monocytogenes* Growth in a Ready-to-Eat Poultry Product Using a Bacteriophage. *Food Microbiol.* **2011**, *28*, 1448–1452. [CrossRef] [PubMed]
86. Scattolini, S.; D'Angelantonio, D.; Boni, A.; Mangone, I.; Marcacci, M.; Battistelli, N.; D'Agostino, K.; Pomilio, F.; Camma, C.; Migliorati, G.; et al. Characterization and In Vitro Efficacy against *Listeria monocytogenes* of a Newly Isolated Bacteriophage, ΦIZSAM-1. *Microorganisms* **2021**, *9*, 731. [CrossRef]
87. Aprea, G.; D'Angelo, A.R.; Prencipe, V.A.; Migliorati, G. Bacteriophage Morphological Characterization by Using Transmission Electron Microscopy. *J. Life Sci.* **2015**, *9*, 214–220.
88. Carlton, R.M.; Noordman, W.H.; Biswas, B.; de Meester, E.D.; Loessner, M.J. Bacteriophage P100 for Control of *Listeria monocytogenes* in Foods: Genome Sequence, Bioinformatic Analyses, Oral Toxicity Study, and Application. *Regul. Toxicol. Pharmacol.* **2005**, *43*, 301–312. [CrossRef]
89. Oliveira, M.; Viñas, I.; Colàs, P.; Anguera, M.; Usall, J.; Abadias, M. Effectiveness of a Bacteriophage in Reducing *Listeria monocytogenes* on Fresh-Cut Fruits and Fruit Juices. *Food Microbiol.* **2014**, *38*, 137–142. [CrossRef] [PubMed]
90. Iacumin, L.; Manzano, M.; Comi, G. Phage Inactivation of *Listeria monocytogenes* on San Daniele Dry-Cured Ham and Elimination of Biofilms from Equipment and Working Environments. *Microorganisms* **2016**, *4*, 4. [CrossRef]
91. Soni, K.A.; Nannapaneni, R. Bacteriophage Significantly Reduces *Listeria monocytogenes* on Raw Salmon Fillet Tissue. *J. Food Prot.* **2010**, *73*, 32–38. [CrossRef]
92. Soni, K.A.; Nannapaneni, R.; Hagens, S. Reduction of *Listeria monocytogenes* on the Surface of Fresh Channel Catfish Fillets by Bacteriophage Listex P100. *Foodborne Pathog. Dis.* **2010**, *7*, 427–434. [CrossRef]
93. Ellis, D.E.; Whitman, P.A.; Marshall, R.T. Effects of Homologous Bacteriophage on Growth of *Pseudomonas fragi* WY in Milk. *Appl. Microbiol.* **1973**, *25*, 24–25. [CrossRef]
94. Whitman, P.A.; Marshall, R.T. Isolation of Psychrophilic Bacteriophage-Host Systems from Refrigerated Food Products. *Appl. Microbiol.* **1971**, *22*, 220–223. [CrossRef]
95. Whitman, P.A.; Marshall, R.T. Characterization of Two Psychrophilic *Pseudomonas* Bacteriophages Isolated from Ground Beef. *Appl. Microbiol.* **1971**, *22*, 463–468. [CrossRef] [PubMed]
96. Greer, G.G. Homologous Bacteriophage Control of *Pseudomonas* Growth and Beef Spoilage 1, 2. *J. Food Prot.* **1986**, *49*, 104–109. [CrossRef]
97. Greer, G.G. Psychrotrophic Bacteriophages for Beef Spoilage Pseudomonads 1. *J. Food Prot.* **1982**, *45*, 1318–1325. [CrossRef] [PubMed]
98. Tanaka, C.; Yamada, K.; Takeuchi, H.; Inokuchi, Y.; Kashiwagi, A.; Toba, T. A Lytic Bacteriophage for Controlling *Pseudomonas lactis* in Raw Cow's Milk. *Appl. Environ. Microbiol.* **2018**, *84*, e00111-18. [CrossRef] [PubMed]
99. Sillankorva, S.; Neubauer, P.; Azeredo, J. *Pseudomonas fluorescens* Biofilms Subjected to Phage PhiIBB-PF7A. *BMC Biotechnol.* **2008**, *8*, 79. [CrossRef]

100. Sillankorva, S.; Neubauer, P.; Azeredo, J. Isolation and Characterization of a T7-like Lytic Phage for *Pseudomonas fluorescens*. *BMC Biotechnol.* **2008**, *8*, 80. [CrossRef]
101. Kang, H.-W.; Kim, J.-W.; Jung, T.-S.; Woo, G.-J. Wksl3, a New Biocontrol Agent for *Salmonella* Enterica Serovars Enteritidis and Typhimurium in Foods: Characterization, Application, Sequence Analysis, and Oral Acute Toxicity Study. *Appl. Environ. Microbiol.* **2013**, *79*, 1956–1968. [CrossRef]
102. Kim, J.H.; Kim, H.J.; Jung, S.J.; Mizan, M.F.R.; Park, S.H.; Ha, S.-D. Characterization of *Salmonella* Spp.-Specific Bacteriophages and Their Biocontrol Application in Chicken Breast Meat. *J. Food Sci.* **2020**, *85*, 526–534. [CrossRef]
103. Fiorentin, L.; Vieira, N.D.; Barioni, W. Oral Treatment with Bacteriophages Reduces the Concentration of *Salmonella* Enteritidis PT4 in Caecal Contents of Broilers. *Avian Pathol.* **2005**, *34*, 258–263. [CrossRef]
104. Fiorentin, L.; Vieira, N.D.; Barioni, J.W.; Barros, S. In Vitro Characterization and in Vivo Properties of Salmonellae Lytic Bacteriophages Isolated from Free-Range Layers. *Braz. J. Poult. Sci.* **2004**, *6*, 121–128. [CrossRef]
105. Fiorentin, L.; Vieira, N.D.; Barioni, J.W. Use of Lytic Bacteriophages to Reduce *Salmonella* Enteritidis in Experimentally Contaminated Chicken Cuts. *Braz. J. Poult. Sci.* **2005**, *7*, 255–260. [CrossRef]
106. Berchieri, A.; Lovell, M.A.; Barrow, P.A. The Activity in the Chicken Alimentary Tract of Bacteriophages Lytic for *Salmonella* Typhimurium. *Res. Microbiol.* **1991**, *142*, 541–549. [CrossRef]
107. Sonalika, J.; Srujana, A.S.; Akhila, D.S.; Juliet, M.R.; Santhosh, K.S. Application of Bacteriophages to Control *Salmonella* Enteritidis in Raw Eggs. *Iran J. Vet. Res.* **2020**, *21*, 221–225. [PubMed]
108. Islam, M.S.; Zhou, Y.; Liang, L.; Nime, I.; Liu, K.; Yan, T.; Wang, X.; Li, J. Application of a Phage Cocktail for Control of *Salmonella* in Foods and Reducing Biofilms. *Viruses* **2019**, *11*, 841. [CrossRef]
109. Islam, M.S.; Zhou, Y.; Liang, L.; Nime, I.; Yan, T.; Willias, S.P.; Mia, M.Z.; Bei, W.; Connerton, I.F.; Fischetti, V.A.; et al. Application of a Broad Range Lytic Phage LPST94 for Biological Control of *Salmonella* in Foods. *Microorganisms* **2020**, *8*, 247. [CrossRef]
110. Huang, C.; Virk, S.M.; Shi, J.; Zhou, Y.; Willias, S.P.; Morsy, M.K.; Abdelnabby, H.E.; Liu, J.; Wang, X.; Li, J. Isolation, Characterization, and Application of Bacteriophage LPSE1 Against *Salmonella* Enterica in Ready to Eat (RTE) Foods. *Front. Microbiol.* **2018**, *9*, 1046. [CrossRef] [PubMed]
111. Whichard, J.M.; Sriranganathan, N.; Pierson, F.W. Suppression of *Salmonella* Growth by Wild-Type and Large-Plaque Variants of Bacteriophage Felix O1 in Liquid Culture and on Chicken Frankfurters. *J. Food Prot.* **2003**, *66*, 220–225. [CrossRef]
112. Whichard, J.M.; Weigt, L.A.; Borris, D.J.; Li, L.L.; Zhang, Q.; Kapur, V.; Pierson, F.W.; Lingohr, E.J.; She, Y.-M.; Kropinski, A.M.; et al. Complete Genomic Sequence of Bacteriophage Felix O1. *Viruses* **2010**, *2*, 710–730. [CrossRef] [PubMed]
113. Felix, A.; Callow, B.R. Typing of Paratyphoid B Bacilli by Vi Bacteriophage. *Br. Med. J.* **1943**, *2*, 127–130. [CrossRef]
114. Guenther, S.; Herzig, O.; Fieseler, L.; Klumpp, J.; Loessner, M.J. Biocontrol of *Salmonella* Typhimurium in RTE Foods with the Virulent Bacteriophage FO1-E2. *Int. J. Food Microbiol.* **2012**, *154*, 66–72. [CrossRef]
115. Higgins, J.P.; Higgins, S.E.; Guenther, K.L.; Huff, W.; Donoghue, A.M.; Donoghue, D.J.; Hargis, B.M. Use of a Specific Bacteriophage Treatment to Reduce *Salmonella* in Poultry Products. *Poult. Sci.* **2005**, *84*, 1141–1145. [CrossRef] [PubMed]
116. Wall, S.K.; Zhang, J.; Rostagno, M.H.; Ebner, P.D. Phage Therapy to Reduce Preprocessing *Salmonella* Infections in Market-Weight Swine. *Appl. Environ. Microbiol.* **2010**, *76*, 48–53. [CrossRef] [PubMed]
117. Zhang, J.; Hong, Y.; Harman, N.J.; Das, A.; Ebner, P.D. Genome Sequence of a *Salmonella* Phage Used to Control Salmonella Transmission in Swine. *Genome Announc.* **2014**, *2*, e00521-14. [CrossRef] [PubMed]
118. Hong, Y.; Schmidt, K.; Marks, D.; Hatter, S.; Marshall, A.; Albino, L.; Ebner, P. Treatment of *Salmonella*-Contaminated Eggs and Pork with a Broad-Spectrum, Single Bacteriophage: Assessment of Efficacy and Resistance Development. *Foodborne Pathog. Dis.* **2016**, *13*, 679–688. [CrossRef] [PubMed]
119. Zhang, Y.; Ding, Y.; Li, W.; Zhu, W.; Wang, J.; Wang, X. Application of a Novel Lytic Podoviridae Phage Pu20 for Biological Control of Drug-Resistant *Salmonella* in Liquid Eggs. *Pathogens* **2021**, *10*, 34. [CrossRef] [PubMed]
120. Li, Z.; Ma, W.; Li, W.; Ding, Y.; Zhang, Y.; Yang, Q.; Wang, J.; Wang, X. A Broad-Spectrum Phage Controls Multidrug-Resistant *Salmonella* in Liquid Eggs. *Food Res. Int.* **2020**, *132*, 109011. [CrossRef]
121. Ahn, J.; Kim, S.; Jung, L.-S.; Biswas, D. In Vitro Assessment of the Susceptibility of Planktonic and Attached Cells of Foodborne Pathogens to Bacteriophage P22-Mediated *Salmonella* Lysates. *J. Food Prot.* **2013**, *76*, 2057–2062. [CrossRef]
122. Susskind, M.M.; Botstein, D. Molecular Genetics of Bacteriophage P22. *Microbiol. Rev.* **1978**, *42*, 385–413. [CrossRef]
123. Zorn, G.A.; Gough, M. Morphology of Bacteriophage P22 as Seen in Thin Sections of Pelleted Phage. *Virology* **1976**, *71*, 434–443. [CrossRef]
124. Zinno, P.; Devirgiliis, C.; Ercolini, D.; Ongeng, D.; Mauriello, G. Bacteriophage P22 to Challenge *Salmonella* in Foods. *Int. J. Food Microbiol.* **2014**, *191*, 69–74. [CrossRef]
125. Islam, M.S.; Hu, Y.; Mizan, M.F.R.; Yan, T.; Nime, I.; Zhou, Y.; Li, J. Characterization of *Salmonella* Phage LPST153 That Effectively Targets Most Prevalent *Salmonella* Serovars. *Microorganisms* **2020**, *8*, 1089. [CrossRef]
126. Spricigo, D.A.; Bardina, C.; Cortés, P.; Llagostera, M. Use of a Bacteriophage Cocktail to Control *Salmonella* in Food and the Food Industry. *Int. J. Food Microbiol.* **2013**, *165*, 169–174. [CrossRef] [PubMed]
127. Bardina, C.; Colom, J.; Spricigo, D.A.; Otero, J.; Sánchez-Osuna, M.; Cortés, P.; Llagostera, M. Genomics of Three New Bacteriophages Useful in the Biocontrol of *Salmonella*. *Front. Microbiol.* **2016**, *7*, 545. [CrossRef] [PubMed]
128. Bardina, C.; Spricigo, D.A.; Cortés, P.; Llagostera, M. Significance of the Bacteriophage Treatment Schedule in Reducing *Salmonella* Colonization of Poultry. *Appl. Environ. Microbiol.* **2012**, *78*, 6600–6607. [CrossRef] [PubMed]

129. Augustine, J.; Bhat, S.G. Biocontrol of *Salmonella* Enteritidis in Spiked Chicken Cuts by Lytic Bacteriophages ΦSP-1 and ΦSP-3. *J. Basic Microbiol.* **2015**, *55*, 500–503. [CrossRef] [PubMed]
130. Augustine, J.; Varghese, S.M.; Bhat, S.G. ΦSP-3, a *Salmonella*-Specific Lytic Phage Capable of Infecting Its Host under Nutrient-Deprived States. *Ann. Microbiol.* **2013**, *63*, 381–386. [CrossRef]
131. Modi, R.; Hirvi, Y.; Hill, A.; Griffiths, M.W. Effect of Phage on Survival of *Salmonella* Enteritidis during Manufacture and Storage of Cheddar Cheese Made from Raw and Pasteurized Milk. *J. Food Prot.* **2001**, *64*, 927–933. [CrossRef]
132. Bao, H.; Zhang, P.; Zhang, H.; Zhou, Y.; Zhang, L.; Wang, R. Bio-Control of *Salmonella* Enteritidis in Foods Using Bacteriophages. *Viruses* **2015**, *7*, 4836–4853. [CrossRef]
133. Bao, H.; Zhou, Y.; Shahin, K.; Zhang, H.; Cao, F.; Pang, M.; Zhang, X.; Zhu, S.; Olaniran, A.; Schmidt, S.; et al. The Complete Genome of Lytic *Salmonella* Phage VB_SenM-PA13076 and Therapeutic Potency in the Treatment of Lethal Salmonella Enteritidis Infections in Mice. *Microbiol. Res.* **2020**, *237*, 126471. [CrossRef] [PubMed]
134. Duc, H.M.; Son, H.M.; Yi, H.P.S.; Sato, J.; Ngan, P.H.; Masuda, Y.; Honjoh, K.-I.; Miyamoto, T. Isolation, Characterization and Application of a Polyvalent Phage Capable of Controlling *Salmonella* and *Escherichia coli* O157:H7 in Different Food Matrices. *Food Res. Int.* **2020**, *131*, 108977. [CrossRef]
135. Kocharunchitt, C.; Ross, T.; McNeil, D.L. Use of Bacteriophages as Biocontrol Agents to Control *Salmonella* Associated with Seed Sprouts. *Int. J. Food Microbiol.* **2009**, *128*, 453–459. [CrossRef] [PubMed]
136. Pao, S.; Rolph, S.P.; Westbrook, E.W.; Shen, H. Use of Bacteriophages to Control *Salmonella* in Experimentally Contaminated Sprout Seeds. *J. Food Sci.* **2006**, *69*, 127–130. [CrossRef]
137. Li, J.; Li, Y.; Ding, Y.; Huang, C.; Zhang, Y.; Wang, J.; Wang, X. Characterization of a Novel Siphoviridae *Salmonella* Bacteriophage T156 and Its Microencapsulation Application in Food Matrix. *Food Res. Int.* **2021**, *140*, 110004. [CrossRef] [PubMed]
138. Kelly, D.; McAuliffe, O.; Ross, R.P.; Coffey, A. Prevention of *Staphylococcus aureus* Biofilm Formation and Reduction in Established Biofilm Density Using a Combination of Phage K and Modified Derivatives. *Lett. Appl. Microbiol.* **2012**, *54*, 286–291. [CrossRef]
139. Gill, J.J. Revised Genome Sequence of *Staphylococcus aureus* Bacteriophage K. *Genome Announc.* **2014**, *2*, e01173-13. [CrossRef]
140. Bueno, E.; García, P.; Martínez, B.; Rodríguez, A. Phage Inactivation of *Staphylococcus aureus* in Fresh and Hard-Type Cheeses. *Int. J. Food Microbiol.* **2012**, *158*, 23–27. [CrossRef]
141. Garcia, P.; Madera, C.; Martinez, B.; Rodriguez, A. Biocontrol of *Staphylococcus aureus* in Curd Manufacturing Processes Using Bacteriophages. *Int. Dairy J.* **2007**, *17*, 1232–1239. [CrossRef]
142. García, P.; Martínez, B.; Obeso, J.M.; Lavigne, R.; Lurz, R.; Rodríguez, A. Functional Genomic Analysis of Two *Staphylococcus aureus* Phages Isolated from the Dairy Environment. *Appl. Environ. Microbiol.* **2009**, *75*, 7663–7673. [CrossRef] [PubMed]
143. Duc, H.M.; Son, H.M.; Ngan, P.H.; Sato, J.; Masuda, Y.; Honjoh, K.-I.; Miyamoto, T. Isolation and Application of Bacteriophages Alone or in Combination with Nisin against Planktonic and Biofilm Cells of *Staphylococcus aureus*. *Appl. Microbiol. Biotechnol.* **2020**, *104*, 5145–5158. [CrossRef] [PubMed]
144. Chang, Y.; Bai, J.; Lee, J.-H.; Ryu, S. Mutation of a *Staphylococcus aureus* Temperate Bacteriophage to a Virulent One and Evaluation of Its Application. *Food Microbiol.* **2019**, *82*, 523–532. [CrossRef] [PubMed]
145. Yang, Z.-Q.; Tao, X.-Y.; Zhang, H.; Rao, S.-Q.; Gao, L.; Pan, Z.-M.; Jiao, X.-A. Isolation and Characterization of Virulent Phages Infecting *Shewanella baltica* and *Shewanella putrefaciens*, and Their Application for Biopreservation of Chilled Channel Catfish (Ictalurus Punctatus). *Int. J. Food Microbiol.* **2019**, *292*, 107–117. [CrossRef] [PubMed]
146. Zhang, H.; Wang, R.; Bao, H. Phage Inactivation of Foodborne *Shigella* on Ready-to-Eat Spiced Chicken. *Poult. Sci.* **2013**, *92*, 211–217. [CrossRef]
147. Zhang, H.; Yang, Z.; Zhou, Y.; Bao, H.; Wang, R.; Li, T.; Pang, M.; Sun, L.; Zhou, X. Application of a Phage in Decontaminating *Vibrio parahaemolyticus* in Oysters. *Int. J. Food Microbiol.* **2018**, *275*, 24–31. [CrossRef]
148. Cooper, I.R. A Review of Current Methods Using Bacteriophages in Live Animals, Food and Animal Products Intended for Human Consumption. *J. Microbiol. Methods* **2016**, *130*, 38–47. [CrossRef] [PubMed]
149. Gray, J.A.; Chandry, P.S.; Kaur, M.; Kocharunchitt, C.; Bowman, J.P.; Fox, E.M. Novel Biocontrol Methods for *Listeria monocytogenes* Biofilms in Food Production Facilities. *Front. Microbiol.* **2018**, *9*, 605. [CrossRef]
150. Hsu, F.C.; Shieh, Y.S.C.; Sobsey, M.D. Enteric Bacteriophages as Potential Fecal Indicators in Ground Beef and Poultry Meat. *J. Food Prot.* **2002**, *65*, 93–99. [CrossRef] [PubMed]
151. Wongsuntornpoj, S.; Moreno Switt, A.I.; Bergholz, P.; Wiedmann, M.; Chaturongakul, S. *Salmonella* Phages Isolated from Dairy Farms in Thailand Show Wider Host Range than a Comparable Set of Phages Isolated from U.S. Dairy Farms. *Vet. Microbiol.* **2014**, *172*, 345–352. [CrossRef] [PubMed]
152. Muniesa, M.; Jofre, J. Abundance in Sewage of Bacteriophages Infecting *Escherichia coli* O157:H7. *Methods Mol. Biol.* **2004**, *268*, 79–88. [CrossRef]
153. DePaola, A.; Motes, M.L.; Chan, A.M.; Suttle, C.A. Phages Infecting *Vibrio vulnificus* Are Abundant and Diverse in Oysters (*Crassostrea virginica*) Collected from the Gulf of Mexico. *Appl. Environ. Microbiol.* **1998**, *64*, 346–351. [CrossRef]
154. Muniain-Mujika, I.; Calvo, M.; Lucena, F.; Girones, R. Comparative Analysis of Viral Pathogens and Potential Indicators in Shellfish. *Int. J. Food Microbiol.* **2003**, *83*, 75–85. [CrossRef]
155. Leclerc, H.; Edberg, S.; Pierzo, V.; Delattre, J.M. Bacteriophages as Indicators of Enteric Viruses and Public Health Risk in Groundwaters. *J. Appl. Microbiol.* **2000**, *88*, 5–21. [CrossRef]

156. Luo, E.; Aylward, F.O.; Mende, D.R.; DeLong, E.F. Bacteriophage Distributions and Temporal Variability in the Ocean's Interior. *mBio* **2017**, *8*, e01903-17. [CrossRef]
157. Suttle, C.A. Marine Viruses–Major Players in the Global Ecosystem. *Nat. Rev. Microbiol.* **2007**, *5*, 801–812. [CrossRef]
158. Kennedy, J.E.; Wei, C.I.; Oblinger, J.L. Distribution of Coliphages in Various Foods. *J. Food Prot.* **1986**, *49*, 944–951. [CrossRef] [PubMed]
159. Kennedy, J.E.; Oblinger, J.L.; Bitton, G. Recovery of Coliphages from Chicken, Pork Sausage and Delicatessen Meats. *J. Food Prot.* **1984**, *47*, 623–626. [CrossRef]
160. Kennedy, J.E.; Wei, C.I.; Oblinger, J.L. Methodology for Enumeration of Coliphages in Foods. *Appl. Environ. Microbiol.* **1986**, *51*, 956–962. [CrossRef] [PubMed]
161. Gautier, M.; Rouault, A.; Sommer, P.; Briandet, R. Occurrence of *Propionibacterium freudenreichii* Bacteriophages in Swiss Cheese. *Appl. Environ. Microbiol.* **1995**, *61*, 2572–2576. [CrossRef] [PubMed]
162. Hagens, S.; Loessner, M.J. Bacteriophage for Biocontrol of Foodborne Pathogens: Calculations and Considerations. *Curr. Pharm. Biotechnol.* **2010**, *11*, 58–68. [CrossRef]
163. Ackermann, H.-W. Sad State of Phage Electron Microscopy. Please Shoot the Messenger. *Microorganisms* **2013**, *2*, 1–10. [CrossRef]
164. Available online: https://talk.ictvonline.org (accessed on 1 August 2021).
165. Ackermann, H.W. Classification of Bacteriophages. In *The Bacteriophages*, 2nd ed.; Calendar, R., Abedon, S.T., Eds.; Oxford University Press: Oxford, UK, 2006.
166. Ackermann, H.-W. Phage Classification and Characterization. *Methods Mol. Biol.* **2009**, *501*, 127–140. [CrossRef]
167. Ackermann, H.-W. 5500 Phages Examined in the Electron Microscope. *Arch. Virol.* **2007**, *152*, 227–243. [CrossRef]
168. Adriaenssens, E.M.; Edwards, R.; Nash, J.H.E.; Mahadevan, P.; Seto, D.; Ackermann, H.-W.; Lavigne, R.; Kropinski, A.M. Integration of Genomic and Proteomic Analyses in the Classification of the *Siphoviridae* Family. *Virology* **2015**, *477*, 144–154. [CrossRef]
169. Lavigne, R.; Darius, P.; Summer, E.J.; Seto, D.; Mahadevan, P.; Nilsson, A.S.; Ackermann, H.W.; Kropinski, A.M. Classification of *Myoviridae* Bacteriophages Using Protein Sequence Similarity. *BMC Microbiol.* **2009**, *9*, 224. [CrossRef]
170. Lavigne, R.; Seto, D.; Mahadevan, P.; Ackermann, H.-W.; Kropinski, A.M. Unifying Classical and Molecular Taxonomic Classification: Analysis of the *Podoviridae* Using BLASTP-Based Tools. *Res. Microbiol.* **2008**, *159*, 406–414. [CrossRef]
171. Dion, M.B.; Oechslin, F.; Moineau, S. Phage Diversity, Genomics and Phylogeny. *Nat. Rev. Microbiol.* **2020**, *18*, 125–138. [CrossRef]
172. Fokine, A.; Rossmann, M.G. Molecular Architecture of Tailed Double-Stranded DNA Phages. *Bacteriophage* **2014**, *4*, e28281. [CrossRef] [PubMed]
173. Casjens, S.R. Comparative Genomics and Evolution of the Tailed-Bacteriophages. *Curr. Opin. Microbiol.* **2005**, *8*, 451–458. [CrossRef] [PubMed]
174. Callewaert, L.; Michiels, C.W. Lysozymes in the Animal Kingdom. *J. Biosci.* **2010**, *35*, 127–160. [CrossRef] [PubMed]
175. Vollmer, W.; Joris, B.; Charlier, P.; Foster, S. Bacterial Peptidoglycan (Murein) Hydrolases. *FEMS Microbiol. Rev.* **2008**, *32*, 259–286. [CrossRef] [PubMed]
176. Vermassen, A.; Leroy, S.; Talon, R.; Provot, C.; Popowska, M.; Desvaux, M. Cell Wall Hydrolases in Bacteria: Insight on the Diversity of Cell Wall Amidases, Glycosidases and Peptidases Toward Peptidoglycan. *Front. Microbiol.* **2019**, *10*, 331. [CrossRef]
177. Young, I.; Wang, I.; Roof, W.D. Phages Will out: Strategies of Host Cell Lysis. *Trends Microbiol.* **2000**, *8*, 120–128. [CrossRef]
178. Loessner, M.J. Bacteriophage Endolysins–Current State of Research and Applications. *Curr. Opin. Microbiol.* **2005**, *8*, 480–487. [CrossRef]
179. Abedon, S.T. Lysis from Without. *Bacteriophage* **2011**, *1*, 46–49. [CrossRef]
180. Catalão, M.J.; Gil, F.; Moniz-Pereira, J.; São-José, C.; Pimentel, M. Diversity in Bacterial Lysis Systems: Bacteriophages Show the Way. *FEMS Microbiol. Rev.* **2013**, *37*, 554–571. [CrossRef] [PubMed]
181. Gutiérrez, D.; Fernández, L.; Rodríguez, A.; García, P. Are Phage Lytic Proteins the Secret Weapon To Kill *Staphylococcus aureus*? *mBio* **2018**, *9*, e01923-17. [CrossRef] [PubMed]
182. Rodríguez-Rubio, L.; Martínez, B.; Donovan, D.M.; Rodríguez, A.; García, P. Bacteriophage Virion-Associated Peptidoglycan Hydrolases: Potential New Enzybiotics. *Crit. Rev. Microbiol.* **2013**, *39*, 427–434. [CrossRef] [PubMed]
183. Oliveira, H.; São-José, C.; Azeredo, J. Phage-Derived Peptidoglycan Degrading Enzymes: Challenges and Future Prospects for In Vivo Therapy. *Viruses* **2018**, *10*, 292. [CrossRef]
184. Latka, A.; Maciejewska, B.; Majkowska-Skrobek, G.; Briers, Y.; Drulis-Kawa, Z. Bacteriophage-Encoded Virion-Associated Enzymes to Overcome the Carbohydrate Barriers during the Infection Process. *Appl. Microbiol. Biotechnol.* **2017**, *101*, 3103–3119. [CrossRef] [PubMed]
185. Tan, J.-X.; Dao, F.-Y.; Lv, H.; Feng, P.-M.; Ding, H. Identifying Phage Virion Proteins by Using Two-Step Feature Selection Methods. *Molecules* **2018**, *23*, 2000. [CrossRef] [PubMed]
186. Rodríguez, L.; Martínez, B.; Zhou, Y.; Rodríguez, A.; Donovan, D.M.; García, P. Lytic Activity of the Virion-Associated Peptidoglycan Hydrolase HydH5 of *Staphylococcus aureus* Bacteriophage VB_SauS-PhiIPLA88. *BMC Microbiol.* **2011**, *11*, 138. [CrossRef]
187. Pires, D.P.; Oliveira, H.; Melo, L.D.R.; Sillankorva, S.; Azeredo, J. Bacteriophage-Encoded Depolymerases: Their Diversity and Biotechnological Applications. *Appl. Microbiol. Biotechnol.* **2016**, *100*, 2141–2151. [CrossRef] [PubMed]
188. Yan, J.; Mao, J.; Mao, J.; Xie, J. Bacteriophage Polysaccharide Depolymerases and Biomedical Applications. *BioDrugs* **2014**, *28*, 265–274. [CrossRef] [PubMed]

189. Fenton, M.; Keary, R.; McAuliffe, O.; Ross, R.P.; O'Mahony, J.; Coffey, A. Bacteriophage-Derived Peptidase CHAP(K) Eliminates and Prevents Staphylococcal Biofilms. *Int. J. Microbiol.* **2013**, *2013*, 625341. [CrossRef]
190. Gutiérrez, D.; Ruas-Madiedo, P.; Martínez, B.; Rodríguez, A.; García, P. Effective Removal of Staphylococcal Biofilms by the Endolysin LysH5. *PLoS ONE* **2014**, *9*, e107307. [CrossRef]
191. Sass, P.; Bierbaum, G. Lytic Activity of Recombinant Bacteriophage Phi11 and Phi12 Endolysins on Whole Cells and Biofilms of *Staphylococcus aureus*. *Appl. Environ. Microbiol.* **2007**, *73*, 347–352. [CrossRef] [PubMed]
192. Son, J.-S.; Lee, S.-J.; Jun, S.Y.; Yoon, S.J.; Kang, S.H.; Paik, H.R.; Kang, J.O.; Choi, Y.-J. Antibacterial and Biofilm Removal Activity of a *Podoviridae Staphylococcus aureus* Bacteriophage SAP-2 and a Derived Recombinant Cell-Wall-Degrading Enzyme. *Appl. Microbiol. Biotechnol.* **2010**, *86*, 1439–1449. [CrossRef] [PubMed]
193. Gerstmans, H.; Criel, B.; Briers, Y. Synthetic Biology of Modular Endolysins. *Biotechnol. Adv.* **2018**, *36*, 624–640. [CrossRef]
194. Yang, H.; Yu, J.; Wei, H. Engineered Bacteriophage Lysins as Novel Anti-Infectives. *Front. Microbiol.* **2014**, *5*, 542. [CrossRef]
195. Son, B.; Kong, M.; Cha, Y.; Bai, J.; Ryu, S. Simultaneous Control of *Staphylococcus aureus* and *Bacillus cereus* Using a Hybrid Endolysin LysB4EAD-LysSA11. *Antibiotics* **2020**, *9*, 906. [CrossRef]
196. Son, B.; Yun, J.; Lim, J.-A.; Shin, H.; Heu, S.; Ryu, S. Characterization of LysB4, an Endolysin from the *Bacillus cereus*-Infecting Bacteriophage B4. *BMC Microbiol.* **2012**, *12*, 33. [CrossRef] [PubMed]
197. Garde, S.; Calzada, J.; Sánchez, C.; Gaya, P.; Narbad, A.; Meijers, R.; Mayer, M.J.; Ávila, M. Effect of *Lactococcus lactis* Expressing Phage Endolysin on the Late Blowing Defect of Cheese Caused by *Clostridium tyrobutyricum*. *Int. J. Food Microbiol.* **2020**, *329*, 108686. [CrossRef]
198. Mayer, M.J.; Gasson, M.J.; Narbad, A. Genomic Sequence of Bacteriophage ATCC 8074-B1 and Activity of Its Endolysin and Engineered Variants against *Clostridium sporogenes*. *Appl. Environ. Microbiol.* **2012**, *78*, 3685–3692. [CrossRef]
199. Zimmer, M.; Vukov, N.; Scherer, S.; Loessner, M.J. The Murein Hydrolase of the Bacteriophage Phi3626 Dual Lysis System Is Active against All Tested *Clostridium perfringens* Strains. *Appl. Environ. Microbiol.* **2002**, *68*, 5311–5317. [CrossRef]
200. Cho, J.-H.; Kwon, J.-G.; O'Sullivan, D.J.; Ryu, S.; Lee, J.-H. Development of an Endolysin Enzyme and Its Cell Wall-Binding Domain Protein and Their Applications for Biocontrol and Rapid Detection of *Clostridium perfringens* in Food. *Food Chem.* **2021**, *345*, 128562. [CrossRef] [PubMed]
201. Loessner, M.J.; Wendlinger, G.; Scherer, S. Heterogeneous Endolysins in *Listeria monocytogenes* Bacteriophages: A New Class of Enzymes and Evidence for Conserved Holin Genes within the Siphoviral Lysis Cassettes. *Mol. Microbiol.* **1995**, *16*, 1231–1241. [CrossRef] [PubMed]
202. Gaeng, S.; Scherer, S.; Neve, H.; Loessner, M.J. Gene Cloning and Expression and Secretion of *Listeria monocytogenes* Bacteriophage-Lytic Enzymes in *Lactococcus lactis*. *Appl. Environ. Microbiol.* **2000**, *66*, 2951–2958. [CrossRef]
203. Turner, M.S.; Waldherr, F.; Loessner, M.J.; Giffard, P.M. Antimicrobial Activity of Lysostaphin and a *Listeria monocytogenes* Bacteriophage Endolysin Produced and Secreted by Lactic Acid Bacteria. *Syst. Appl. Microbiol.* **2007**, *30*, 58–67. [CrossRef] [PubMed]
204. Xu, S.; Campisi, E.; Li, J.; Fischetti, V.A. Decontamination of *Escherichia coli* O157:H7 on Fresh Romaine Lettuce Using a Novel Bacteriophage Lysin. *Int. J. Food Microbiol.* **2021**, *341*, 109068. [CrossRef] [PubMed]
205. Ribelles, P.; Rodríguez, I.; Suárez, J.E. LysA2, the *Lactobacillus casei* Bacteriophage A2 Lysin Is an Endopeptidase Active on a Wide Spectrum of Lactic Acid Bacteria. *Appl. Microbiol. Biotechnol.* **2012**, *94*, 101–110. [CrossRef]
206. Van Tassell, M.L.; Ibarra-Sánchez, L.A.; Hoepker, G.P.; Miller, M.J. Hot Topic: Antilisterial Activity by Endolysin PlyP100 in Fresh Cheese. *J. Dairy Sci.* **2017**, *100*, 2482–2487. [CrossRef]
207. Zhang, H.; Bao, H.; Billington, C.; Hudson, J.A.; Wang, R. Isolation and Lytic Activity of the *Listeria* Bacteriophage Endolysin LysZ5 against *Listeria monocytogenes* in Soya Milk. *Food Microbiol.* **2012**, *31*, 133–136. [CrossRef] [PubMed]
208. Simmons, M.; Morales, C.A.; Oakley, B.B.; Seal, B.S. Recombinant Expression of a Putative Amidase Cloned from the Genome of *Listeria monocytogenes* That Lyses the Bacterium and Its Monolayer in Conjunction with a Protease. *Probiotics Antimicrob. Proteins* **2012**, *4*, 1–10. [CrossRef] [PubMed]
209. Schmelcher, M.; Waldherr, F.; Loessner, M.J. *Listeria* Bacteriophage Peptidoglycan Hydrolases Feature High Thermoresistance and Reveal Increased Activity after Divalent Metal Cation Substitution. *Appl. Microbiol. Biotechnol.* **2012**, *93*, 633–643. [CrossRef]
210. Dorscht, J.; Klumpp, J.; Bielmann, R.; Schmelcher, M.; Born, Y.; Zimmer, M.; Calendar, R.; Loessner, M.J. Comparative Genome Analysis of Listeria Bacteriophages Reveals Extensive Mosaicism, Programmed Translational Frameshifting, and a Novel Prophage Insertion Site. *J. Bacteriol.* **2009**, *191*, 7206–7215. [CrossRef] [PubMed]
211. Yan, J.; Yang, R.; Yu, S.; Zhao, W. The Application of the Lytic Domain of Endolysin from *Staphylococcus aureus* Bacteriophage in Milk. *J. Dairy Sci.* **2021**, *104*, 2641–2653. [CrossRef]
212. Gu, J.; Xu, W.; Lei, L.; Huang, J.; Feng, X.; Sun, C.; Du, C.; Zuo, J.; Li, Y.; Du, T.; et al. LysGH15, a Novel Bacteriophage Lysin, Protects a Murine Bacteremia Model Efficiently against Lethal Methicillin-Resistant *Staphylococcus aureus* Infection. *J. Clin. Microbiol.* **2011**, *49*, 111–117. [CrossRef]
213. Chang, Y.; Kim, M.; Ryu, S. Characterization of a Novel Endolysin LysSA11 and Its Utility as a Potent Biocontrol Agent against *Staphylococcus aureus* on Food and Utensils. *Food Microbiol.* **2017**, *68*, 112–120. [CrossRef] [PubMed]
214. García, P.; Martínez, B.; Rodríguez, L.; Rodríguez, A. Synergy between the Phage Endolysin LysH5 and Nisin to Kill *Staphylococcus aureus* in Pasteurized Milk. *Int. J. Food Microbiol.* **2010**, *141*, 151–155. [CrossRef] [PubMed]

215. Duarte, A.C.; Fernández, L.; De Maesschalck, V.; Gutiérrez, D.; Campelo, A.B.; Briers, Y.; Lavigne, R.; Rodríguez, A.; García, P. Synergistic Action of Phage PhiIPLA-RODI and Lytic Protein CHAPSH3b: A Combination Strategy to Target *Staphylococcus aureus* Biofilms. *NPJ Biofilms Microbiomes* **2021**, *7*, 39. [CrossRef]
216. Rodríguez-Rubio, L.; Martínez, B.; Donovan, D.M.; García, P.; Rodríguez, A. Potential of the Virion-Associated Peptidoglycan Hydrolase HydH5 and Its Derivative Fusion Proteins in Milk Biopreservation. *PLoS ONE* **2013**, *8*, e54828. [CrossRef]
217. Schmelcher, M.; Powell, A.M.; Camp, M.J.; Pohl, C.S.; Donovan, D.M. Synergistic Streptococcal Phage ΛSA2 and B30 Endolysins Kill Streptococci in Cow Milk and in a Mouse Model of Mastitis. *Appl. Microbiol. Biotechnol.* **2015**, *99*, 8475–8486. [CrossRef]
218. Pritchard, D.G.; Dong, S.; Baker, J.R.; Engler, J.A. The Bifunctional Peptidoglycan Lysin of *Streptococcus agalactiae* Bacteriophage B30. *Microbiology* **2004**, *150*, 2079–2087. [CrossRef]
219. Zhang, Y.; Huang, H.-H.; Duc, H.M.; Masuda, Y.; Honjoh, K.-I.; Miyamoto, T. Endolysin LysSTG2: Characterization and Application to Control *Salmonella* Typhimurium Biofilm Alone and in Combination with Slightly Acidic Hypochlorous Water. *Food Microbiol.* **2021**, *98*, 103791. [CrossRef]
220. Ding, Y.; Zhang, Y.; Huang, C.; Wang, J.; Wang, X. An Endolysin LysSE24 by Bacteriophage LPSE1 Confers Specific Bactericidal Activity against Multidrug-Resistant *Salmonella* Strains. *Microorganisms* **2020**, *8*, 737. [CrossRef]
221. Oliveira, H.; Thiagarajan, V.; Walmagh, M.; Sillankorva, S.; Lavigne, R.; Neves-Petersen, M.T.; Kluskens, L.D.; Azeredo, J. A Thermostable *Salmonella* Phage Endolysin, Lys68, with Broad Bactericidal Properties against Gram-Negative Pathogens in Presence of Weak Acids. *PLoS ONE* **2014**, *9*, e108376. [CrossRef] [PubMed]
222. Bai, J.; Lee, S.; Ryu, S. Identification and in Vitro Characterization of a Novel Phage Endolysin That Targets Gram-Negative Bacteria. *Microorganisms* **2020**, *8*, 447. [CrossRef]
223. Skorynina, A.V.; Piligrimova, E.G.; Kazantseva, O.A.; Kulyabin, V.A.; Baicher, S.D.; Ryabova, N.A.; Shadrin, A.M. Bacillus-Infecting Bacteriophage Izhevsk Harbors Thermostable Endolysin with Broad Range Specificity. *PLoS ONE* **2020**, *15*, e0242657. [CrossRef] [PubMed]
224. Ha, E.; Son, B.; Ryu, S. *Clostridium perfringens* Virulent Bacteriophage CPS2 and Its Thermostable Endolysin LysCPS2. *Viruses* **2018**, *10*, 251. [CrossRef]
225. Stentz, R.; Bongaerts, R.J.; Gunning, A.P.; Gasson, M.; Shearman, C. Controlled Release of Protein from Viable *Lactococcus lactis* Cells. *Appl. Environ. Microbiol.* **2010**, *76*, 3026–3031. [CrossRef] [PubMed]
226. Singh, B.K.; Tiwari, S.; Dubey, N.K. Essential Oils and Their Nanoformulations as Green Preservatives to Boost Food Safety against Mycotoxin Contamination of Food Commodities: A Review. *J. Sci. Food Agric.* **2021**, *101*, 4879–4890. [CrossRef]
227. Tavakoli, S.; Regenstein, J.M.; Daneshvar, E.; Bhatnagar, A.; Luo, Y.; Hong, H. Recent Advances in the Application of Microalgae and Its Derivatives for Preservation, Quality Improvement, and Shelf-Life Extension of Seafood. *Crit. Rev. Food Sci. Nutr.* **2021**, 1–14. [CrossRef] [PubMed]
228. Beya, M.M.; Netzel, M.E.; Sultanbawa, Y.; Smyth, H.; Hoffman, L.C. Plant-Based Phenolic Molecules as Natural Preservatives in Comminuted Meats: A Review. *Antioxidants* **2021**, *10*, 263. [CrossRef] [PubMed]
229. Davies, C.R.; Wohlgemuth, F.; Young, T.; Violet, J.; Dickinson, M.; Sanders, J.-W.; Vallieres, C.; Avery, S.V. Evolving Challenges and Strategies for Fungal Control in the Food Supply Chain. *Fungal Biol. Rev.* **2021**, *36*, 15–26. [CrossRef] [PubMed]
230. Ritota, M.; Manzi, P. Natural Preservatives from Plant in Cheese Making. *Animals* **2020**, *10*, 749. [CrossRef] [PubMed]
231. Dykes, G.A.; Moorhead, S.M. Combined Antimicrobial Effect of Nisin and a Listeriophage against *Listeria monocytogenes* in Broth but Not in Buffer or on Raw Beef. *Int. J. Food Microbiol.* **2002**, *73*, 71–81. [CrossRef]
232. Leverentz, B.; Conway, W.S.; Camp, M.J.; Janisiewicz, W.J.; Abuladze, T.; Yang, M.; Saftner, R.; Sulakvelidze, A. Biocontrol of *Listeria monocytogenes* on Fresh-Cut Produce by Treatment with Lytic Bacteriophages and a Bacteriocin. *Appl. Environ. Microbiol.* **2003**, *69*, 4519–4526. [CrossRef] [PubMed]
233. Rohde, C.; Wittmann, J. Phage Diversity for Research and Application. *Antibiotics* **2020**, *9*, 734. [CrossRef]
234. Sulakvelidze, A. Safety by Nature: Potential Bacteriophage Applications. *Microbe* **2011**, *6*, 122–126. [CrossRef]
235. Naanwaab, C.; Yeboah, O.-A.; Ofori Kyei, F.; Sulakvelidze, A.; Goktepe, I. Evaluation of Consumers' Perception and Willingness to Pay for Bacteriophage Treated Fresh Produce. *Bacteriophage* **2014**, *4*, e979662. [CrossRef]
236. Dalmasso, M.; Hill, C.; Ross, R.P. Exploiting Gut Bacteriophages for Human Health. *Trends Microbiol.* **2014**, *22*, 399–405. [CrossRef] [PubMed]
237. Moye, Z.D.; Woolston, J.; Sulakvelidze, A. Bacteriophage Applications for Food Production and Processing. *Viruses* **2018**, *10*, 205. [CrossRef] [PubMed]

molecules

MDPI

Article

Physicochemical Properties, Antioxidant Capacity, Prebiotic Activity and Anticancer Potential in Human Cells of Jackfruit (*Artocarpus heterophyllus*) Seed Flour

Ibna Suli Trejo Rodríguez [1,2], Luz Eugenia Alcántara Quintana [2], Paola Algara Suarez [2], Miguel Angel Ruiz Cabrera [1] and Alicia Grajales Lagunes [1,*]

1 Faculty of Chemical Sciences, Autonomous University of San Luis Potosí, Universitaria, San Luis Potosí 78210, Mexico; ibna.trejo@uaslp.mx (I.S.T.R.); mruiz@uaslp.mx (M.A.R.C.)

2 Faculty of Nursing and Nutrition, Autonomous University of San Luis Potosí, San Luis Potosí 78240, Mexico; luz.alcantara@uaslp.mx (L.E.A.Q.); paola.algara@uaslp.mx (P.A.S.)

* Correspondence: grajales@uaslp.mx

Abstract: The general aim of this study was to evaluate physicochemical properties, prebiotic activity and anticancer potential of jackfruit (*Artocarpus heterophyllus*) seed flour. The drying processes of jackfruit seeds were performed at 50, 60 and 70 °C in order to choose the optimal temperature for obtaining the flour based on drying time, polyphenol content and antioxidant capacity. The experimental values of the moisture ratio during jackfruit seed drying at different temperatures were obtained using Page's equation to establish the drying time for the required moisture between 5 and 7% in the flour. The temperature of 60 °C was considered adequate for obtaining good flour and for performing its characterization. The chemical composition, total dietary fiber, functional properties and antioxidant capacity were then examined in the flour. The seed flour contains carbohydrates (73.87 g/100 g), dietary fiber (31 g/100 g), protein (14 g/100 g) and lipids (1 g/100 g). The lipid profile showed that the flour contained monounsaturated (4 g/100 g) and polyunsaturated (46 g 100 g) fatty acids. Sucrose, glucose, and fructose were found to be the predominant soluble sugars, and non-digestible oligosaccharides like 1-kestose were also found. The total polyphenol content was 2.42 mg of gallic acid/g of the sample; furthermore, the antioxidant capacity obtained by ferric reducing antioxidant power (FRAP) and 2,2-diphenyl-1-picrylhydrazyl (DPPH) was 901.45 μmol Trolox/100 g and 1607.87 μmol Trolox/100 g, respectively. The obtained flour exhibited good functional properties, such as water and oil absorption capacity, swelling power and emulsifier capacity. Additionally, this flour had a protective and preventive effect which is associated with the potential prebiotic activity in *Lactobacillus casei* and *Bifidobacterium longum*. These results demonstrate that jackfruit seed flour has good nutritional value and antioxidant and prebiotic activity, as well as potential protective effects and functional properties, making it an attractive food or ingredient in developing innovative functional products.

Keywords: jackfruit seeds flour; functional properties; antioxidant activity; dietary fiber; prebiotic; anticancer activity

Citation: Trejo Rodríguez, I.S.; Alcántara Quintana, L.E.; Algara Suarez, P.; Ruiz Cabrera, M.A.; Grajales Lagunes, A. Physicochemical Properties, Antioxidant Capacity, Prebiotic Activity and Anticancer Potential in Human Cells of Jackfruit (*Artocarpus heterophyllus*) Seed Flour. *Molecules* **2021**, *26*, 4854. https://doi.org/10.3390/molecules26164854

Academic Editors: Manuela Pintado, Ezequiel Coscueta and María Emilia Brassesco

Received: 13 June 2021
Accepted: 20 July 2021
Published: 11 August 2021

1. Introduction

The jackfruit (*Artocarpus heterophyllus*) is highly produced and consumed in the southeast of Asia and Brazil [1]. In Mexico, the production and consumption of jackfruit is recent and has increased in the last years because of its flavor and nutritional properties. The seeds however, have no use and are considered waste products, although they are rich in starch, proteins, crude fiber [2,3] and phytonutrients, such as lignans and isoflavones, which have wide-ranging health benefits [4]. In some countries to extend shelf life seeds are boiled or roasted are eaten as a snack [5]. Obtaining flour from these seeds has been considered another alternative for their use, but the conditions of the drying process for obtaining it are very variable [1,2,6–8]. The nutrients composition and some functional properties of seeds flour of jackfruit have been reported [1,2,8]. These studies, however, have only reported the global chemical composition of the jackfruit seeds flour, but the type of carbohydrates, lipids and total fiber have not been reported. Additionally, the antioxidant capacity in the flour has not been reported either. Regarding to biological activity as prebiotic effect and colon anti-cancer activity of jackfruits flour there are no report in the literature. The prebiotics component in jackfruits seeds has been reported, yet its effect has not been evaluated [9]. The prebiotic activity plays an important role in the health of intestinal microbiota which is associated with the prevention of the gastrointestinal diseases including colon cancer. Gastrointestinal diseases are a global health problem that has been increasing in recent years [10], and prebiotic consumption can active immune components from the gut lumen, and with intraepithelial lymphocytes, which modulate the innate immune barrier by secretion of IgA [11–13]. Additionally, it has also been demonstrated that prebiotic fermentation by anaerobic bacteria produces short-chain fatty acids (SCFAs), mainly acetic, propionic and butyric acids, whose increase antagonizes the growth of some pathogenic bacterial strains and favors mucin production in the colon [11]. For this reason, the objectives of this work were (1) to perform the drying kinetics of jackfruit seeds at different temperatures and to use a mathematical model to calculate the time required for obtaining a flour with a desired moisture content based on drying time, polyphenol content and antioxidant capacity, (2) to determine chemical composition and techno-functional properties and (3) in vitro evaluation of prebiotic effect and anti-cancer activity on human colon epithelial cell.

2. Materials and Methods

2.1. *Jackfruit Seeds*

The seeds were obtained in the Huasteca Potosina region (Huichihuayán, in the state of San Luis Potosí, Mexico) from 15 jackfruits. The weight of each seeds was obtained, and the seeds were then cleaned and stored in resealable bags at −20 °C until later use and analysis.

2.2. *Drying Kinetics of Jackfruit Seeds*

Because of the variability that exists in seed drying to obtain the flour, drying kinetics studies were performed to control the drying process. Seed samples were stored overnight at 4 °C before starting the experiments. At the start of each experiment, the seeds were cut into eight pieces, including the shell or aryl, with dimensions of 10 × 7 × 5 mm. About 50 g of these samples were spread in a square tray in a single layer. Drying experiments were performed using three replicates at the temperatures of 50, 60 and 70 °C in a forced air oven (model ULM 500, Memmert, Schwabach, Germany) with an air velocity of 1 m/s in all the experiments. The weight reduction of the samples during the drying process was recorded every 30 min with an electronic balance (±0.0001 g, Explorer model E12140, Ohaus, NJ, USA). Drying was stopped when consecutive weight measurements gave constant values. Average data are reported in drying curves.

When the drying was complete, the samples were ground separately in a food processor (NutriBullet Rx) and then placed in Tyler sieves in order to obtain flour with a particle size of ≤250 μm. The flour was stored in airtight, amber bottles at room temperature until use.

The best temperature of drying was selected based on average drying time, polyphenol content and antioxidant capacity.

2.3. Mathematical Modeling of Drying Curves

Moisture content (X) was transformed to a dimensionless form referred to as the moisture ratio (X*) and was then related to the time for deriving the drying curves. The drying curves were then analyzed with Page's model given by the following equation:

$$X^* = \frac{X_t - X_e}{X_0 - X_e} = exp^{(-kt)^n} \quad (1)$$

where $X\ (t)$, X_0 and Xe are the moisture content at any time, at $t = 0$ and $t = \infty$ respectively; k is the drying constant (min^{-1}), which is a phenomenological property representative of mass and heat transport phenomena; t is the drying time (min); and n is the dimensionless parameter of the Page equation. The goodness of fit was estimated by the coefficient of determination (r^2).

2.4. Analysis Performed in Jackfruit Seeds Flour

Polyphenols content and antioxidant activity tests were performed on flour obtained at the three temperatures with an average moisture content of 6.5 g/100 g. Chemical composition, techno-functional properties, prebiotic and anticancer activity assays were carried out only on the flour obtained at 60 °C because at this temperature, it was possible to achieve a short drying time without affecting the polyphenol content and antioxidant capacity.

2.5. Chemical Composition

Proximate analyses of moisture, proteins, lipids, ash, total carbohydrates and reducing sugars were performed based on established techniques [14]. The total dietary fibers (TDF) was determined using the enzymatic-gravimetric method [985.29 of the AOAC]. This methodology consists of the simulation of the digestion of the sample with the use of enzymes. One gram of defatted flour was gelatinized with α-amylase (at 95 °C, pH 6, for 15 min) and afterward was digested with protease at 60 °C, pH 7.5 for 30 min. Finally, the sample was incubated with amyloglucosidase (at 60 °C, pH 4.5, for 30 min). After the enzymatic treatment, the sample was filtered and simultaneously washed with 95% ethanol and acetone. The recovered filtrate was dried, and the protein and ash content were determined for calculating the TDF content. The TDF was determined in quadruplicate.

2.6. FTIR Infrared Spectroscopy

In order to determine the main functional groups in the flour, a Nicolet iS 10 FTIR spectrometer (Thermo Scientific, Madison, WI, USA) equipped with a universal attenuated total reflection (ATR) detector was used. The spectrum of the sample and its duplicate were recorded in transmittance mode with a 4 cm^{-1} resolution in the region of 4000–300 cm^{-1} at room temperature [15].

2.7. Carbohydrates Characterization by HPLC

Carbohydrates characterization was performed using the methodology reported by [16]. 20 μL of sample were analyzed using an 1100 Series HPLC chromatography system (Agilent Technologies, Santa Clara, CA, USA). The equipment includes a degassing device, a quaternary pump and a refractive index detector (Waters 410, Waters Corporation, Milford, MA, USA). An ion exchange column was used as a stationary phase (model HPX-87C, 7.8 mm internal diameter × 300 mm, Aminex, Hercules, CA, USA). The mobile phase was deionized HPLC grade water with a flow rate of 0.5 mL/min at 75 °C. The program Empower QuickStart 5.0 was used as a control and data analysis system. Arabinose, fructose, galactose, glucose, lactose, maltose, mannose, ribose, sucrose and xylose, as well as chicory inulin, all with purities greater than 98% and from Sigma-Aldrich (St. Louis, MO, USA) were used as standards. The resolution time was 20 min, and carbohydrates were identified by comparing the retention times with the corresponding references standards. Quantification was obtained according to the reference carbohydrate calibration curve ($r = 0.99$).

2.8. Lipids Characterization

The lipids were extracted using the Soxhlet extraction method with analytical grade ether as the reflux solvent for 4 h. At the end of the extraction, the oil was recovered from the mixture by distillation on a rotary evaporator. Sodium sulfate was added to remove moisture from the extract which was stored at 4 °C until use.

2.9. Sample Derivatization and Fatty Acid Profiles

Fatty acid methyl esters (FAME) were prepared according to the AOAC 969.33 method. A methanol solution was prepared with NaOH (0.1 g) in methanol (5 mL), and the mixture was heated to dissolve the sodium hydroxide. The extract was placed in a test tube (900 mg) after, 0.8 mL of the methanol solution was added and the mixture subjected to a microwave cycle method using a Discover System Model 908005 (CEM, Watertown, MA, USA) with the following conditions: 100 °C, 150 W, 290 Psi, with a running time of 3 min and a warm-up time of 10 min, with high agitation. Subsequently, 1 mL of boron trifluoride in methanol was added, and a microwave cycle was applied to the same conditions. Finally, 1 mL of trimethylpentane was added, and again a microwave cycle was applied. The sample obtained had two phases, and the organic phase was used for the injection into the chromatograph.

The FAME were analyzed on a 6890N gas chromatograph of an Agilent Technologies model 5973 system, equipped with a ZB-WAX column (30 m in length, 320 µm internal diameter and 0.50 µm of thickness) and an FID detector. The carrier gas was helium. The temperature of the column was initially 100 °C (for 5 min) and was gradually increased to 240 °C for 4 °C/min. This temperature was maintained for 20 min. The injector temperature was 240 °C, and 1 µL of the organic phase was injected twice. The identification was performed by comparing the retention times of the FAME standards, which consist of a mixture of 37 fatty acid methyl ester.

2.10. Total Polyphenols Content and Antioxidant Capacity

The total polyphenol content was evaluated in the flour obtained from the seeds. For extraction, the sample was allowed to stand for 24 h in the methanol (80%) at 4 °C. The solution was filtered for evaluating the total phenolic content following the Folin-Ciocalteau reagent assay [17].

The antioxidant capacity was measured using two methods. The first was 2,2-diphenyl-1-picrylhydrazyl (DPPH) radical assay following the protocol from Santos-Zea [18], and the results were expressed in µmoles of Trolox/100 g. The second was the ferric reducing antioxidant power (FRAP) assay following the method [17] and the results were expressed as µmoles of Trolox/100 g.

2.11. Techno-Functional Properties

2.11.1. Water and Oil Absorption Capacity

One g of sample was dispersed in 10 mL of deionized water or in 6 mL of soybean oil in a centrifuge tube. After stirring, the sample for water and oil absorption was centrifuged at 4280× g for 60 min and 3077× g for 25 min, respectively, the measurements were performed at room temperature (25 °C). The supernatant was decanted, and the water and oil absorption were calculated as a percentage [19].

2.11.2. Emulsifying Capacity

One g of flour was suspended in 20 mL of deionized water, and then 7 mL of soybean oil was added to it. The mixture was emulsified with a T 25 digital ULTRA-TURRAX (IKA, Wilmington, NC, USA) at 192× g for 1 min. The emulsion obtained was centrifuged at 1731× g for 60 min. the measurements were performed at room temperature (25 °C). The emulsifying capacity was calculated as a percentage of the height of emulsified layer/the height of the whole layer in the tube) × 100 [20].

2.11.3. Swelling Power and Water Solubility Capacity

The 500 mg of flour (W_1) were placed in a tube with 20 mL of distilled water and weighed (W_2). The sample was heated in a water bath at 40, 50, 60 and 70 °C for 30 min, then a centrifugation was performed at 2292× *g* for 10 min. The supernatant was separated and used to determine the solubility, and the tube and its contents were weighed (W_3). For solubility, 10 mL of each supernatant (Vs) were taken and placed in a crucible at constant weight (W_4). The crucible with the sample was dried in a forced air oven at 105 °C for 12 h [8]. After constant weight (W_5) of the sample, these properties were calculated with the following equations:

$$\text{Swelling power (g/g)} = \frac{W_3 - W_2}{W_1} \quad (2)$$

$$\text{Solubility (\%)} = \frac{W_5 - W_4}{Vs} \times \frac{100}{W_1} \quad (3)$$

2.11.4. Viscosity Evaluation

To evaluate the viscosity of the flour, a RheoPlus/32 rheometer (Anton Paar, Ostfildern-Scharnhausen, Germany) was used, which is equipped with a parallel stainless steel plates (θ = 60 mm) geometry. The temperature was increased from 30 to 80 °C. A gap of 0.5 mm between the plates was used. The concentrations evaluated were 1, 3 and 5 g/100 mL, and the amount of sample used was 1.5 mL. The viscosity Pa·s was determined as a function of the increase in the shear rate (γ) from 10 to 600 s^{-1}.

2.12. Prebiotic Extracted from Jackfruit Seed

The prebiotic extraction was carried out according to a published procedure [9] with some modifications. Fresh jackfruit seeds were cut and mixed with 80% ethanol and then ground in a processor to obtain smaller particles. This mixture was allowed to stand 24 h at 4 °C. The sample was then filtered and centrifuged at 1000 rpm/1 h at 4 °C. The supernatant was concentrated on a heating plate at 60 °C with shaking until the content was reduced to half. The concentrate was dehydrated in a silicone tray at 60 °C for 24 h in a forced air oven.

2.13. Prebiotic Effect of Jackfruits Seed Flour and Jackfruits Seeds Extracted

The prebiotic effect of flour and prebiotic extracted was performed using the methodology [21] with some modifications. The growth of *Lactobacillus casei* and *Bifidubacterium longum* was evaluated using a MRS culture medium (Difco, San José, CA, USA) with the same composition but replacing the carbohydrates source. Different sources of carbohydrates such as *Agave salmiana* fructans, commercial inulin and whole wheat flour were used because they are already recognized as prebiotics and we consider it important to compare them with the prebiotic effect of jackfruit flour. The growth of these bacterial was also evaluated in commercial MRS media and MRS without carbohydrates source and were used as control. All carbohydrates source including jackfruit flour were evaluated to 20 g/L of concentration. The broth MRS for *B. longum* were supplemented with 0.05 g/L of L-cysteine. *L. casei* was inoculated was 0.06% (*v*/*v*) corresponding to approximately 1×10^8 CFU/mL in all culture media (25 mL) and incubated at 37 °C/24 h. The concentration of each inoculum was verified through plate count with MRS agar using the Miles and Misra methods [22] the bacteria were incubated anaerobically at 37 °C for 48 h and the growth was expressed as CFU/mL.

2.14. Anticancer Activity in Human Colon Epithelial Cells

2.14.1. Cell Cultures

The human colon epithelial cell lines (normal) CRL1831, (cancerous, grade 1–2) HT29 and (cancerous, grade 3–4) SW480 were transferred to 25-cm^2 culture flasks and grown in Dulbecco's modified Eagle medium (DMEM, Gibco, Invitrogen Corp., Carlsbad, CA, USA) supplemented with 10 % of fetal bovine serum (FBS) (Gibco) and antibiotics (100 U/mL penicillin, 100 μg/mL streptomycin, 0.25 μg/mL amphotericin). The cells were cultured in a humidified incubator with 5 % CO_2 and 95 % air atmosphere at 37 °C. When the cells grew to 80 to 90 % confluence, they were washed with PBS and were trypsinized with 0.5% trypsin-EDTA (Gibco). In addition, the cells were used in subsequent experiments.

2.14.2. Treatments

The HT29, and SW480 cells were seeded in 96-well plates at a density of 1×10^5/well with PBS. The PBS was replaced with DMEM supplemented only for the control for the treatments, PBS was replaced with DMEM-supplemented with jackfruit flour at a concentration of 50 μg/mL, 100 μg/mL, 500 μg/mL and 1 mg/mL. The cells were incubated in a humidified incubator with 5% CO_2 at 37 °C for 24 h.

2.14.3. Cytotoxicity Assay

Viability was assessed with a thiazolyl blue tetrazolium bromide (MTT) assay at 5 mg/mL, at 24, 48 and 72 h, measuring absorbance with a spectrophotometer at a wavelength of 570 nm.

2.14.4. Antioxidant Assay

The CRL1831, HT-29, SW480 colon epithelial cells stored in PBS (100 μL) were homogenized adding proteinase k (2 μg/μL, Sigma-Aldrich, placed in a water bath for 24 h at 25 °C. After this period, proteinase k was inactivated, the samples were centrifuged at 3000× *g* at 4 °C for 5 min. The cell-free supernatants from all the treatments and controls were collected in new sterile tubes. The activities of superoxide dismutase (SOD) and catalase (CAT) enzymes were measured using a colorimetric assay kit according to the manufacturer's protocol: SOD (EC 1.15.11) Assay kit item no. 19160, CAT (EC 1.11.1.6) Assay kit item no. CAT100 (Sigma-Aldrich), using a plate reader.

2.15. Statistical Analysis

All analyses were carried out in triplicate, and the data are given as the mean ± standard deviation. An analysis of variance (ANOVA) and Fisher' least significant difference (LSD) test were performed at a level of $p < 0.05$ and $p < 0.01$ for all response and colon cell respectively using the GraphPad Prism 9.1 software (GraphPad, San Diego, CA, USA).

3. Results and Discussion

3.1. Jackfruit Seeds

The individual weight of each seed was between 6 and 8 g, and these seeds represent 8.5% of the total weight (0.56 ± 0.3 kg) of jackfruit, which is an attractive percentage for use in flour manufacturing. This value was similar to that reported by [23] who indicated that the percentage of seed weight in jackfruit is between 8 and 15%.

3.2. Drying Kinetics of Seeds

The experimental values of the moisture ratio during the drying of jackfruit seeds at different temperatures are shown in Figure 1. The shape of these curves is typical for food products with a drying process that is controlled by internal water transfer. Significant differences ($p < 0.05$) were observed between samples dried at 50, 60 and 70 °C to reach the equilibrium moisture. In Figure 1, the curves predicted with Page's equation are also shown. The values of r^2, n and k (min^{-1}) calculated with Page's equation are given in Table 1. A good correlation between experimental and predicted values of the moisture ratio with r^2 values higher than 0.99 was observed. Therefore, Page's equation is a good model for describing the drying of jackfruit seeds and can be used to calculate the required or desired moisture content of samples at 50, 60 and 70 °C. These results suggest that using any of these temperatures is not required long term to obtain a stable product with a moisture content of less than 10% of the flour. For example, from this figure the time required to obtain a flour with 6.5% of moisture at 60 °C was 8.5 h.

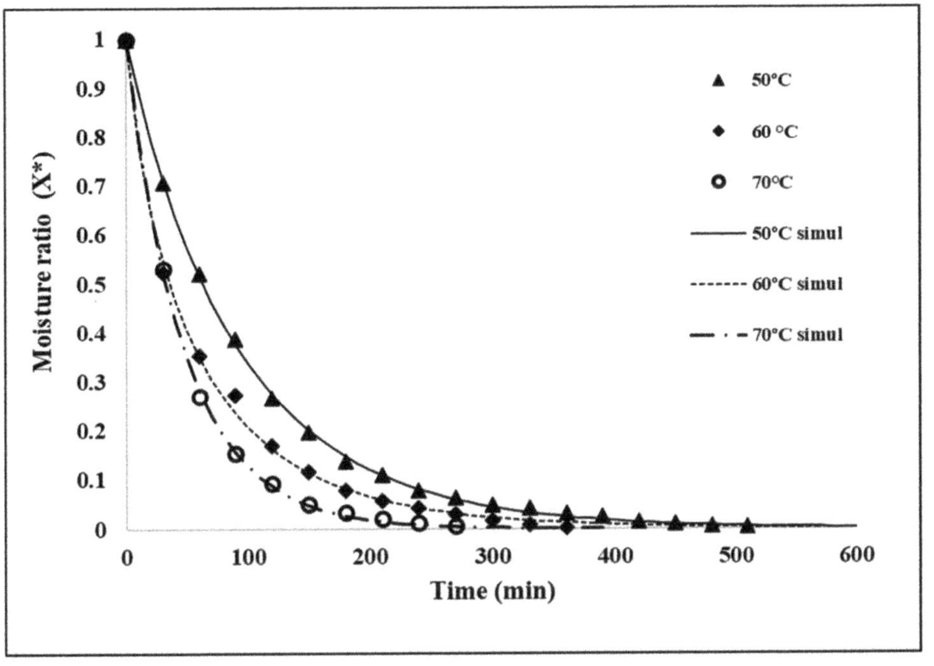

Figure 1. Experimental evolution and the predicted moisture ratio of jackfruit seeds' flour with drying times at 50, 60 and 70 °C.

Table 1. Coefficients obtained from the Page model according to experimental data evaluated at different temperatures.

Model	50 °C		60 °C		70 °C	
	Coefficients	r^2	Coefficients	r^2	Coefficients	r^2
Page	k = 0.0129 n = 0.9618	0.999	k = 0.0402 n = 0.7956	0.996	k = 0.0248 n = 0.9593	0.999

3.3. Chemical Composition

The chemical composition values of flour are presented in Table 2. The mean values of moisture and total ash content were of 6.54 g/100 g and 3.25 g/100 g, respectively. With this moisture content, the flour was stable for 12 months at 25 °C and a relative humidity of 50%. This confirms that a moisture content of less than 10% is required for stable powdered products. With respect to the other biomolecules, the carbohydrates were the major highest component of jackfruit seeds' flour (73.8 ± 1.06 g/100 g), which coincides with other studies reported [1,2,7]. The values of this study, however, were lower than those reported by Ocloo et al. [2] (79.3 g/100 g). The protein content of the flour was 14 ± 0.15 g/100 g; this value is similar to [2,7]) but [1] reported higher values. According to those reported by Lima et al. [1], the protein content of this flour is usually higher than that obtained from wheat flour with an approximate protein content of 10 g/100 g. This value qualifies jackfruit flour as a protein-rich product, and it could be used as an ingredient to prevent malnutrition. The lipids content was the lowest in the flour; the mean value was 1.3 g/100 g and is in the range of [2,8]. The differences observed between the values of chemical composition reported in the literature and those of this work can be attributed to the drying process and jackfruit variability.

Table 2. Proximate composition of jackfruit seed flour in dry basis.

Component g/100 g	Means ± Standard Deviation
Moisture	6.54 ± 0.03
Ash	4.19 ± 0.13
Total carbohydrates	73.87 ± 1.06
Reducing sugars	6.11 ± 0.04
Total dietary fiber	31.59 ± 11.14
Proteins **	14.07 ± 0.15
Lipids	1.3 ± 0.24

** Quantified by the Kjeldahl method. Conversion factor of $n = 6.25$.

A high content of total dietary fiber (TDF) was found in the flour (31.5 g/100 g), which suggests that this flour is an ingredient with a high fiber content. The TDF of this flour has not been reported by other authors; only crude fiber was reported (3.19 and 3.72 g/100 g) by [2,20], respectively. The TDF value is greater than for whole wheat flour that has a value ranging from 11 to 14 g/100 g or for flours obtained from quinoa, oats and corn that have values of 4 to 5 g/100 g in fiber [24].

3.4. FTIR Analysis

Figure 2 shows the FTIR spectra signal and the chemical bonds identified in jackfruit seed flour. The infrared analysis confirmed the presence of carbohydrates and proteins in flour samples made from jackfruit seeds. A typical peak appears in the region of 1013 cm^{-1}, which reflects the presence of carbohydrates. This value is in the range of 1189–953 cm^{-1} reported by [1], and this band is the result of vibrations, stretching and deformation of bonds C-C, C-O and C-H, respectively [25]. The value of 1633 cm^{-1} and 1533 cm^{-1} have been related to amide group characteristics of proteins [1]. The range of 2500–2000 cm^{-1} the bands correspond to double and/or triple bonds and aromatic amino acids [26], which suggests that these types of component are present in the flour. The spectral region between 3000 and 3600 cm^{-1} corresponds to OH stretching vibration, which has been associated with moisture content [13]. No signals were identified in this region because the moisture content of the flour was below 10%.

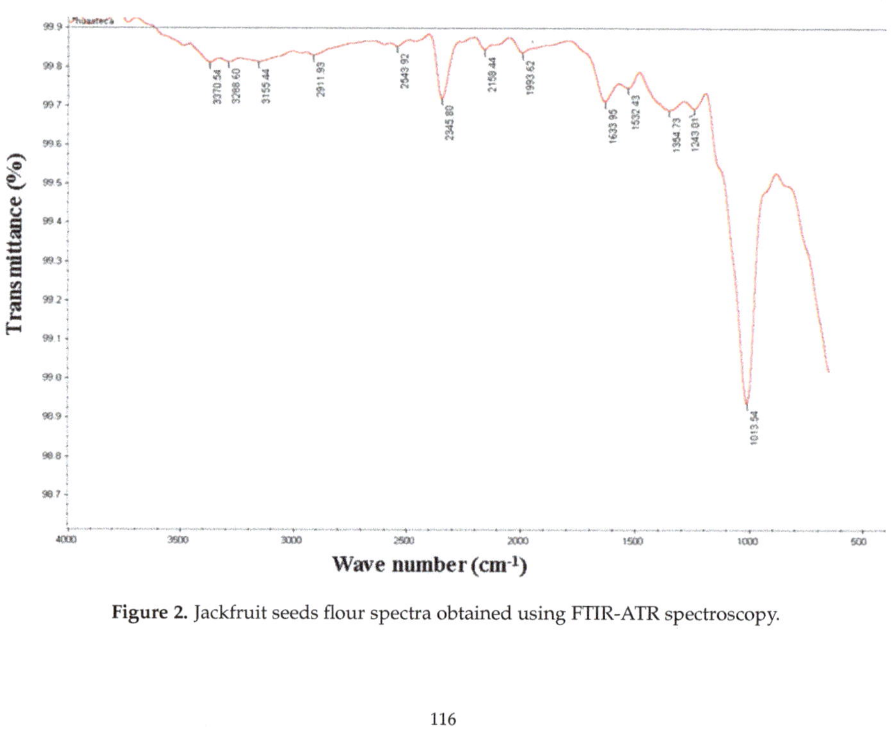

Figure 2. Jackfruit seeds flour spectra obtained using FTIR-ATR spectroscopy.

3.5. Carbohydrates Characterization by HPLC

According to the HPLC analysis, five types of sugars were identified and quantified in jackfruit seeds flour (Table 3). The main carbohydrates identified were mono- and disaccharides like sucrose (35 mg/g), glucose (29 mg/g) and fructose (23 mg/g). According to the standard used, 20 mg/g corresponds to carbohydrates with a polymerization degree greater than 5, which according to the literature [27] could correspond to fructans, oligosaccharides, inulin, resistant starch, cellulose, hemicellulose, gums and pectin. Thus, it will be interesting to identify these compounds using the corresponding standard. In a lower content, 1-kestose (4 mg/g) was identified, which has been described as a non-digestible oligosaccharide considered a prebiotic because of its structure [28]. These types of compounds are relevant because they are not absorbed into the gastrointestinal tract and are not hydrolyzed by human enzymes and thus act as a substrate for endogenous bacteria of the intestine [28].

Table 3. Carbohydrate composition of jackfruit seed flour.

Carbohydrate	Content (mg/g)
Compounds greater than 5 degrees of polymerization	20.0 ± 1.3
1-Kestose	3.8 ± 1.43
Sucrose	35.5 ± 5.4
Glucose	29.4 ± 2.7
Fructose	22.9 ± 6.2

3.6. Lipids Characterization

Table 4 shows the percentages obtained for each fatty acid identified. The average of saturated fatty acids (SFAs) was 49.13 g/100 g, and the main components of SFAs were palmitic acid, tricosanoic acid and stearic acid. With respect to monounsaturated fatty acids (MUFAs), only oleic acid was identified with a total percentage of 4.15 g/100 g. The average percentage of polyunsaturated fatty acids (PUFAs) was 46.72 g/100 g. Linoleic acid, which is an omega-6 fatty acid, was the most dominant, with a relative percentage of 35.11 g/100 g. Furthermore, linolenic acid, arachidonic acid, eicosapentaenoic acid (EPA) and docosadienoic acid were also found. The total of monounsaturated and polyunsaturated fatty acids was 50.87 g/100 g, which is slightly higher than the SFAs. Some of these fatty acids were previously reported in jackfruit seeds [29], others SFAs such as dodecanoic, tridecanoic, tetradecanoic, pentadecanoic and heptadecanoic were also reported by these authors. However, the PUFAs such as arachidonic, eicosapentaeoic and docosadienoic were not reported by these authors. The high content of these fatty acids is relevant because they participate in the control and prevention of cardiovascular diseases, in the control and prevention of rheumatoid arthritis and in the metabolism of HDL lipoproteins [30]. These results suggest that for this flour the unsaturated/saturated fatty acids ratio was 1:1, which could be considered a good balance between both fatty acid type. The relation of ω6/ω3 was 5:1, although some studies report that the ideal ratio in a diet of ω6/ω3 to prevent cardiovascular risk is 1:1 or 1:2 [31]. In contrast, other authors suggest that for the secondary prevention of cardiovascular diseases a ratio of 4:1 has been associated with a 70% reduction in total mortality [32]. German-Austrian-Swiss recommendations stated that ω6 and ω3 fatty acids together should contribute 7 to 10% of the total energy intake with a ratio of linoleic acid (ω6) to α-linolenic acid (ω3) of 5:1 [33]. Additionally, this ratio was also similar to the value obtained for maize pollen (ZP 5557 Lady Fingers, (Maize Research Institute in Zemun Polje. Belgrade, Serbia) which was considered a good balance between ω6 and ω3 fatty acids [34].

Table 4. Fatty acid composition of jackfruit seed flour.

Fatty Acid	Content (g/100 g)
Palmitic acid (C16:0)	36.20 ± 2.68
Stearic acid (18:0)	3.54 ± 0.50
Tricosanoate acid (C23:0)	9.39 ± 0.51
Oleic acid (C18:1)	4.15 ± 1.38
Linoleic acid (C18:2)	35.11 ± 0.59
Linolenic acid (C18:3)	2.82 ± 0.25
Arachidonic acid (C20:4)	3.82 ± 0.23
Eicosapentaenoic acid (20:5 (n-3))	4.01 ± 0.65
Docosadienoic acid (C22:2 (n-6))	0.97 ± 0.02
Total saturated fatty acids	49.13 ± 4.90
Total monounsaturated fatty acids	4.15 ± 2.30
Total polyunsaturated fatty acids	46.72 ± 2.12
ω6/ω3	5:1

3.7. Total Polyphenol Content and Antioxidant Capacity

The total polyphenol content (TPC) and antioxidant activity were determined in the three flours obtained at different temperatures. In Table 5, these results are presented, and the temperature used for the production of flour showed no significant difference ($p > 0.05$) in polyphenol content. It was observed, however, that the polyphenol content decreases when the temperature increases. For this reason and for the process time, we considered 60 °C as the average temperature adequate for obtaining the flour. Information on polyphenol content in flour obtained from jackfruit seeds is scarce. The TPC has only been reported in jackfruit seeds (2.12 ± 0.009 µg gallic acid/mg extract) [35], which is similar to the value reported in this work. [29], however, reported values between 1.31 and 213.41 mg/kg, which were dependent on the solvent and extraction method. The solvent and extraction method used for the isolation of antioxidants is an important factor that can affect the polyphenol content value. The polyphenol profile of jackfruit seeds was also reported by [29], where 12 types of polyphenols were found. Furthermore, 5-caffeoylquinic acid was the main component, and the presence of carotenoids was also indicated. Additionally, [4] also showed that jackfruits seeds contain polyphenols such as lignans and flavones. These compounds have benefits in the uptake of free radicals, protect different cell organelles, can prevent the oxidation of various cellular compounds and protect the human body against damage by reactive oxygen species [17] therefore, they could then provide beneficial effects in the prevention of coronary heart disease.

Table 5. Polyphenol content and antioxidant activity of jackfruit seed flour at the three evaluated temperatures.

Temperature (°C)	Polyphenol Content (mg of Gallic Acid/g of Sample)	DPPH Radical (µmol Trolox/100 g)	FRAP (µmol Trolox/100 g)
50	2.65 ± 0.86 [a]	1579.51 ± 1.91 [a]	1717.20 ± 3.9 [a]
60	2.42 ± 0.74 [a]	1607.87 ± 2.14 [b]	901.45 ± 5.84 [b]
70	2.39 ± 0.92 [a]	1617.48 ± 0.72 [b]	489.77 ± 2.87 [c]
Fisher	0.539	0.0211	<0.0001

Means within a column followed by different letters are significantly different ($p < 0.05$).

The antioxidant capacity of ingredients derived from vegetables has been a relevant field in the last years. The antioxidant capacity was determined by DPPH radical and FRAP. DPPH radical is a compound that is capable of generating free radicals and has been widely used to evaluate the ability to capture free radicals in compounds with antioxidant activity. The values of DPPH radical ranged from 1579 to 1617 μmol Trolox/100 g are shown in Table 5. These results revealed an increasing trend in radical scavenging activity with an increasing drying temperature. This high antioxidant capacity can be explained because a high temperature may lead to the formation of new compounds with higher antioxidant capacity [36]. These authors indicated that the increase in antioxidant capacity in plums dried at 60 and 85 °C was the result of an increase in hydroxymethylfurfural, which is an intermediate in Maillard reaction products (MRPs) production. The increase in DPPH radical value was also reported in hot air drying of pumpkin flour [37] and cowpea seed thermally processed [38]. The DPPH radical values of this study were higher than okra flour (0.309 mM Trolox/100 g) [39] and that in some drinks such as tea (631 μmol Trolox/100 mL) and orange juice (249 μmol Trolox/100 mL) but lower than the fiber-rich fraction of chia (488 μmol Trolox/g) [40]. Higher phenolic compounds and DPPH radical values were obtained in this study in comparisons to rye flour (0.08 to 0.58 mg gallic Acid/g d.m. and 235 μg Trolox/100 g d.m, respectively) [41].

The FRAP is a method used to evaluate the ferric reducing capacity by electron transfer. Antioxidant compounds cause the reduction of the ferric (Fe^{3+}) to the ferrous (Fe^{2+}) because of their reductive power [42]. The FRAP values are shown in Table 5 and were significantly ($p < 0.05$) reduced in the flours obtained at 60 and 70 °C, which is correlated with the trend observed in the polyphenol content. The FRAP and DPPH radical scavenging activity assays, however, determine antioxidant capacities through a different mechanism; therefore, these cannot be correlated [38]. It has been reported that thermal processing reduces TPC and FRAP and increases the DPPH value [38]. The antioxidant power of this flour is a relevant fact since it could be used to counteract reactive oxygen species, which trigger a cascade of diseases associated with oxidative stress that can damage human health.

3.8. *Techno-Functional Properties of Flour*

The techno-functional properties are important in food systems for the development or implementation of new products.

3.8.1. Water Absorption Capacity (WAC)

The WAC is referred to as the ability of a material to retain water, such as linked, hydrodynamic and physically trapped water, under centrifugation conditions [40]. The WAC value of the flour was 463 mL/100 g ± 1.06. This result suggests that the flour has a good WAC, which can be associated with the increase of porosity, as well as the exposition of amino acid residues because of protein denaturation, starch gelatinization and raw fiber swelling during drying [20,43]. Fiber structure and the high proportions of hemicellulose and lignin may also augment WAC [40]. WAC is an important parameter in food processing; for example, a high WAC helps to preserve the freshness of bread, cakes and sausage and can be used as a soup thickener [2,8,19]. However, reported lower values than this study, this difference could be due to process conditions for obtaining the flour or to the effect that in our study the whole seed was used, and in the two studies the thin brown spermoderm was removed.

3.8.2. Oil Absorption Capacity (OAC)

A good oil absorption capacity (OAC) of 34.83 ± 0.001 g/100 mL was obtained in this flour, which is higher than that reported for chia flour [40]. The high OAC is the result of protein denaturation during drying because it has been reported that protein hydrophobicity plays an important role in oil absorption [17], and can also depend on surface characteristics, total charge density and the hydrophobicity of fiber particles [42,44]. Additionally, it was reported that dietary fiber has the ability to retain fat and has been related to the capacity to decrease serum cholesterol levels and remove excess fat from the human body [42,45]. This property is important because fat can improve the flavor of foods and then, this flour could be a high fat and flavor retainer and may therefore find useful application in food systems such as ground meat formulations [2].

Bifidobacterium spp. and *Lactobacillus* spp. in vivo. For commercial inulin, however, the same ($p > 0.05$) behavior was observed in both bacteria. The growth of *B. longum* was higher in all tested prebiotics than the controls. The differences observed in bacterial growth can be associated with the ability that these bacteria have for fermenting non digestible carbohydrates. According to our results, we can say that *B. longum* is capable of degrading prebiotic compounds that are more complex than *L. casei* because jackfruit flour has a variety of nutrients like dietary fiber, polyphenols, resistant starch, and ketose. Additionally, it has been reported that *L. casei* preferentially degrades fructans with a low degree of polymerization [21]. These results suggest that jackfruit seed flour and the prebiotic extract from seeds induce the growth of probiotic bacteria equal to or better than other sources of prebiotics.

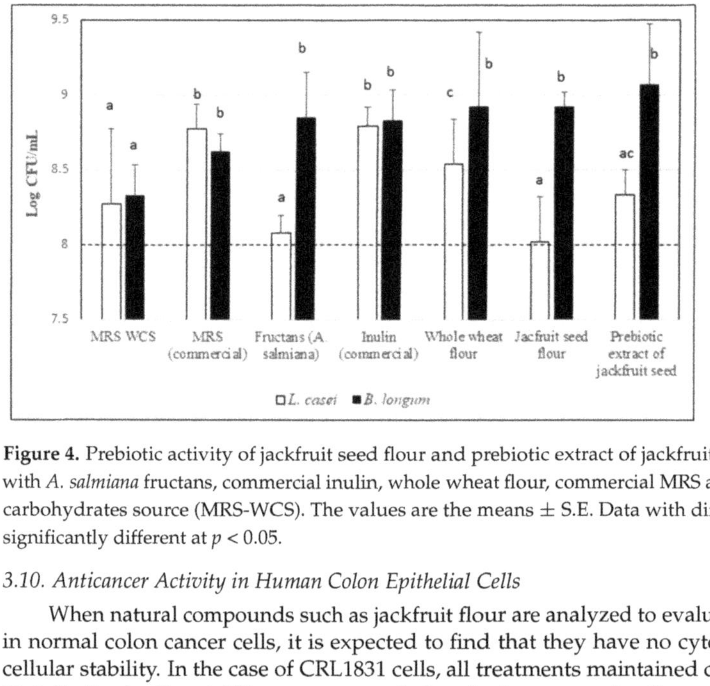

Figure 4. Prebiotic activity of jackfruit seed flour and prebiotic extract of jackfruit seed, compared with *A. salmiana* fructans, commercial inulin, whole wheat flour, commercial MRS and MRS without carbohydrates source (MRS-WCS). The values are the means ± S.E. Data with different letters are significantly different at $p < 0.05$.

3.10. Anticancer Activity in Human Colon Epithelial Cells

When natural compounds such as jackfruit flour are analyzed to evaluate cytotoxicity in normal colon cancer cells, it is expected to find that they have no cytotoxic effect on cellular stability. In the case of CRL1831 cells, all treatments maintained cellular balance. Statistically significant differences were only found when comparing the control (50 μg/mL and 100 μg/mL) against the 1 mg/mL concentration at 72 h, where a slight increase in absorbance can be observed, which means a greater amount of MTT (3-(4,5-dimethylthiazol-2-yl)-2,5-diphenyltetrazolium bromide) was transformed by the cell or greater cell viability (Figure 5A). When these same compounds were analyzed in colon cancer cell lines, however, as is the case of the HT29 cell line, we found that jackfruit flour has an anticancer effect at 50 μg/mL, 100 μg/mL, 500 μg/mL and 1 mg/mL at 48 h. Nevertheless, if 500 μg/mL is added and we wait 24 h more, the effect is the opposite (72 h), higher cell viability (Figure 5B). In the case of the SW480 cell line, there was no significant difference when comparing the groups because the advanced stage of this cancer, the cells cannot sense the extracts of the jackfruit flour. For this reason, they have an exacerbated viability. During biological processes, chemical species known as free radicals are generated, which are characterized by having an unpaired electron and by being very reactive. Of all the radicals, oxygen-derived reactive species (ROS) are of great interest because of the dual radical structure of these molecules and the large number of processes that generate them and in which they can be involved. The main ROS are superoxide anion (O_2), hydroxyl radical (OH^+), singlet oxygen, and hydrogen peroxide (H_2O_2). These radical species are involved in damage cell in such a way that oxidative stresses can lead to carcinogenesis, inflammatory diseases, cellular senescence, and neurodegenerative diseases, among other pathological processes. In the organism, there is an antioxidant protection system formed by enzymes and low molecular weight compounds. Two of the enzymes involved in the protection and consequently the maintenance of the oxidant/antioxidant balance are catalase (CAT) and superoxide dismutase (SOD).

Through the different treatments with jackfruit flour, it was observed that the control epithelial cells (CRL 1831 from a normal colon) presented a significant difference when 100 µg/mL, 500 µg/mL, and 1 mg/mL of jackfruit flour was added at 24 h, as well as 100 µg/mL and 1 mg/mL of the jackfruit flour at 48 h, and finally 500 µg/mL of proteins of jackfruit flour at 72 h (Figure 6A). No significant differences were found, however, for HT29 (Figure 6B) and SW480 (Figure 6C) cells; both lines are colon cancer but at a different stage: stage 1–2 and stage 3–4, respectively. CAT levels in SW480 cells are tenfold higher than even in normal and HT29 cells because of the excessive stress that stage 3–4 cells are under.

For the CRL 1831 cells, in the case of SOD, it was found that the highest concentration of jackfruit flour (1 mg/mL) at the maximum of 72 h showed significant differences when compared with the control and even with each treatment (Figure 7A). In the case of the HT29 cell line, significant differences were found at 100 µg/mL at 72 h, when compared with the other treatments and even with the control. Similarly, for the concentration of 1 mg/mL at 72 h, differences were found with the four other treatments. (Figure 7B). In the case of the SW 480 cell line, significant differences were found when comparing 1 mg/mL at 72 h with the four treatments: 50 µg/mL and 500 µg/mL at 48 h, and 100 ug/mL and 1 mg/mL at 24 h. For that, we consider that the best time to see the effect is at 72 h using 1 mg/mL of concentration (Figure 7C). If we compare only the controls for SOD, we can see an increase in the SW480 cell line for stage 3–4. Therefore, these colon cancer tumor cells are under great oxidative stress, which can generate continuous genetic changes that are manifested as an increase in chromosomal abnormalities and mutations and in turn can lead to tumor genesis and spreading (metastasis). Colon cancer is one of the most prevalent malignant diseases and a major health problem throughout the world, causing a high rate of mortality [53]. Strategies to prevent colon cancer consisting of improved diet components have become an important tool and need much more understanding for its proper use. The main finding of this work is that jackfruit flour shows antioxidant and antiproliferative effects on a colonic cancer cell line. These effects depend on the stage of cancer progression and were not observed in more advanced stages; they were also milder in the control epithelial cell line, which shows the potential for jackfruit flour as a preventive nutraceutical.

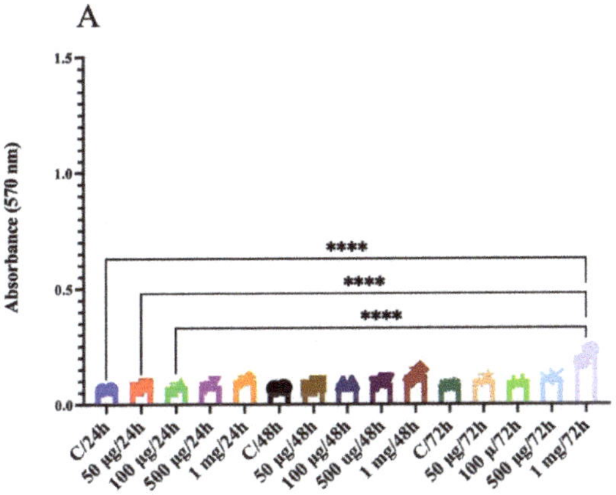

Figure 5. *Cont.*

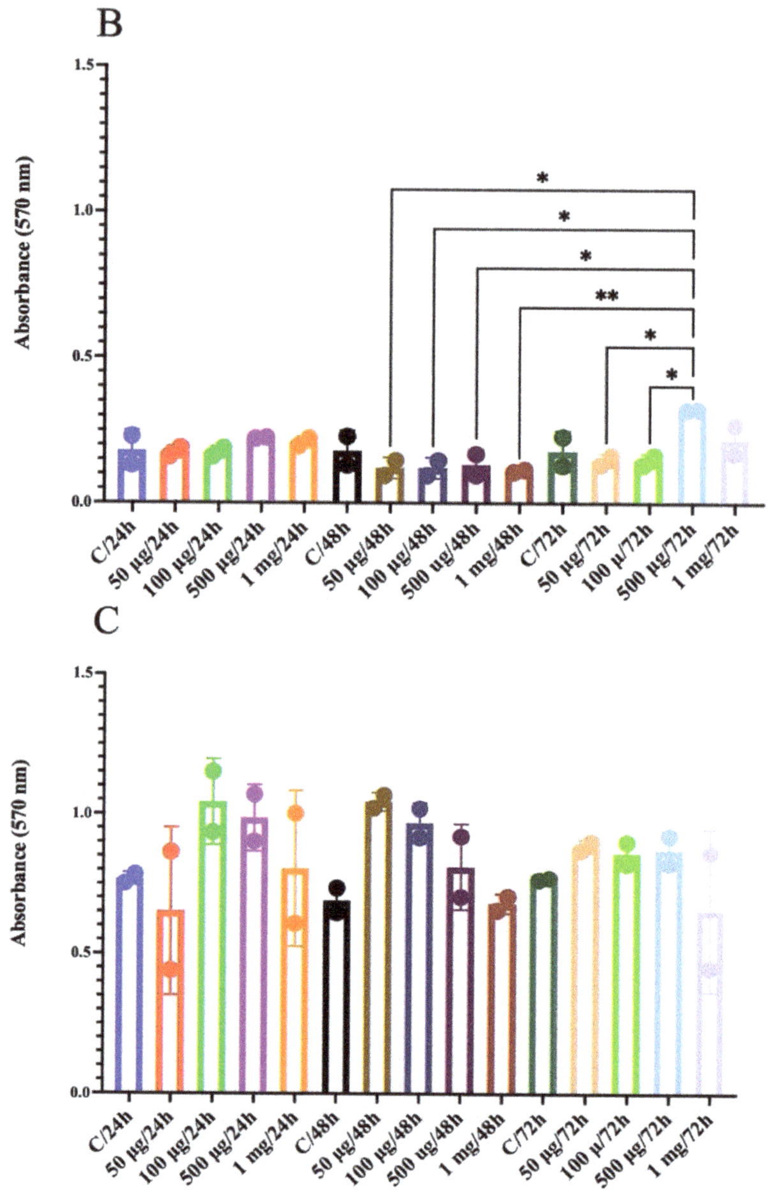

Figure 5. Cytotoxicity assay, was assessed with an MTT (thiazolyl blue tetrazolium bromide), at 24, 48 and 72 h. (**A**) CRL1831, (**B**) HT29, (**C**) SW480. (* $p < 0.05$, ** $p < 0.01$, **** $p < 0.0001$).

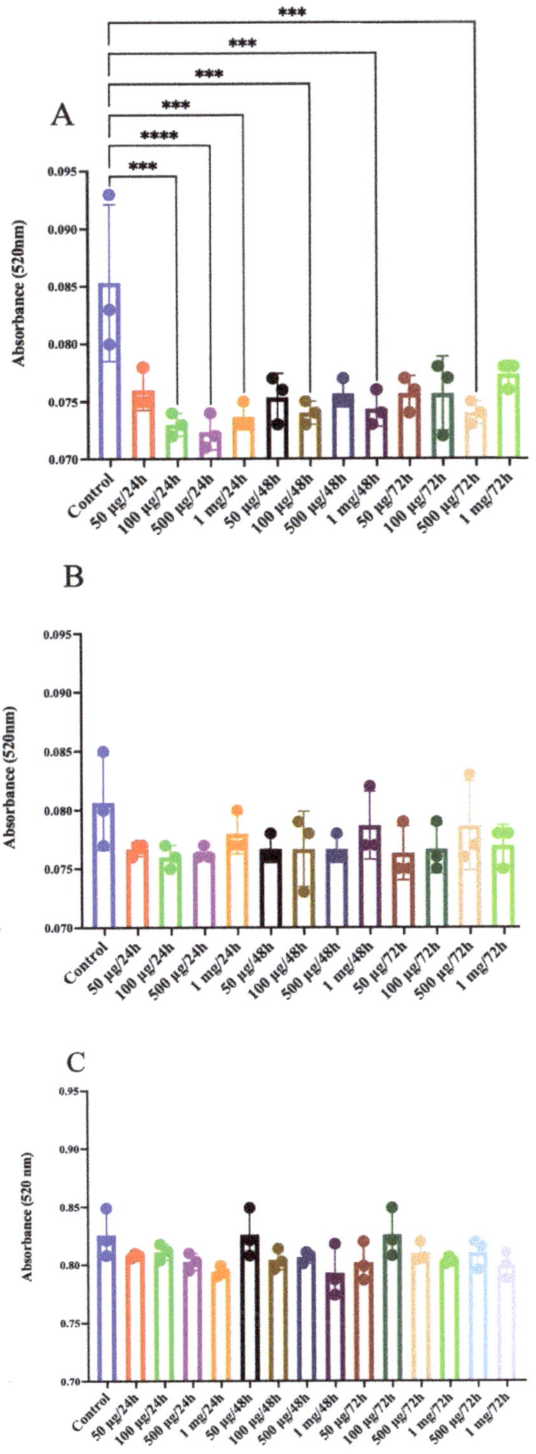

Figure 6. Catalase assay, (**A**) CRL1831, (**B**) HT29, (**C**) SW480. (*** $p < 0.005$, **** $p < 0.0001$).

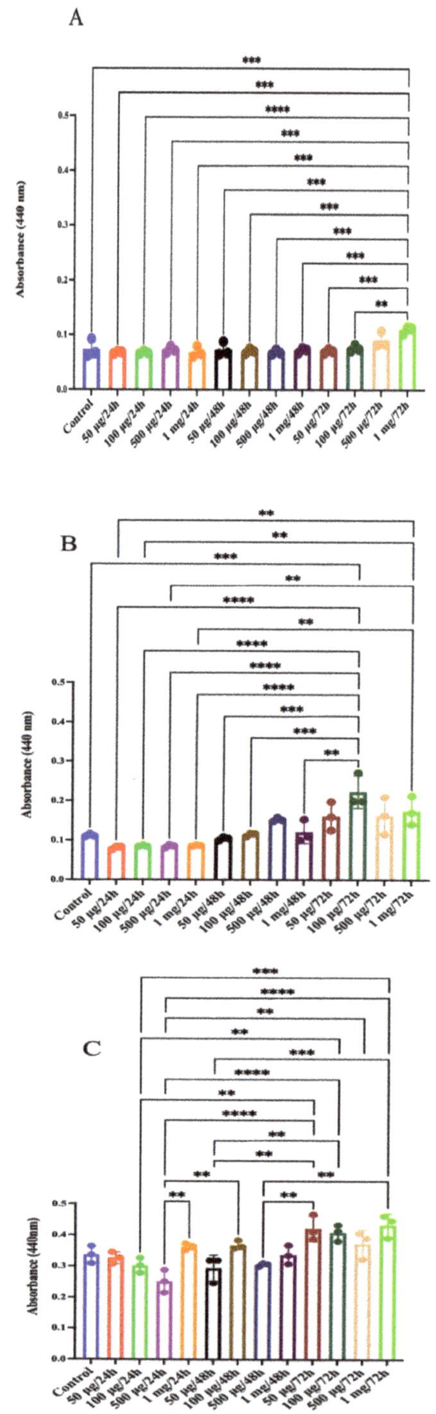

Figure 7. SOD assay. (**A**) CRL1831, (**B**) HT29, (**C**) SW480. (** $p < 0.01$, *** $p < 0.005$, **** $p < 0.0001$).

Tumorigenesis, as the hallmark of cancer, consists of a dysregulation of cellular proliferation, leading to uncontrolled cell growth. Therefore in this work we tested the effect of jackfruit flour on cell viability in a control colonic epithelial cell line and in two colonic cancer cell lines. Jackfruit flour did not alter cell viability in a controlled colonic epithelial cell line, either by showing cytotoxicity nor by inducing proliferation. It has been reported that artocarpin, a phenolic compound found in the wood of jackfruit decreased cell viability in five different colon adenocarcinoma cell lines (DLD1, HCT15, HCT116, HT29, and SW480) but not in normal colon fibroblast cells (CCD-18Co). It induced apoptosis and autophagy, inducing G1 phase cell cycle arrest by targeting Akt1 and Akt 2 kinase activity, and reduced tumorigenesis in vivo [54].

On the other hand, the flour decreased cell viability in the HT29 cell line, which is a cancerous cell line, and thus showed anticancer effects. Reference [55] recently reported that water-soluble polysaccharides obtained from jackfruit did not affect viability in HT29 nor SW620 colonic cancer cell lines after 24 or 48 h as tested by a neutral red assay. Nevertheless, an MTT assay of these same cell lines showed a gradual decrease in cell viability with increasing concentrations of polysaccharide extracts. The anti-proliferative effect was also supported by a high antioxidant activity of these polysaccharide extracts.

Viability in the SW480 cell line was very high in basal conditions, which is expected since it is a cell line from advanced cancer stages characterized by Broders' classification as grade 4, and shows multiple mitosis and high ROS production [56]. Jackfruit flour did not alter viability in these cells at any concentration or incubation time. A possible explanation for this is that this cell line shows a very high basal proliferation rate related to mutations and a loss of apoptosis, which has been well characterized in this cell line [57]. A very interesting feature of the jackfruit flour treatment was its antioxidant activity in colonic cells. Oxidative stress consists mainly of the excessive formation of oxygen-containing chemical species, such as superoxide, hydroxyl and hydrogen peroxide by enzymatic and non-enzymatic cellular pathways. Oxidative stress in a healthy background is naturally counterbalanced by specific enzymes (catalase, SOD, etc.) that transform ROS into harmless chemical species, such as oxygen and water. This process is deeply involved in the initiation and progression of colorectal cancer [58], given that the colon is chronically exposed to oxidizing compounds, toxins and iron. An oxidative environment promotes chromosome damage and thus mutations that lead to increased proliferation and tumour formation [59]. It is not surprising then that antioxidant strategies have been widely proposed as a strong preventive tool against cancer. In this work we reported many compounds found in jackfruit extracts such as polysaccharides and polyphenols which have antioxidants activity. In this work we measured catalase and SOD activity as a means of evaluating the effect of jackfruit flour on oxidative stress response in colonic cell lines. We observed that control epithelial cells showed basal catalase activity, which significantly decreased after incubation with jackfruit flour at different doses after 24, 48 and 72 h. Nevertheless, SOD activity did not decrease, which might mean that catalase activity is sufficient in controlling ROS formation. Regarding cancerous cells, in HT-29 cells there was a slight decrease in catalase activity, which was not significant, whereas SOD activity decreased at 24 h of incubation but increased over longer periods of time. This might be explained by increased oxidative stress because of the incubation in media for longer periods of time. Regarding SW480 cells, catalase and SOD activities were high under all conditions. These high levels of ROS have been previously reported for this cell line [60], which is highly proliferative and thus very active metabolically, leading to a high production of ROS that could not be scavenged efficiently by jackfruit flour.

4. Conclusions

Drying curves of jackfruit seeds were satisfactorily described by mathematical models based on Page's equation and can be used to calculate the required or desired moisture content of samples at 50, 60 and 70 °C. Jackfruit seeds flour obtained at 60 °C is a nutritional food because of the presence of proteins, monosaccharides, oligosaccharides and polysaccharides, as well as omega-3 and omega-6 fatty acids. This flour also provides bioactive ingredients, such as total dietary fiber, polyphenols and antioxidant properties. Additionally, this flour exhibits good functional properties, such as water and fat absorption capacity, swelling powder and emulsifier capacity. The flour and prebiotic extract of jackfruits seed showed a potential prebiotic effect compared with other prebiotic sources. Moreover, jackfruit flour has a protective and preventive effect in colon cancer cells. Additional research, however, is needed to optimize the application of this flour, to assess its effect as a texture adjuvant in food matrices like meat products, bread, beverages or to develop new food products with functional ingredients that may prevent intestinal diseases.

Author Contributions: I.S.T.R., A.G.L. designed the research plan. L.E.A.Q. contributed with her expertise on cell analyses, she discussed and wrote the results. I.S.T.R. performed the experiments. I.S.T.R., A.G.L. and P.A.S. analyzed and discussed the results and wrote the document. M.A.R.C. contributed to the flour drying, mathematical modeling and revised and corrected the English version. All authors revised the manuscript, contributed to the article. All authors have read and agreed to the published version of the manuscript.

Funding: This work was supported by Research Support Fund of the Autonomous University of San Luis Potosi (C16-FAI-09-09-09).

Institutional Review Board Statement: Not applicable.

Informed Consent Statement: Not applicable.

Data Availability Statement: Data used to support the findings of this study are available from the corresponding author upon request.

Acknowledgments: I.S.T.R. thanks the Consejo Nacional de Ciencia y Tecnología (CONACyT, Mexico) for the scholarship granted to her. Thank you to Yolanda Terán for lending us the microbiology laboratory facilities.

Conflicts of Interest: The authors declare that there is no conflict of interest.

References

1. Lima, B.N.B.; Lima, F.F.; Tavares, M.I.B.; Costa, A.M.M.; Pierucci, A.P.T. Determination of the centesimal composition and characterization of flours from fruit seeds. *Food Chem.* **2014**, *151*, 293–299. [CrossRef] [PubMed]
2. Ocloo, F.C.K.; Bansa, D.; Boatin, R.; Adom, T.; Agbemavor, W.S. Physico-chemical, functional and pasting characteristics of flour produced from jackfruits (*Artocarpus heterophyllus*) seeds. *Agric. Biol. J. N. Am.* **2010**, *1*, 903–908. [CrossRef]
3. Butool, S.; Butool, M. Nutritional quality on value addition to jack fruit seed, flour. *Int. J. Sci. Res.* **2015**, *4*, 2406–2411.
4. Umesh, J.B.; Panaskar-Shrimant, N.; Bapat, V.A. Evaluation of antioxidant capacity and phenol content in jackfruit (*Artocarpus heterophyllus*) fruit pulp. *Plant Foods Hum. Nutr.* **2010**, *65*, 99–104.
5. Madruga, M.S.; de Alburquerque, F.S.M.; Silva, I.R.A.; Do Amaral, D.S.; Magnani, M.; Neto, V.Q. Chemical morphological and functional properties of Brazilian jackfruit (*Artocarpus heterophyllus* L.) seeds starch. *Food Chem.* **2014**, *143*, 440–445. [CrossRef] [PubMed]
6. Reza, F.; Tajul-Aris, Y.; Wan-Nadiah, W.A.; Wahidu, Z. Effects of incorporation of jackfruit rind power on chemical and functional properties of bread. *Trop. Life Sci. Res.* **2018**, *29*, 113–126.
7. Eke-Ejiofor, J.; Beleya, E.A.; Onyenorah, N.I. The effect of processing methods on the functional and compositional properties of jackfruit seed flour. *Int. J. Nutr. Food Sci.* **2014**, *3*, 166–173. [CrossRef]
8. Islam, S.; Begum, R.; Khatun, M.; Dey, K.C. A study on nutritional and functional properties analysis of Jackfruit seed flour and value addition to biscuits. *Int. J. Eng. Res. Technol.* **2015**, *4*, 139–147.
9. Prasertsit, K.; Thitipong, R.; Chetpattananondh, P. Possible prebiotic and gallic acid separation from jackfruit seed extract. *Songklanakarin J. Sci. Technol.* **2015**, *37*, 353–359.
10. Faghfoori, Z.; Gargari, B.P.; Gharamaleki, A.S.; Bagherpour, H.; Khosroushahi, A.Y. Cellular and molecular mechanisms of probiotics effects on colorectal cancer. *J. Funct. Foods* **2015**, *18*, 463–472. [CrossRef]
11. Franco-Robles, E.; López, M.G. Implication of fructans in health: Immunomodulatory and antioxidant mechanisms. *Sci. World J.* **2015**, *2015*, 289267. [CrossRef] [PubMed]
12. Ito, H.; Takemura, N.; Sonoyama, K.; Kawagishi, H.; Topping, D.L.; Conlon, M.A.; Morita, T. Degree of polymerization of inulin-type fructans differentially affects number of lactic acid bacteria, intestinal immune functions, and immunoglobulin A Secretion in the rat cecum. *J. Agric. Food Chem.* **2011**, *59*, 5771–5778. [CrossRef]
13. Vogt, L.; Meyer, D.; Pullens, G.; Faas, M.; Smelt, M.; Venema, K.; Ramasamy, U.; Schols, H.A.; De Vos, P. Immunological properties of inulin-type fructans. *Crit. Rev. Food Sci. Nutr.* **2016**, *55*, 414–436. [CrossRef] [PubMed]
14. AOAC. *Official Methods of Analysis of AOAC*, 18th ed.; Association of Official Analytical Chemists: Gathersburg, MD, USA, 2005.

15. González-Muñoz, A.; Montero, B.; Enrione, J.; Mariacevich, S. Rapid prediction of moisture content of quinoa (*Chenopodium quinoa* Willd.) flour by Fourier transform infrared (FTIR) spectroscopy. *J. Cereal Sci.* **2016**, *71*, 246–249. [CrossRef]
16. Castillo-Andrade, A.I.; Rivera-Bautista, C.; Soria Guerra, R.E.; Ruiz-Cabrera, M.A.; Fuentes-Ahumada, C.; Garcia-Chávez, E.; Grajales-Lagunes, A. *Agave fructans* as gut health promoters: Prebiotic activity and inflammatory response in Wistar healthy rats. *Int. J. Biol. Macromol.* **2019**, *136*, 785–795. [CrossRef] [PubMed]
17. Soong, Y.; Barlow, P. Antioxidant activity and phenolic content of select fruit seeds. *Food Chem.* **2004**, *88*, 411–417. [CrossRef]
18. Santos-Zea, L.; Gutiérrez-Uribe, J.A.; Serna-Saldivar, S. Comparative analyses of total phenols, antioxidant activity, and flavonol glycoside profile of cladode flours from different varieties of *Opuntia* spp. *J. Agric. Food Chem.* **2011**, *59*, 7054–7061. [CrossRef] [PubMed]
19. Falade, K.O.; Okafor, C.A. Physical, functional, and pasting properties of flours from corms of two Cocoyam (*Colocasia esculenta* and *Xanthosoma sagittifolium*) cultivars. *J. Food Sci. Technol.* **2015**, *52*, 3440–3448. [CrossRef] [PubMed]
20. Juárez-Barrientos, J.M.; Hernández-Santos, B.; Herman-Lara, E.; Martínez-Sánchez, C.E.; Torruco-Uco, G.J.; Ramírez-Rivera, E.J.; Pineda-Pineda, J.M.; Rodríguez-Miranda, J. Effects of boiling on the functional, thermal and compositional properties of the Mexican jackfruit (*Artocarpus heterophyllus*) seed jackfruit seed meal properties. *Emir. J. Food Agric.* **2017**, *29*, 1–9. [CrossRef]
21. Moreno-Vilet, L.; García-Hernández, M.H.; Delgado-Portales, R.E.; Corral-Fernández, N.M.; Cortez-Espinoza, N.; Ruiz-Cabrera, M.A.; Portales-Pérez, D.P. In vitro assessment of *Agave fructans* (*Agave salmiana*) as prebiotics and immune system activators. *Int. J. Biol. Macromol.* **2014**, *63*, 181–187. [CrossRef] [PubMed]
22. ICMSF (International Commission on Microbiological Specifications for Food of the International Union of Microbiological Societies). *Microorganism in Foods 1: Their Significance and Methods and Enumeration Microorganism in Foods*, 2nd ed.; University of Toronto Press: Toronto, ON, Canada, 1978.
23. Amornrat, M.; Kamontip, S. Physico-chemical properties of flour and starch from jackfruit seeds (*Artocarpus heterophyllus* Lam.) compared with modified starches. *Int. J. Food Sci. Technol.* **2004**, *39*, 271–276.
24. Hager, A.; Wolter, A.; Jacob, F.; Zannini, E.; Arendt, E. Nutritional properties and ultra-structure of commercial gluten free flours from different botanical sources compared to wheat flours. *J. Cereal Sci.* **2012**, *56*, 239–247. [CrossRef]
25. Liu, H.; Chaudhary, D.; Shin-Ichi, Y.T.; Moses, O. Glycerol/starch/Na^+ montmorillonite nanocomposites: An XRD, FTIR, DSC and 1H NMR study. *Carbohydr. Polym.* **2011**, *83*, 1591–1597. [CrossRef]
26. Salunkhe, D.K.; Chavan, J.K.; Kadam, S.S.; Reddy, N.R. Pigeon pea as an important food source. *Crit. Rev. Food Sci. Nutr.* **1986**, *23*, 103–145. [CrossRef] [PubMed]
27. Davani-Davari, D.; Negahdaripour, M.; Karimzadeh, I.; Seifan, M.; Mohkam, M.; Jalil-Masoumi, S.; Berenjian, A.; Ghasemi, Y. Prebiotics: Definition, types, sources, mechanism and clinical applications. *Foods* **2019**, *8*, 92. [CrossRef] [PubMed]
28. Shirai, T.; Suzuki, Y.; Kamikado, K.; Koga, Y.; Aoki, R. Kestose, a prebiotic fructooligosaccharide, enhances intercelular tight junction recovery via a rho-associated knase-dependent mechanism in intestinal CaCo-2-Cells. *Int. J. Probiotics Prebiotics* **2013**, *8*, 53–60.
29. Fernandes, F.; Ferreres, F.; Gil-Izquierdo, A.; Oliveira, A.; Valentão, P.; Andrade, P.B. Accumulation of primary and secondary metabolites in edible jackfruit seed tissues and scavenging of reactive nitrogen species. *Food Chem.* **2017**, *233*, 85–95. [CrossRef] [PubMed]
30. Jin, H.S.; Park, Y. Effect of conjugated linoleic acid on bone formation and rheumatoid arthritis. *Eur. J. Pharmacol.* **2007**, *568*, 16–24.
31. Granados, S.; Quiles, J.L.; Gil, A.; Ramírez-Tortosa, M.C. Lípidos de la dieta y cáncer. *Nutr. Hosp.* **2006**, *21*, 44–54.
32. Simopoulos, A.P. Omega-6/omega-3 essential fatty acids: Biological effects. *World Rev. Nutr. Diet.* **2009**, *99*, 1–16.
33. EFSA Panel on Dietetic Products, Nutrition and Allergies (NDA). Scientific Opinion on Dietary reference values for fat, including saturated fatty acids, polyunsaturated fatty acids, monounsaturated fatty acids, trans fatty acids and cholesterol. *EFSA J.* **2010**, *8*, 1461. [CrossRef]
34. Kostic, A.Z.; Macucanovic-Jocic, M.P.; Spirovic-Trifunovic, B.D.; Vukasinovic, I.Z.; Pavlovic, V.B.; Pesic, M.V. Fatty acids of mize pollen—Quantification, nutritional and morphological evaluation. *J. Cereal Sci.* **2017**, *77*, 180–185. [CrossRef]
35. Gupta, D.; Mann, S.; Sood, A.; Gupta, R.K. Phytochemical, nutritional and antioxidant activity evaluation of seeds of jackfruit (*Artocarpus heterophyllus* Lam.). *Int. J. Pharm. Biol. Sci.* **2011**, *2*, 336–345.
36. Piga, A.; Del Caro, A.; Corda, G. From plums to prunes: Influence of drying parameters on polyphenols and antioxidant activity. *J. Agric. Food Chem.* **2003**, *51*, 3675–3681. [CrossRef] [PubMed]
37. Que, F.; Mao, L.; Fang, X.; Wu, T. Comparison of hot air-drying and freeze-drying on the physicochemical properties and antioxidant activities of pumpkin (*Cucurbita moschata* Duch.) flours. *Int. J. Food Sci. Technol.* **2008**, *43*, 1195–1201. [CrossRef]
38. Yadav, N.; Kaur, D.; Malaviya, R.; Singh, M.; Fatima, M.; Singh, L. Effect of thermal and non-thermal processing on antioxidant potential of cowpea seeds. *Int. J. Food Prop.* **2018**, *21*, 437–451. [CrossRef]
39. Guimarães, R.M.; Ida, E.I.; Falcão, H.G.; de Rezende, T.A.M.; de Santana Silva, J.; Fernandes Alves, C.C.; Pereira da Silva, M.A.; Buranelo-Egea, M. Evaluating technological quality of okara flours obtained by different drying processes. *LWT Food Sci. Technol.* **2020**, *123*, 1–28. [CrossRef]
40. Vázquez-Ovando, A.; Rosado-Rubio, G.; Chel-Guerrero, L.; Betancur-Ancona, D. Physicochemical properties of a fibrous fraction from chia (*Salvia hispanica* L.). *LWT Food Sci. Technol.* **2009**, *42*, 168–173.
41. Warechowska, M.; Warechowski, J.; Tyburski, J.; Siemianowska, E.; Nawrocka, A.; Mis, A.; Skrajda-Brdak, M. Evaluation of physicochemical properties, antioxidant potential and baking quality of grain and flour of primitive rye (*Secale cereale* var. *Multicaule*). *J. Food Sci. Technol.* **2019**, *56*, 3422–3430. [CrossRef] [PubMed]
42. Meng, X.; Liu, F.; Xiao, Y.; Cao, J.; Wang, M.; Duan, X. Alterations in physicochemical and functional properties of buckwheat straw insoluble dietary fiber by alkaline hydrogen peroxide treatment. *Food Chem. X* **2019**, *3*, 1–10. [CrossRef]

43. Ma, Z.; Boye, J.I.; Simpson, B.K.; Prasher, S.O.; Monpetit, D.; Malcolmson, L. Thermal processing effects on the functional properties and microstructure of lentil, chickpea, and pea flours. *Food Res. Int.* **2011**, *44*, 2534–2544. [CrossRef]
44. Zheng, Y.; Li, Y. Physicochemical and functional properties of coconut (*Cocos nucifera* L.) cake dietary fibres: Effects of cellulase hydrolysis, acid treatment and particle size distribution. *Food Chem.* **2018**, *257*, 135–142. [CrossRef] [PubMed]
45. Chen, J.; Zhao, Q.; Wang, L.; Zha, S.; Zhang, L.; Zhao, B. Physicochemical and functional properties of dietary fiber from maca (*Lepidium meyenii* Walp.) liquor residue. *Carbohydr. Polym.* **2015**, *132*, 509–512. [CrossRef] [PubMed]
46. Kumuro, A.C.; Widiyanti, M.; Ratnawati, R.; Retnowati, D.S. Nutritional and functional properties changes during facultative submerged fermentation of gadung (*Dioscorea hispida* Dennst) tuber flour using *Lactobacillus plantarum*. *Heliyon* **2020**, *6*, 1–7. [CrossRef] [PubMed]
47. Kusumayanti, H.; Handayani, N.A.; Santosa, H. Swelling power and water solubility of cassava and sweet potatoes flour. *Procedia Environ. Sci.* **2015**, *23*, 164–167. [CrossRef]
48. Motta-Romero, H.; Zhang, Y. Physicochemical properties and rheological behavior of flour and starches from four bean varieties for gluten-free pasta formulation. *J. Agric. Food Res.* **2019**, *1*, 100001. [CrossRef]
49. Yu, J.; Ahmednay, M.; Goktepe, I. Peanut protein concentrate: Production and functional properties as affected by processing. *Food Chem.* **2007**, *103*, 121–129. [CrossRef]
50. Regand, A.Z.; Chowdhury, S.M.; Tosh, T.M.; Wolever, S.; Wood, P. The molecular weight, solubility and viscosity of oat beta-glucan affect human glycemic response by modifying starch digestibility. *Food Chem.* **2011**, *129*, 297–304. [CrossRef] [PubMed]
51. Lima-Batista, V.; da Silva, T.F.; Lima de Jesús, L.C.; Coelho-Rocha, N.D.; Lima Barroso, F.A.; Tavares, L.M.; Azevedo, A.; Mancha-Agresti, M.; Martins-Drumond, M. Probiotics, prebiotic and synbiotic and paraprobiotics as a therapeutic alternative for intestinal mucositis. *Front. Microbiol.* **2020**, *11*, 544490. [CrossRef]
52. Massa, L.N.M.; Menezesa, F.N.D.D.; de Albuquerquea, T.M.R.; de Oliveira, S.P.A.; dos Santos Lima, M.; Magnanic, M.; de Souzaa, E.L. Effects of digested jabuticaba (*Myrciaria jaboticaba* (Vell.) Berg) by-product on growth and metabolism of Lactobacillus and Bifidobacterium indicate prebiotic properties. *LWT Food Sci. Technol.* **2020**, *131*, 1–8.
53. IARC, OMS. The Global Cancer Observatory: Colorectal Cancer Factsheet, 2020. Available online: https://gco.iarc.fr/today/data/factsheets/cancers/8-Colon-fact-sheet.pdf (accessed on 25 July 2021).
54. Sun, G.; Zheng, Z.; Lee, M.H.; Xu, Y.; Kang, S.; Dong, Z.; Wang, M.; Gu, Z.; Li, H.; Chen, W. Chemoprevention of colorectal cancer by artocarpin, a dietary phytochemical from *Artocarpus heterophyllus*. *J. Agric. Food Chem.* **2017**, *65*, 3474–3480. [CrossRef] [PubMed]
55. Wiater, A.; Paduch, R.; Trojnar, S.; Choma, A.; Pleszczy, M.; Adamczyk, P.; Pi, M.; Próchniak, K.; Szczodrak, J.; Strawa, J.; et al. The effect of water-soluble polysaccharide from jackfruit (*Artocarpus heterophyllus* Lam.) on human colon carcinoma cells cultured *in vitro*. *Plants* **2020**, *9*, 103. [CrossRef] [PubMed]
56. Leibovitz, A.; Stinson, J.C.; McCombs, W.B.; Ill McCoy, C.E.; Mazur, K.C.; Mabry, N.D. Classification of human colorectal adenocarcinoma cell lines. *Cancer Res.* **1976**, *36*, 4562–4569. [PubMed]
57. Tomita, N.; Jiang, W.; Hibshoosh, H.; Warburton, D.; Kahn, S.M.; Weinstein, I.B. Isolation and characterization of a highly malignant variant of the SW480 human colon cancer cell line. *Cancer Res.* **1992**, *52*, 6840–6847. [PubMed]
58. Basak, D.; Uddin, M.N.; Hancock, J. The role of oxidative stress and its counteractive utility in colorectal cancer (CRC). *Cancers* **2020**, *12*, 3336. [CrossRef] [PubMed]
59. Hayes, J.D.; Dinkova-Kostova, A.T.; Tew, K.D. Oxidative stress in cancer. *Cancer Cell* **2020**, *38*, 167–197. [CrossRef] [PubMed]
60. Aykin-Burns, N.; Ahmad, I.M.; Zhu, Y.; Oberley, L.W.; Spitz, D.R. Increased levels of superoxide and H_2O_2 mediate the differential susceptibility of cancer cells versus normal cells to glucose deprivation. *Biochem. J.* **2009**, *418*, 29–37. [CrossRef] [PubMed]

molecules

Review

Role of Honey in Advanced Wound Care

Hana Scepankova [1,†], Patricia Combarros-Fuertes [2,†], José María Fresno [2], María Eugenia Tornadijo [2], Miguel Sousa Dias [3], Carlos A. Pinto [1], Jorge A. Saraiva [1] and Letícia M. Estevinho [3,*]

1 LAQV-REQUIMTE, Department of Chemistry, University of Aveiro, 3810-193 Aveiro, Portugal; hanka.scepankova@gmail.com (H.S.); carlospinto@ua.pt (C.A.P.); jorgesaraiva@ua.pt (J.A.S.)
2 Department of Food Hygiene and Technology, Faculty of Veterinary Science, Campus de Vegazana, University of León, 24071 León, Spain; pcomf@unileon.es (P.C.-F.); jmfreb@unileon.es (J.M.F.); metorr@unileon.es (M.E.T.)
3 CIMO, Mountain Research Center, Polytechnic Institute of Bragança, Campus Santa Apolónia, 5301-855 Bragança, Portugal; miglsdias@gmail.com
* Correspondence: leticia@ipb.pt; Tel.: +351-273303342
† These authors contributed equally to this work.

Abstract: Honey is a natural product rich in several phenolic compounds, enzymes, and sugars with antioxidant, anticarcinogenic, anti-inflammatory, and antimicrobial potential. Indeed, the development of honey-based adhesives for wound care and other biomedical applications are topics being widely investigated over the years. Some of the advantages of the use of honey for wound-healing solutions are the acceleration of dermal repair and epithelialization, angiogenesis promotion, immune response promotion and the reduction in healing-related infections with pathogenic microorganisms. This paper reviews the main role of honey on the development of wound-healing-based applications, the main compounds responsible for the healing capacity, how the honey origin can influence the healing properties, also highlighting promising results in in vitro and in vivo trials. The challenges in the use of honey for wound healing are also covered and discussed. The delivery methodology (direct application, incorporated in fibrous membranes and hydrogels) is also presented and discussed.

Keywords: honey; wound-healing; antioxidant; antimicrobial; hydrogels; dermal repair; hydrogel

Citation: Scepankova, H.; Combarros-Fuertes, P.; Fresno, J.M.; Tornadijo, M.E.; Dias, M.S.; Pinto, C.A.; Saraiva, J.A.; Estevinho, L.M. Role of Honey in Advanced Wound Care. *Molecules* **2021**, *26*, 4784. https://doi.org/10.3390/molecules26164784

Academic Editor: Juraj Majtan

Received: 13 July 2021
Accepted: 29 July 2021
Published: 7 August 2021

1. Introduction

The skin is composed of three layers (epidermis, dermis, and hypodermis), and is considered the first physical barrier against external infectious agents. Wounds are defined as the disruption in the continuity of the skin, induced by mechanical, chemical, or thermal harms, and resulting in the loss of the defensive functions of this tissue [1].

The wound-healing process has the purpose of recovering the integrity of the damaged tissue and the regeneration of the epithelium that was lost, and it is a dynamic and complex process that globally occurs in four overlapping steps: hemostasis, inflammation, tissue proliferation, and regeneration [1,2] (Figure 1).

The cascade of initial vasoconstriction of blood vessels and platelet aggregation play a key role in stopping the loss of blood. The initial vasoconstriction is followed by a vasodilation, which allows an influx of a variety of inflammatory cells which release several types of mediators and cytokines to promote thrombosis, angiogenesis, and re-epithelialization. In addition, the fibroblasts release extracellular components which initiate the formation of collagen fibers that will serve as scaffolding [1,3]. During the inflammatory phase the hemostasis, chemotaxis, and the increased vascular permeability limit further damage, close the wound, remove cellular debris and bacteria, and promote cellular migration [3]. Afterwards, the proliferative phase implicates the formation of granulation tissue, re-epithelialization, and neovascularization, a process that can last from several weeks until some months, or even more time in the case of the existence of some co-morbidities or particular patient situations [4]. In the end, during the maturation period, the new tissues are remodeled, the excess of collagen is reduced, and the wound contracts and reaches the maximum tensile strength [1].

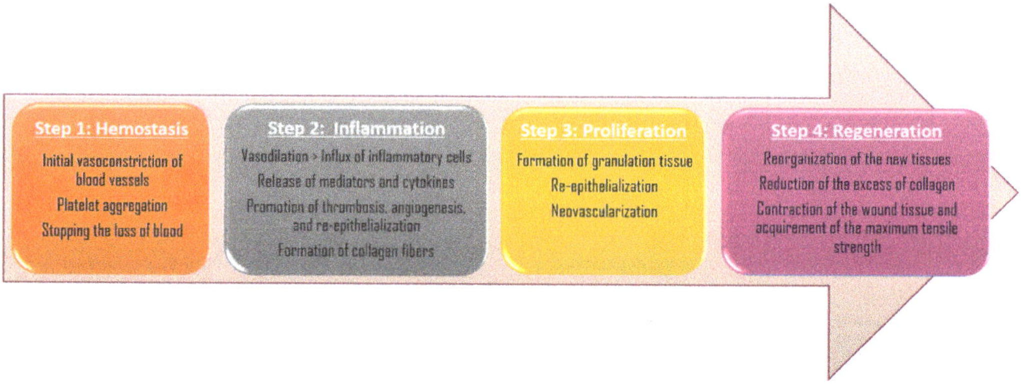

Figure 1. Stages of wound-healing process.

Acute wounds derived from unexpected accidents or surgical injury commonly heal within a predictable period depending on the size, depth, and extent of damage. Nevertheless, deficiencies in the wound-healing process cause more than 38 million patients with chronic wounds worldwide, which reaches epidemic proportions and causes a large economic weight on healthcare systems [5].

Prolonged chronicity of wounds is normally related to a bacterial injured-tissue colonization, which can progress into a bacterial resistance to topical and systemic antimicrobial agents, or into biofilm development, which complicates, in both cases, their treatment [6]. In the end, this type of wound can cause sepsis and inflammation in organs and lead to increased morbidity and mortality.

The clinical considerations in wound-healing management include preventing and controlling the infection and/or contamination, maintaining the adequate moisture environment, treating edema, and preventing further injury. Conventional chronic wound care involves debridement to remove non-viable tissue and bacterial biofilms, followed by wound dressing. The common wound dressings consist of a standard cotton bandage or highly absorbent dressings, such as collagen and alginate, or hydrocolloids; however, this procedure of wound care is often ineffective. Due to this need for new, efficient, and improved therapies, there has been a revived interest in alternative treatment approaches, such as honey. The management of wound healing has become the primary field of therapeutic application of this natural product [7,8].

Honey has been used for wound healing since ancient times, mainly due to its antimicrobial activity. In addition to the broad spectrum of antibacterial activity against common wound-infecting microorganisms, honey has been demonstrated to be effective against antibiotic-resistant bacteria and was able to restore the efficacy of some antibiotics against bacteria with previously acquired resistance [8–10]. Furthermore, due to its several antimicrobial components and its different antibacterial action mechanisms, the development of bacterial resistance to honey is unlikely [8,11,12]. Moreover, the wound-healing ability of honey is also related to its anti-inflammatory and antioxidant activity, as well as its capacity to promote re-epithelialization and angiogenesis and stimulate skin and immune cells [13,14]. All these mechanisms act together favoring the regeneration process of the damaged tissue.

Several case studies and randomized controlled trials provide considerable evidence of the effectiveness of honey in healing different types of wounds, such as amputation wounds, burns, skin grafting sites, skin lesions, or skin ulcers including leg, varicose, malignant, diabetic, and sickle cell ulcers [15–18].

The resolution time of wounds using honey or honey-derivates varies from some days to several months depending on the type, the depth, the anatomical location, and the chronicity of the wound [17,19]. Due to its osmotic effect, honey creates a liquid layer between the dressing and the wound bed. This liquid layer is an advantage in the removal and change of wound-dressings by reducing or eliminating the pain of this process and avoiding damage of the newly grown tissue, reducing the healing time [20].

The use of honey obtained a remarkable improvement of recalcitrant wounds, and has demonstrated even more efficacy than conventional treatments using commercial wound dressings or antibiotics (systemic and topical) [19,21,22]. Honey rapidly replaces sloughs with granulation tissue and promotes a quick epithelialization and absorption of edema from around the ulcer margins, significantly reducing the healing time [19,23]. In addition, honey-based products showed excellent cytocompatibility with tissue cell cultures when compared with conventional treatments, such as silver dressings [24].

Moreover, some studies demonstrated that the combination of honey with other compounds or in combination with conventional treatment can be beneficial for diabetic foot ulcer healing, reducing the wound-resolution time, the cost of the hospital stay, and the rate of amputation when compared to other conventional treatment [17].

However, the use of honey by itself might present some limitations which are being overcome with the development of different honey formulations and honey wound dressings.

This review aims to highlight the mechanism of honey's action in wound healing and gather the literature available regarding the use of honey and modern engineering templates for promoting modern solutions for wound and skin healing and regeneration.

2. The Mechanisms of Honey in Wound and Burn Healing

Honey is a natural and greatly complex substance with hundreds of compounds in its composition [8]. Honey bioactivity, and in consequence, its wound-healing potential, will be influenced by its composition, which depends mainly on the floral source and other factors, such as seasonal, environmental, as well as processing, manipulation, packaging, and storage conditions [25,26]. In addition to some inherent characteristics, such as the acidity and the osmotic pressure, the healing properties of honey in diverse types of wounds and burns have also been attributed to other components which act through different mechanisms that work together to restore the structural integrity of the damaged tissues [6,18,27] (Figure 2).

2.1. Antibacterial Effects

Honey has been traditionally used in the prevention and treatment of wound infections [28,29]. However, with the arrival of antibiotics, the use of honey gradually decreased. Nowadays, microbial drug-resistance has become an increasingly common concern, and honey has regained the scientific interest as an antibacterial agent [8,12,13]. Intrinsic characteristics of honey, such as high osmolarity, low water activity, and acidity, as well as some compounds, such as hydrogen peroxide, phenolic compounds, methylglyoxal, or bee defensin-1 peptide, directly affect the bacterial growth and survival [30–32]. In addition, honey shows an indirect antimicrobial action which involves the promotion of lymphocyte and antibody production, cytokines and immunomodulation, and nitric oxide (NO) [6,33–35].

Non-healing wounds, as well as burns, present an elevated risk of infection, which might increase morbidity and mortality derived from sepsis and inflammation in organs [13]. In addition, drug-resistant infections and wounds with biofilms are particularly difficult to treat, since bacteria do not respond to the therapy or are protected by a self-produced matrix of polysaccharide material [36].

Several studies have demonstrated, in vitro and in vivo, the efficacy of different varieties of honey against a broad spectrum of bacteria, including those that commonly caused wound and burn infections, such as *Staphylococcus aureus*, *Pseudomonas aeruginosa*, *Escherichia coli*, *Acinetobacter baumannii*, or *Staphylococcus epidermidis* [6,18,37–40]. In addition, honey has also been demonstrated to be effective against antibiotic-resistant bacteria [19,41–43], as well as against biofilms by preventing the formation and the development of the biofilm [31,44–47], by reducing the metabolic activity of already formed biofilms [44,48], or by altering the gene expression of different genes related to the formation and the development of biofilms [48,49], and is related to the bacterial quorum sensing [49,50].

Moreover, some studies demonstrated that manuka honey acts synergistically with several antibiotics, reducing the doses required to inhibit bacterial growth or reverting the antibiotic resistance previously acquired [9,10,51,52]. These results suggest a potential application of a combined therapy of honey and antibiotics.

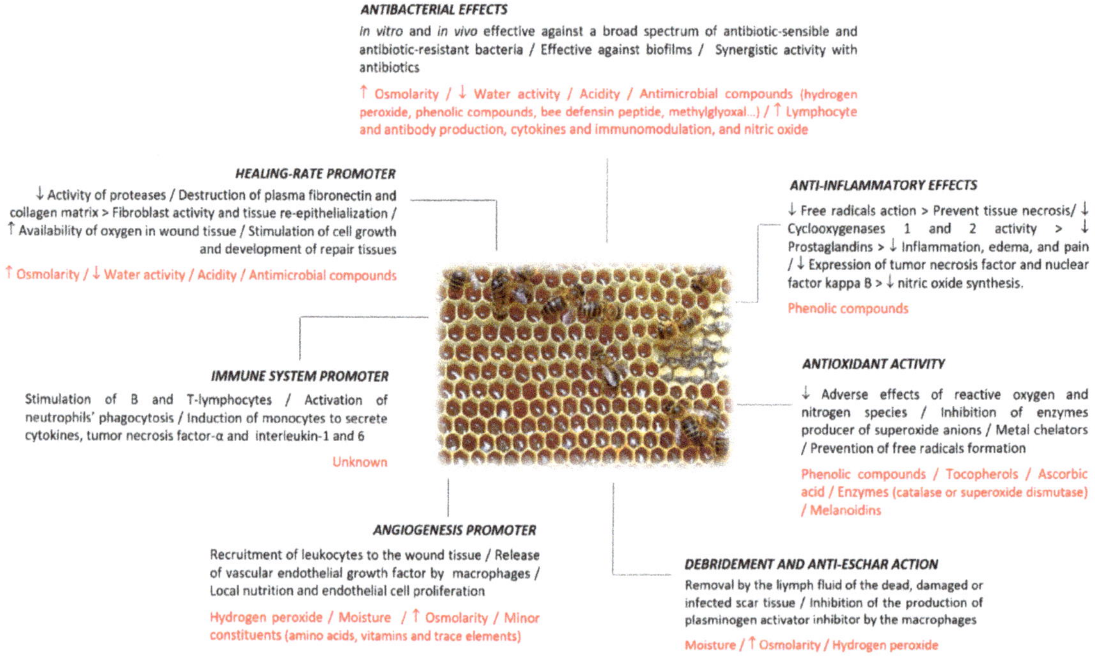

Figure 2. Wound healing mechanisms described for honey, different effects related to each mechanism, and the compound(s) or honey characteristics associated with the mechanism.

2.2. Anti-Inflammatory Effects

Inflammation is the response of a living tissue to a local injury and plays a fundamental role as a defense and protection mechanism to avoid infections and to repair the affected tissue. The inflammatory phase is a necessary part of wound healing; however, when this response is not adequate, an overproduction of inflammatory mediators by immune cells, which do not respond to initial triggers, might be produced, becoming a problem for wound resolution [53]. The anti-inflammatory activity of honey is a consequence of different mechanisms.

During the inflammatory phase, the affected tissues release a high concentration of free radicals. The antioxidant compounds in honey act synergistically and can reduce the damage caused by these radicals, and therefore prevent tissue necrosis [14].

In addition, in vitro and in vivo studies have demonstrated that honey reduces the activity of cyclooxygenases 1 and 2 (COX1 and COX2) that intervene in the synthesis of prostaglandins [54,55]. Prostaglandins participate in the inflammatory response by producing vasodilation, increasing the permeability of blood vessels and allowing the passage of leukocytes, acting as an antiplatelet agent, and stimulating the nerve endings of pain. The reduction in prostaglandin concentration in plasma may induce a diminution of inflammation, edema, and pain [56].

Moreover, honey can inhibit the expression of tumor necrosis factor (TNF-α) and reduce the concentration of pro-inflammatory cytokines through the attenuation of nuclear factor kappa B (NF-κB) [54]. Furthermore, NF-κB is involved in the activation of the inducible NO synthase enzyme (iNO). During inflammation, iNO is induced by cytokines, TNF-α, interleukins, and bacterial endotoxins, producing NO.

NO is a free radical that acts as a mediator in acute and chronic inflammation and favors the healing process of tissues. However, an excess of NO or an overproduction at the wrong time can be detrimental and contribute to the development of pathologies related to inflammation [56].

Another advantage of the anti-inflammatory action of honey is the decrease in edema, thus reducing the pressure on the microvasculature of wound tissue that allows the availability of oxygen and nutrients required for growth of tissue and wound repair [20]. This effect also allows the control of the wounds' exudate with an appropriate moisture balance, which is still a constant challenge in the healing processes [39].

The anti-inflammatory activity of honey has been mainly attributed to phenolic compounds [53,57]. However, until now, no correlation was found between the level of anti-inflammatory activity in different honey samples and the phenolic compound content [58], which might be due to the distinct types of interactions that can occur among these compounds and other compounds present in honey.

2.3. *Antioxidant Activity*

The antioxidant activity of honey is due to a wide variety of compounds, such as flavonoids, phenolic acids, tocopherols, ascorbic acid, and enzymes including catalase or superoxide dismutase [14,59,60]. In addition, melanoidins, products of the Maillard reaction, were described as the main components responsible for the radical-scavenging capacity of honey [61,62]. These substances reduce the adverse effects of reactive oxygen species (ROS) and reactive nitrogen species (RNS), inhibit the enzymes responsible for producing superoxide anions, act as metal chelators, and interfere in the chain reactions of free radicals and can play a preventative role in the process of their formation [63]. Through these antioxidant mechanisms, honey contributes to wound and burn healing by interfering with abnormal inflammatory response [6].

ROS act as messengers to give feedback amplification of the inflammatory response [20] and mediate TNF-α induced cytotoxicity [64]. Moreover, in chronic wounds, neutrophils and macrophages liberate high levels of ROS against invading bacteria [64]. The extended exposure to ROS causes cell damage of the tissue and might delay wound healing. In addition, the ROS formed in the inflammatory phase of wound healing stimulate the activity of the fibroblasts which produce the collagen fibers of scar tissue. If the inflammatory phase prolongates, it could induce hyper-granulation and fibrosis, so honey minimizes or prevents hypertrophic scarring [14,64]. In addition, flavonoids protect tissue against RNS, such as NO and peroxy-nitrite [65].

2.4. *Debridement and Anti-Eschar Action*

Wound debridement is essential in producing the functional process of tissue reparation. The conventional procedure is the surgical remotion of dead tissue, which is painful, may cause infections, and produces toxins that can destroy the surrounding tissues [6,14,27].

The moist environment produced by honey facilitates the wounds' autolytic debridement process. The high osmotic pressure pulls out lymphatic fluid from the deeper zones, which automatically remove dead, damaged, or infected scar tissue [27,39]. In addition, lymph is a rich source of proteases that activated by the hydrogen peroxide produced when honey is diluted and assists in the debriding activity [6].

Additionally, honey inhibits the production of plasminogen activator inhibitor (PAI) by the macrophages derived from its anti-inflammatory activity [39]. PAI blocks the transformation of plasminogen, the enzymically inactive precursor of plasmin, into active plasmin. Plasmin is an enzyme that specifically digests fibrin attached to the wound surface, but does not digest the collagen matrix, which is necessary for tissue reparation, thus preventing eschar formation [20,66]. Inflammation increases the generation of PAI, so the mechanism through which honey decreases the production of PAI is probably related to its anti-inflammatory activity [20].

2.5. Angiogenesis Promoter

Angiogenesis occurs in the proliferative phase of wound healing. The development of new blood vessels from pre-existing ones supplies the required oxygen in the wound, which is an important stage in the healing process. This dynamic process is strongly regulated by signals from serum and the surrounding extracellular matrix environment [67]. Stimulation of angiogenesis by honey was demonstrated in an in vitro study with analogues of angiogenesis and an endothelial proliferation assay [68], and more recently, in another in vivo-model study [69].

Hydrogen peroxide (generated from glucose by the action of the enzyme glucose oxidase present in raw honey) induces the recruitment of leukocytes to wounds through a concentration gradient mechanism. Due to an oxidant induct, macrophages release vascular endothelial growth factor (VEGF), which stimulates angiogenesis [67]. In addition, the high concentration of sugars present in honey, as well as other minor constituents, such as amino acids, vitamins and trace elements, provide, in a moist environment, a local cellular energy source, which may improve local nutrition and endothelial cell proliferation [70–72].

On the contrary, another study has demonstrated the anti-angiogenic activity of honey is mediated by the modulation of prostaglandin E_2 and VEGF production [73]. This disparity among studies might be explained by the honey concentration tested, since the highest pro-angiogenic effect was found in a low concentration of honey, whereas higher concentrations demonstrated anti-angiogenic activity [68].

2.6. Immune System Promoter

Some studies have also demonstrated the activity of honey in stimulating some immune system mediators. Honey can stimulate B- lymphocytes and T-lymphocytes and activate neutrophil phagocytosis in cell culture [20,74]. In addition, honey induces monocytes (MM6 cells) to secrete cytokines, tumor necrosis factor-α (TNF-α), interleukin-1 (IL-1), and interleukin-6 (IL-6), which activate the immune response to infection [33–35].

Moreover, honey stimulates antibody production during primary and secondary immune responses against thymus-dependent and thymus-independent antigens [75] and increases humoral immunity by the intrinsic NO, which activates specific signal transduction pathways in monocytes in a concentration-dependent manner [76,77].

2.7. Healing-Rate Promoter

In addition to all the honey effects previously described, honey acts in the regeneration of the new formed tissue, an essential step in the wound-healing process [78].

The acidification of the wound environment favors the action of macrophages, limits bacterial growth, and neutralizes the ammonia produced by bacterial metabolisms that could damage tissues [14]. However, the acidic pH of honey also limits the activity of proteases; these enzymes might inactivate the tissue growth factors and destroy the plasma fibronectin and the collagen matrix, which are necessary for fibroblast activity and tissue re-epithelialization [14,42]. In addition, the diminution of pH in the wound bed makes more oxygen available from hemoglobin in the blood [42].

Furthermore, all the nutritious components present in honey (sugars, amino acids, vitamins, and other trace elements) stimulate cell growth and the development of repair tissues [70–72,79,80]. In addition, re-epithelialization would also be promoted by the increment of TNF-α and interleukin-1β (IL-1β) levels. A concentration of 1% of honey has been found to stimulate the release of the cytokine TNF-α, IL-1β and IL-6 from monocytes [81], which induce the keratinocyte migration and proliferation, which are the major cellular components that are involved in the intricate mechanisms of initiation, maintenance, and completion of wound healing, and may induce collagen synthesis by fibroblasts [14,68,82].

3. Safety of Honey Used for Topical Treatment

The extensive scientific evidence proves that honey may offer distinct advantages over the chemotherapeutic substances currently used in the wound- and burn- healing processes. However, this natural product shows a series of limitations, and is not completely free from adverse effects.

The composition of honey is rather variable, depending primarily on the botanical origin, and secondarily on other factors such as geographical origin, or harvesting, processing, and storage conditions [38,39]. This variability determines its bioactive properties, and consequently influences the therapeutic efficacy of the wound treatment [39]. In addition, the absence of standardization and the incomplete knowledge of the active components, and the mechanisms through which they interact and act in wound healing, are the major limitations for the application of honey in medicine. For this reason, is essential to select the more appropriate varieties of honey, and it is recommended to carry out a previous screening [38].

In addition, other considerations must be considered before honey application in wounds. The low pH, derived from the presence of organic acids in honey, may contribute to a stinging or burning sensation when it is applied to a damaged tissue [83]. Besides this unpleasant sensation, it is necessary to consider that, although minor, there is a risk of wound infection, mainly related to the presence of clostridial spores which have occasionally been found in honey [39,84]. This risk can be reduced by using gamma-irradiation, which inactivates the spores without modifying the original biological activity [85,86]. Nevertheless, no cases of wound infection due to clostridial spores related to the use of non-irradiated honey on wounds have been reported to date.

Furthermore, the honey used for medical purposes must be free of any chemical contamination, such as pesticides, herbicides, or heavy metals. In this sense, to guarantee the maximum purity, honey should be collected in areas that meet the requirements for organic production, as well as following rigorous quality, processing and storage standards [8,19]. In addition, is necessary to consider that some varieties of honey might present toxic active compounds originating from the nectar of species such as rhododendron, oleanders, mountain and sheep laurels, or azaleas. However, these effects have been described by honey ingestion [87,88].

4. Biomedical Application of Honey in Advanced Wound Care

4.1. Medical Grade Honey and Honey Ointments for Topical Application

The safety threats previously described in wound treatment with honey are overcome by medical-grade honey approved for wound care [18,39]. Indeed, the medical-grade honeys are sterilized by gamma irradiation with the aim to kill Clostridium spores, produced under rigorous standards of hygiene, without pollutants or contaminating pesticides in its composition, and standardized under different defined criteria [8]. They have potent in vitro bactericidal activity against antibiotic-resistant bacteria and are approved for application in wound management. Having reproducible antibacterial activity, these honeys are produced under controlled conditions in greenhouses (i.e., Revamil source honey); medical grade Leptospermum-derived Manuka honeys are analyzed individually by each batch to assess the Unique Manuka Factor (UMF) that gives a number based on its bactericidal activity [89]. Unlike other varieties, the antibacterial activity of Manuka honey is based on its non-peroxide activity related to compounds which are mainly present in this variety, such as methylglyoxal, leptosperin, or methyl syringate [38].

Medical-grade honey can be directly applied to the wound bed and then covered with conventional dressing. However, in high-exuding wounds, honey can become less viscous and diluted. The liquid state of honey might complicate its application and its permanence on wound and burn environments, and in the treatment of excessively exudative lesions, honey might be diluted to concentrations that present minimal or no effects in short periods of time [7,39]. Nevertheless, according to [6], even when honey is heavily diluted by wound exudate, it will still have potent enough antibacterial activity to inhibit the growth of bacteria (the MIC values were found to be below 11%).

In clinical practice today, several honey-based wound-care commercial preparations in the form of gels, ointment, and dressings, are approved by the US Food and Drug Administration (FDA) and registered as medical devices [90]. Most of them are formulated with medical-grade Manuka honey, since it is one of the most studied varieties of honey in the world, and it was the first honey type to obtain the status of medical-grade honey. However, there are also alternatives that use other types of honey, such as buckwheat, multifloral, and Revamil source honey, among others [8]. For instance, a honey-based gel formula can be prepared with 100 % medical-grade honey without the addition of other ingredient (Table 1) or can be mixed with other agents(s) such as natural emollients (e.g., lanolin, polyethylene glycol, glycerine, myristil myristrate) or different plant waxes and gelling agents [90].

Table 1. Commercially available medical-grade honey and honey-based gels and ointments for wound healing.

Product Name	Product Type	Composition
Manuka Guard® Medical Grade Manuka Honey	Honey	100% Manuka honey
Manuka Fill®	Honey	100% Manuka honey
Ectocare® Manuka Fill™	Honey	100% Manuka honey
ManukaDress–T	Honey	100% Manuka honey
Activon® Tube	Paste formula	100% Manuka honey
Manuka Health® Wound Gel	Gel formula	94% Manuka honey with natural gelling agents
Medihoney® Barrier Cream	Cream formula	30% Manuka honey, other non-described components
Medihoney® Gel Wound & Burn Dressing	Gel formula	100% Manuka honey in a hydrocolloidal suspension
Melladerm® Plus	Gel formula	45% medical-grade multi-flower honey, other non-described components
Melloxy®	Gel formula	40% medical-grade multi-flower honey, 11% ozonated vegetable olive oil, other non-described components
MANUKApli®	Gel formula	100% Manuka honey
L-Mesitran® Soft	Gel formula	40% medical-grade honey (not Manuka) with lanolin, polyethylene glycol, and vitamins C and E
L-Mesitran® Ointment	Gel formula	48% medical-grade honey (not Manuka), lanolin, cod liver oil, sunflower oil, calendula, aloe vera, zinc oxide, and vitamins C and E
Revamil Gel®	Gel formula	100% medical-grade honey (not Manuka)
Revamil Balm®	Balm formula	25% medical-grade honey (not Manuka), arachis oleum, cera alba, glyceryl oleate, aqua.
Surgihoney™RO®	Gel formula	Mixture of medical-grade honey from various sites/floral sources engineered to produce hydrogen peroxide and reactive oxygen species when diluted in water
Therahoney® Gel	Gel formula	100% Manuka honey

The FDA-approved honey-based devices are indicated in the treatment of different types of wounds, such as low and moderate-to-heavy exuding wounds, diabetic foot ulcers, leg ulcers, pressure ulcers, burns, traumatic wounds, surgical wounds, chronic wounds, or colonized acute wounds, among other indications [18]. Despite the availability of these products, their use in medical practice is still limited, probably due to the misconception that there is no evidence to support the use of honey with therapeutic purposes, as well as the scarce promotion and diffusion of honey products for wound care [20].

Nevertheless, effectiveness of the Medihoney^{TM} Antibacterial Wound Gel has been evaluated in eight post-coronary artery bypass graft (CABG) patients. The gel was selected as the active primary product of choice for all the graft wounds (seven of the eight patients had wound infections). The honey-based gel was applied directly onto the wound bed and covered with an adhesive-bordered non-adherent gauze dressing. The wounds all reduced in size and there was a significant reduction in pain, odor, and exudate. Moreover, the wound gel had reduced the bioburden of the wounds enabling them to progress to healing. Finally, the use of medical honey became a regular dressing choice within the authors' cardiothoracic unit [91]. Moreover, the combined treatment with honey-based gels (L-Mesitran Soft) and ointment (L-Mesitran Ointment) has been successfully applied for treatment of infected ulcers in diabetics patients [19]. Indeed, the use of ointments help to entrap water, keep the skin moist, and provide an emollient protective film, which are all crucial elements for wound healing [19,90], while the layer of gel fights infections and optimizes wound healing [19,90].

The medical-grade honey, as well as the ointments and gels, is applied to the wound bed and requires a secondary conventional dressing (e.g., cotton wool bandage) to contain the honey in the wound bed environment, which on removal cause pain [92].

Therefore, recent research focuses on the development of different materials or matrices to convey honey, control its delivery, and act as absorbent secondary dressings [7,39,93–95].

4.2. Honey-Based Advanced Wound Care Products

Many years ago, wound management was based on covering the wound using conventional dressings (i.e., gauzes, absorbent cotton, bandages). However, they are limited in terms of influencing/accelerating wound healing and preventing/treating infections. Currently, wound management has been updated due to a greater understanding of the molecular and cellular processes involved in wound healing. Additionally, with advancements in technology, the design and functionality of wound dressings has advanced in the direction of multi-functionality [96]. The modern dressings are designed to maintain the moist wound environment and promote healing [97]. Moreover, the critical necessities of modern wound dressings include biocompatibility, no cytotoxic effects, a rate of biodegradability directly proportional with the rate of formation of new tissue, a release of incorporated bioactive compounds (drugs), and the control of possible infections [96].

Tissue engineering has recently introduced wound dressings/scaffolds as an alternative treatment of wounds with advanced properties, suitable for keeping a moist environment while absorbing exudates, creating a barrier against pathogens, and facilitating drug delivery systems [96].

The recent in vitro and in vivo research demonstrated that honey is a valuable addition to many tissue-engineering templates in eliminating bacterial infection, aiding in inflammation resolution and improving tissue integration with the template (Tables 2 and 3) [28]. Currently, hydrogels and electrospun nanofibers are the most researched types of honey-incorporated scaffolds [98].

Table 2. Examples of honey hydrogel-based wound dressings, methods of formation, evaluated models and main findings.

Honey Hydrogel Formulation			In Vivo Wound and Burn Model				
Incorporated Honey	**Hydrogel Matrix**	**Method of Hydrogel Formation**	**Type of Mice**	**Location of Lesson**	**Type of Wound/Burn**	**Findings**	**References**
6% (*v*/*v*) Gelam (*Melaleuca Apis melifera*) honey (Malaysia)	PVP 15% (*w*/*v*), PEG 1% (*v*/*v*), protein-free agar solution 1% (*w*/*v*).	Electron beam irradiation (25 kGy)	96 male Sprague Dawley rats	Dorsum of rat	Deep partial thickness burn wounds	Good transparency; slightly acidic (pH 4.3); high capabilities in absorbing fluid. Significant acceleration of dermal repair and advanced re-epithelialization. Modulation of proinflammatory cytokines in wound healing. Synergistic effect of hydrogel matrix and incorporated honey.	[99]
Up to 3.5 % (*w*/*w*) Iran honey	PVA 10 % (*w*/*w*), CM-chitosan up to 3.5 % (*w*/*w*), water 85 % (*w*/*w*).	^{60}Co Gamma-ray (radiation method) up to 40 kGy; ^{60}Co Gamma-ray followed by 3 cycles of freeze–thawing (combinational method).	male NMRI mice	Dorsum of mice	0.7 cm × 0.7 cm wound	Acceptable swelling degree, transparency and inhibition of the growth of *E. coli* bacteria. The hydrogel containing more honey in its formulation has a more effective action in the wound healing process of the mouse. The mechanical strength of the hydrogels prepared by the combinational method was higher than by radiation method.	[100]
70% honey-based alginate hydrogel	Alginate hydrogel	-	20 male Wistar rats	Side of vertebral column between the ears	Full-thickness wound (1 cm × 1 cm)	Epiderm growth (after 21 days) and collagen synthesis (after 14 days). Wound-healing influences were attributed to the synergistic effect of the alginate hydrogel and the incorporated honey.	[101]
15 wt% PVA/chitosan nanoclay hydrogel	PVA 10% (*w*/*w*), Chitosan 2% (*w*/*w*), TPP (chitosan crosslinking agent), Montmorillonite up to 3 wt%, Acetic acid solution 2% (*v*/*v*)	Freeze–thawing method (freezing at −15 °C/24 h and subsequently thawing at room temperature for 24 h).	15 female mice	Dorsum	Full-thickness wound (1 cm × 1 cm)	Honey-loaded hydrogel nanocomposite wound dressings (PCMH) had better wound-healing ability than nanoclay hydrogel without honey (PCM) and hydrogel without nanoclay and honey (control group). The wound size reduction at the third post-operation day was: 39.62% (control), 39.62% (PCM), and 39.62% (PCMH); at the sixth day: 55.23% (control), 58.38% (PCM), and 72.60% (PCMH).	[102]
10% and 20% (*v*/*v*) Chicory honey (Iran)	Chitosan, gelatin 5% (*w*/*v*), PVA 10% (*w*/*v*), acetic acid 3% (*w*/*v*); Ratio of 2:1:1 (*v*/*v*/*v*) chitosan, PVA, gelatin solution	3 cycles of freeze–thawing (freezing at −20 °C for 20 h, and subsequently thawing at 25 °C for 4 h)	18 male Wistar rats	Back of rats	Full thickness excisional wounds (2 cm × 2 cm)	The higher concentration of honey in the hydrogel facilitated the wound-healing process from inflammation to proliferation, and finally, to the maturation phase. Almost 50% wound closure was observed after 4 days (20% *v*/*v* honey); and 95% after 12 days (10 and 20% *v*/*v* honey).	[103]

Table 2. *Cont.*

Honey Hydrogel Formulation			In Vivo Wound and Burn Model			Findings	References
Incorporated Honey	Hydrogel Matrix	Method of Hydrogel Formation	Type of Mice	Location of Lesson	Type of Wound/Burn		
20 g of diluted (50:50 *w/w*) raw sunflower honey (China)	Chitosan, gelatin, honey (ratio of 0.5:20:20, *w/w/w*). Distilled water up to 100 g (final volume).	Standing and cooling to room temperature. Sterilization of hydrogel sheets with UV rays for 45 min.	4 male rabbits	Dorsum	Second degree burn wound (3 cm × 3 cm)	On day 12, the burns treated with honey hydrogel sheets (HS) were completely healed with intact epidermis and topical proliferation of hair follicles. In contrast, MEBO-treated burns (commercial ointment) and non-treated burns presented 15% and63% unclosed wound area, respectively.	[104]
Solution (1:1 *v/v*) of liquid Manuka honey (New Zealand)	pectin	Hot-air-dried at 40 °C and conditioned in air drier at 25 °C for 5 days. Sterilization by gamma-irradiation at 25 kGray	36 male Sprague Dawley rats	Dorsum	Full thickness excisional wounds (2 cm × 2 cm)	Topical administration of pectin and pectin-honey hydrogels accelerate wound healing in rats. On the 23rd day, the entire surface of the lesion treated with the dressing was covered with hair follicles and matured fibrous tissue.	[105]
Honey (Egypt) Up to 75% (*w/w*)	Carbopol 934, Chitosan, Methyl paraben, TEA, GAA, purified water.	Cold mechanical method (placed in refrigerator)	10 albino mice	Dorsum	Third-degree burn type with a focal wide area of necrosis in the epidermis	Honey 75%-chitosan formula showed the best healing properties (regeneration of the epidermis tissue and the formation of new blood capillaries) compared to the pure honey and the commercial product tested (silver sulphadiazine).	[106]
Manuka honey 80% *w/w*)	PVA, borax (crosslinking agent)	Solution was molded in Petri dishes and kept at 50 °C overnight.	-	-	-	A wound dressing hybrid hydrogel with sustained release of honey over 24 h and with progressively low adhesion to the wound bed that protects new epithelialization and promotes cell proliferation. Antibacterial activity observed against the tested S. aureus.	[107]

Table 3. Examples of honey electrospun nanofibrous scaffolds, obtention methods and main findings.

	Composition		Crosslinking of Fibres		Findings			
Nanofibrous Honey Scaffold	**Honey**	**Material**	**Method**	**Conditions**		**Wound Healing Properties**	**Characteristics**	**References**
Honey/PVA/ Chitosan								
High concentration honey chitosan electrospun nanofibers (HP-chitosan)	Honey (20–40%)	chitosan (1.5% to 5.5%), PVA.	Chemical crosslink Physical crosslink-heating and freeze/thawing	GA vapors heating (under vacuum in an oven, up to 110° /up to 24 h) and freezing (in liquid nitrogen)/thawing (at room temperature.	In vitro	HP–chitosan; (30%:7%:3.5%) enhanced antibacterial activity against *S. aureus* (complete inhibition after 48 h with 30%:7%:5.5 %), poor antibacterial activity against *E. coli*.	HP–chitosan (30%:7%:3.5%) upon aging for more than 2 days acquired the optimum viscosity required for easy spinning and formation of uniform nanofiber. Effective biocompatible wound dressing.	[108]
Honey /PVA/chitosan nanofibers	Honey (10–30%)	chitosan (3.5%), PVA (7%), acetic acid (1%).	Chemical crosslink Heating	GA vapors 40 °C	In vitro	Enhanced antibacterial activity against Gram-positive *S. aureus* over the Gram-negative *E.coli*	Increase in fiber diameter; Large pore diameter reaching 140 µm (10, 30% honey); Degradation decreased with crosslinking of the fibre mats.	[109]
Honey-PVA-chitosan nanofibers green wound dressing (bio-compatible apitherapeutic nanofibers international patent (2006.01)	Honey (25–50%)	PVA, chitosan (1.5–11%) bee venom, propolis, garlic (2–30%), bacteriophage			In vitro	Honey-PVA-chitosan nanofibers loaded with bee venom/bacteriophage exhibited potent antibacterial activity against Gram-positive and Gram-negative strains and achieved nearly complete killing of multidrug-resistant *Pseudomonase aeruginosa*. Enhanced wound healing and improved biocompatibility		[110,111]
Honey/PVA								
Honey, pomegranate peel extract and bee venomnanofibrous wound dressing	Manuka honey (MH) (10–25%), lyophilized multiflora honey powder (25 %)	PVA (up to 12%), bee venom (BV, 0.01%), methanolic pomegranate peel extract (PPP, up to 2.5%).	Chemical crosslink Heating	25% GH vapours in a vacuum oven at 40 °C/24 h	In vivo	MH/PPP/BV/PVA (25%/2.5%/0.01%/9.7%) close resemblance to normal skin at day 10;	**No cytotoxic** (100 % viability, tested on L929 fibroblast cells)	[112]
					In vitro	effective inhibition of bacterial growth for *S. aureus* and *E. coli*.		
PVA/honey nanofibers	Iran-Tabriz honey (up to 40%)	PVA, dexamethasone sodium phosphate (anti-inflammatory drug loaded up to 15%)	Only electro-spinning				Decreased diameter of electrospun fibers caused by increasing honey concentration.	[104]
PVA/honey nanofibrous scaffolds (with low honey concentration for internal tissue regeneration)	Dabur honey (India) (0.2–1% *w*/*v*)	PVA 12% (*w*/*v*)	Chemical crosslink	GA vapours (2 M) for 24 h.	In vitro	Drastically reduced biofilm Nanofiber membranes with 0.5% honey loading can be suggested as optimum concentration	Minimal weight loss of fibers for 10 days.	[113]

Table 3. *Cont.*

	Composition		Crosslinking of Fibres		Findings			
Nanofibrous Honey Scaffold	Honey	Material	Method	Conditions		Wound Healing Properties	Characteristics	References
PVA–DES–honey nanofibers	Acacia honey (China) (5%, *w*/*v*)	PVA (8%, *w*/*v*), DES (5%, *w*/*v*)	Only electro-spinning		In vitro	Possess excellent antimicrobial activity (*E. coli*, *S. aureus*) total bacterial reduction of 37.0% and 37.9% against *E. coli* and *S. aureus*, respectively, after 6 hour incubation in bacterial cultures; excellent cytocompatibility, non-toxic	The nanofiber materials dissolved rapidly in artificial saliva solutions, suggesting potential use of materials for fast-dissolving drug delivery in oral cavities	[114]
					In vivo	PVA–DES–honey nanofibers accelerated the wound healing process, and improved the wound healing rate on rat skin to 85.2% after 6 days of surgery, when compared to the control PVA (68.2%) and PVA–DES (76.3%) nanofibe		
Honey/PICT								
PICT/honey nanofibrous	Pakistan forest honey (10–20%)	PICT					Good elastic behavior and tensile strength (PICT/honey nanofibers containing 15% honey); good releasing efficiency, complete release of honey in 15 min, the maximum release in 10 min (72 mg/L, 56% of honey).	[115]
Honey/silk fibroin								
Honey-silk fibroin (SF) electrospun scaffold	Medical grade Manuka honey (Melita) (5% of 5 and 20 UMF)	Lyophilized SF (5%)			In vitro	Tissue engineered scaffold could be incorporated with MH of any UMF, resulting in the same bactericidal outcome	No significant difference in porosity, bacterial clearance and adhesion, glucose release, or proliferation of cells as effected by the incorporation of 5 versus 20 UMF MH.	[116]
MH/SF composite fibrous matrices manufactured by green electrospinning	Manuka honey (UMF 5+) 0, 10%, 30%, 50%, 70%	SF 20% (*wt*/*v*) and 2% (*wt*/*v*) PEO			In vivo	The addition of MH improved the wound healing rate of the SF fibrous matrices; wound treated with MH (70%)/SF showed a similar healing effect as the AquacelAg dressing.	Excellent biocompatibility m (incorporation of MH could further improve the affinity of SF fibrous matrices for cells)	[117]
					In vitro	Increase in the bacterial inhibition efficacy with increasing the content of MH.		
Honey/PCL								
Manuka honey-PCL nanofiber scaffolds	1, 5, 10, and 20% *v*/*v* Manuka honey solutions	15 wt% of PCL (Polycaprolactone)				Honey scaffolds demonstrated significant clearance in only the Gram-negative *E. coli*	Lower elasticity and strength with honey incorporation, but showed no notable change in material degradation rate with the presence of honey over a 28 day PBS soak.	[118]

4.2.1. Honey-Based Hydrogels

Hydrogels are high-water-content materials prepared from cross-linked polymers, such as chitosan, and can provide sustained, local delivery of a variety of therapeutic agents [106]. Incorporation of honey into the hydrogel system can beneficially affect the water absorption capacity of the polymer (of the hydrogel) and increase the antibacterial activity of the scaffold [99].

Noori, Kokabi, and Hassan (2018) investigated a honey-loaded PVA/chitosan/mont morillonite nanocomposite (PCMH) hydrogel dressing as a drug model for wound healing. Results demonstrated the ability of the PCMH nanocomposite hydrogel to smart release honey against pH and temperature changes. The maximum release of honey from the hydrogel occurred at pH 7, while the minimum was at pH 2. Independently of the pH, increasing temperature caused higher honey release from the hydrogel matrix. However, the addition of the nanoclay (montmorillonite) to the hydrogel decreases the hydrogel swelling and delays the honey release. Nevertheless, the authors suggested that the honey-loaded nanocomposite hydrogel could be used in low exudate wounds to supply optimized humidity in the wound bed [102].

Different honey concentrations have different efficacies in the scavenging of free radicals and promoting epithelial cell proliferation [119]. It has been shown that the release of honey increases when increasing its concentration in hydrogel, regardless of the polymeric hydrogel used in the formulation. For instance, the honey 75%-chitosan formula showed the best healing properties (regeneration of the epidermis tissue and the formation of new blood capillaries) compared to honey hydrogel formulae with a lower concentration of honey (up to 50%, *w*/*w*). Similarly, an 80%-Manuka/PVA wound dressing hybrid hydrogel showed sustained release of honey over 24 h with progressively low adhesion to the wound bed that protects new epithelialization and promotes cell proliferation. Both dressings demonstrate the high value of cell viability and proliferation and promoted antibacterial activity, being suitable for wounds with moderate to relatively high exudate [100].

Further, one of the advantages of using hydrogels instead of conventional designs is their transparency, which allows us to observe the status of the burn or wound without removal of the dressing [103]. In vivo study showed that burns treated with honey hydrogel sheets were completely healed after 12 days with intact epidermis and topical proliferation of hair follicles. In contrast, burns treated with commercial ointment (MEBO-treated burns) and non-treated burns presented 15% and 63% unclosed wound areas, respectively [104].

In addition, hydrogel wound scaffolds containing honey do not function merely as coverage to provide a clean, moist environment for healing, but also directly contribute to enhanced tissue regeneration and recovery [104].

4.2.2. Honey-Based Electrospun Nanofiber Scaffolds

In biomedical applications, the nanofiber membranes prepared by electro-spinning are used in wound dressings, biosensing, tissue engineering scaffolds, artificial organs, and drug delivery [114]. The formulation of the electrospun nanofiber scaffolds contain protein-based polymers, such as collagen, gelatin, and silk, or/and polysaccharide-based polymers such as chitosan, hyaluronic acid, and alginate. Prepared stirred polymeric solution is loaded into a syringe that is attached to a needle of which the tip exhibits voltage [120].

The advantageous properties of nanofibers are a large surface area to volume ratio, high porosity, and a very small pore size, which lead to high exudate absorption, better wound permeation, and prevention of further infection [112].

The fabrication of honey-based electrospun nanofibers increases interest due to the enhanced activity realized upon combining the advantages of the nanofibrous structure, primarily the increased surface-to-volume ratio with the advantageous properties of honey (Table 3) [112].

Wound healing scaffolds are expected to absorb body fluids and maintain hydration, but without increasing infection of the biofilm. Indeed, wounds with biofilm fail to re-epithelialize, show vascular granulation tissue, and consist of recalcitrant microbes. Honey/PVA nanofiber membranes were found to effectively decrease the biofilm formation [113]. Manuka honey is effective against both Gram-positive and Gram-negative bacteria; however, results of honey scaffolds containing 1–20% Manuka/PCL nanofibers indicated that the controlled release of smaller amounts of honey by the scaffold is more effective against Gram-negative-bacteria-infected wounds. Interestingly, opposite results have been found for high-concentration honey chitosan electrospun nanofibers. The honey/PVA/chitosan membrane (30%:7%:3.5%) enhanced antibacterial activity against *S. aureus* (complete inhibition after 48 h with 30%: 7%:5.5 %) and showed poor antibacterial activity against the Gram-negative *E. coli*. Moreover, the scaffold showed high biocompatibility and low cytotoxicity effects [108].

5. Conclusions

The use of honey for biomedical applications has gained special focus over the years, with the development of novel applications for this natural product, taking advantage of its unique chemical characteristics. Due to its characteristics, namely, low pH and water activity, it presents a good microbiological, enzymatic and (bio)chemical stability, which can be lost if not properly processed for safe use by means of keeping both functionality and microbiological safety. Thus, the design of proper honey processing methodologies is of utmost importance for its use. As reviewed, honey presents a very promising potential to be used in wound-healing processes, either by direct application, incorporated in fibrous membranes, or in hydrogel, with very promising results in either in vitro and in vivo trials. Nonetheless, further research is needed to overcome the main challenges on the use of honey for biomedical applications.

Author Contributions: Conceptualization, writing—original draft preparation, H.S., P.C.-F.; writing—review and editing, H.S., P.C.-F., J.M.F., M.E.T., M.S.D., C.A.P., L.M.E.; visualization, C.A.P.; Project administration, supervision, J.A.S., L.M.E. All authors have read and agreed to the published version of the manuscript.

Funding: This research received no external funding.

Institutional Review Board Statement: Not applicable.

Informed Consent Statement: Not applicable.

Data Availability Statement: The data presented in this study are available on request from the corresponding author.

Acknowledgments: Thanks are due to the University of Aveiro and FCT/MCT for the financial support of LAQV-REQUIMTE research Unit (UIDB/50006/2020) through national funds and, where applicable, co-financed by the FEDER within the PT2020 Partnership Agreement, and for financing the PhD grant of Hana Scepankova (SFRH/BD/88133/2012) and Carlos A. Pinto (SFRH/BD/137036/2018).

Conflicts of Interest: The authors declare no conflict of interest.

References

1. Kus, K.J.B.; Ruiz, E.S. Wound Dressings—A Practical Review. *Curr. Dermatol. Rep.* **2020**, *9*, 298–308. [CrossRef]
2. Khaleghverdi, S.; Karimi, A.; Soltani, R.; Zare, R. The Effect of Myrtus, Honey, Aloe Vera and Pseudomonas Phage Treatment on Infected Second Degree Burns: In Vivo Study. *Biointerface Res. Appl. Chem.* **2021**, *11*, 7422–7430. [CrossRef]
3. Bowden, L.G.; Byrne, H.M.; Maini, P.K.; Moulton, D.E. A Morphoelastic Model for Dermal Wound Closure. *Biomech. Model. Mechanobiol.* **2016**, *15*, 663–681. [CrossRef] [PubMed]
4. Anderson, K.; Hamm, R.L. Factors That Impair Wound Healing. *J. Am. Coll. Clin. Wound Spec.* **2012**, *4*, 84–91. [CrossRef] [PubMed]
5. Olsson, M.; Järbrink, K.; Divakar, U.; Bajpai, R.; Upton, Z.; Schmidtchen, A.; Car, J. The Humanistic and Economic Burden of Chronic Wounds: A Systematic Review. *Wound Repair Regen.* **2019**, *27*, 114–125. [CrossRef]
6. Molan, P.; Rhodes, T. Honey: A Biologic Wound Dressing. *Wounds* **2015**, *27*, 141–151. [PubMed]
7. Rossi, M.; Marrazzo, P. The Potential of Honeybee Products for Biomaterial Applications. *Biomimetics* **2021**, *6*, 6. [CrossRef]
8. Combarros-Fuertes, P.; Fresno, J.M.; Estevinho, M.M.; Sousa-Pimenta, M.; Tornadijo, M.E.; Estevinho, L.M. Honey: Another Alternative in the Fight against Antibiotic-Resistant Bacteria? *Antibiotics* **2020**, *9*, 774. [CrossRef]

9. Campeau, M.E.M.; Patel, R. Antibiofilm Activity of Manuka Honey in Combination with Antibiotics. *Int. J. Bacteriol.* **2014**, *2014*, 795281. [CrossRef]
10. Hayes, G.; Wright, N.; Gardner, S.L.; Telzrow, C.L.; Wommack, A.J.; Vigueira, P.A. Manuka Honey and Methylglyoxal Increase the Sensitivity of *Staphylococcus Aureus* to Linezolid. *Lett. Appl. Microbiol.* **2018**, *66*, 491–495. [CrossRef]
11. Lu, J.; Cokcetin, N.N.; Burke, C.M.; Turnbull, L.; Liu, M.; Carter, D.A.; Whitchurch, C.B.; Harry, E.J. Honey Can Inhibit and Eliminate Biofilms Produced by Pseudomonas Aeruginosa. *Sci. Rep.* **2019**, *9*, 18160. [CrossRef]
12. Maddocks, S.E.; Jenkins, R.E. Honey: A Sweet Solution to the Growing Problem of Antimicrobial Resistance? *Future Microbiol.* **2013**, *8*, 1419–1429. [CrossRef]
13. McLoone, P.; Tabys, D.; Fyfe, L. Honey Combination Therapies for Skin and Wound Infections: A Systematic Review of the Literature. *Clin. Cosmet. Investig. Dermatol.* **2020**, *13*, 875–888. [CrossRef] [PubMed]
14. Oryan, A.; Alemzadeh, E.; Moshiri, A. Biological Properties and Therapeutic Activities of Honey in Wound Healing: A Narrative Review and Meta-Analysis. *J. Tissue Viability* **2016**, *25*, 98–118. [CrossRef]
15. Jull, A.; Walker, N.; Parag, V.; Molan, P.; Rodgers, A. Randomized Clinical Trial of Honey-Impregnated Dressings for Venous Leg Ulcers. *Br. J. Surg.* **2008**, *95*, 175–182. [CrossRef]
16. Smaropoulos, E.; Cremers, N.A.J. Treating Severe Wounds in Pediatrics with Medical Grade Honey: A Case Series. *Clin. Case Rep.* **2020**, *8*, 469–476. [CrossRef] [PubMed]
17. Alam, F.; Islam, M.A.; Gan, S.H.; Khalil, M.I. Honey: A Potential Therapeutic Agent for Managing Diabetic Wounds. *Evid. Based Complement. Altern. Med.* **2014**, *2014*, 169130. [CrossRef] [PubMed]
18. Hixon, K.R.; Klein, R.C.; Eberlin, C.T.; Linder, H.R.; Ona, W.J.; Gonzalez, H.; Sell, S.A. A Critical Review and Perspective of Honey in Tissue Engineering and Clinical Wound Healing. *Adv. Wound Care* **2019**, *8*, 403–415. [CrossRef]
19. Nair, H.K.R.; Tatavilis, N.; Pospíšilová, I.; Kučerová, J.; Cremers, N.A.J. Medical-Grade Honey Kills Antibiotic-Resistant Bacteria and Prevents Amputation in Diabetics with Infected Ulcers: A Prospective Case Series. *Antibiotics* **2020**, *9*, 529. [CrossRef]
20. Molan, P.C. The Evidence and the Rationale for the Use of Honey. *Wound Pract. Res.* **2011**, *19*, 204–220.
21. Güneş, Ü.Y.; Eşer, I. Effectiveness of a Honey Dressing for Healing Pressure Ulcers. *J. Wound Ostomy Cont. Nurs.* **2007**, *34*, 184–190. [CrossRef]
22. Kamaratos, A.V.; Tzirogiannis, K.N.; Iraklianou, S.A.; Panoutsopoulos, G.I.; Kanellos, I.E.; Melidonis, A.I. Manuka Honey-Impregnated Dressings in the Treatment of Neuropathic Diabetic Foot Ulcers. *Int. Wound J.* **2012**, *11*, 259–263. [CrossRef] [PubMed]
23. Efem, S.E.E. Clinical Observations on the Wound Healing Properties of Honey. *Br. J. Surg.* **1988**, *75*, 679–681. [CrossRef] [PubMed]
24. Du Toit, D.F.; Page, B.J. An in Vitro Evaluation of the Cell Toxicity of Honey and Silver Dressings. *J. Wound Care* **2009**, *18*, 383–389. [CrossRef] [PubMed]
25. Silvia, P.M.D.; Gauche, C.; Gonzaga, L.V.; Costa, A.C.O.; Fett, R. Honey: Chemical Composition, Stability and Authenticity. *Food Chem.* **2016**, *196*, 309–323. [CrossRef]
26. Combarros-Fuertes, P.; Valencia-Barrera, R.M.; Estevinho, L.M.; Dias, L.G.; Castro, J.M.; Tornadijo, M.E.; Fresno, J.M. Spanish Honeys with Quality Brand: A Multivariate Approach to Physicochemical Parameters, Microbiological Quality, and Floral Origin. *J. Apic. Res.* **2019**, *58*, 92–103. [CrossRef]
27. Yilmaz, A.C.; Aygin, D. Honey dressing in wound treatment: A systematic review. *Complement. Ther. Med.* **2020**, *51*, 102388. [CrossRef]
28. Martinotti, S.; Ranzato, E. Honey, Wound Repair and Regenerative Medicine. *J. Funct. Biomater.* **2018**, *9*, 34. [CrossRef]
29. Mcloone, P.; Warnock, M.; Fyfe, L. Honey: A Realistic Antimicrobial for Disorders of the Skin. *J. Microbiol. Immunol. Infect.* **2016**, *49*, 161–167. [CrossRef]
30. Albaridi, N.A. Antibacterial Potency of Honey. *Int. J. Microbiol.* **2019**, *2019*, 2464507. [CrossRef]
31. Proaño, A.; Coello, D.; Villacrés-Granda, I.; Ballesteros, I.; Debut, A.; Vizuete, K.; Brenciani, A.; Álvarez-Suarez, J.M. The Osmotic Action of Sugar Combined with Hydrogen Peroxide and Bee-Derived Antibacterial Peptide Defensin-1 Is Crucial for the Antibiofilm Activity of Eucalyptus Honey. *LWT* **2021**, *136*, 110379. [CrossRef]
32. Peršurić, Ž.; Pavelić, S.K. Bioactives from Bee Products and Accompanying Extracellular Vesicles as Novel Bioactive Components for Wound Healing. *Molecules* **2021**, *26*, 3770. [CrossRef] [PubMed]
33. Abuharfeil, N.; Al-Oran, R.; Abo-Shehada, M. Food and Agricultural Immunology The Effect of Bee Honey on the Proliferative Activity of Human B-and T-Lymphocytes and the Activity of Phagocytes. *Food Agric. Immunol.* **1999**, *11*, 169–177. [CrossRef]
34. Tonks, A.J.; Cooper, R.A.; Jones, K.P.; Blair, S.; Parton, J.; Tonks, A. Honey Stimulates Inflammatory Cytokine Production from Monocytes. *Cytokine* **2003**, *21*, 242–247. [CrossRef]
35. Tonks, A.; Cooper, R.A.; Price, A.J.; Molan, P.C.; Jones, K.P. Stimulation of TNF-α Release in Monocytes by Honey. *Cytokine* **2001**, *14*, 240–242. [CrossRef]
36. Vestby, L.K.; Grønseth, T.; Simm, R.; Nesse, L.L. Bacterial Biofilm and Its Role in the Pathogenesis of Disease. *Antibiotics* **2020**, *9*, 59. [CrossRef] [PubMed]
37. Molan, P.C. Honey: Antimicrobial Actions and Role in Disease Management. In *New Strategies Combating Bacterial Infection*; Wiley-VCH Verlag GmbH & Co. KGaA: Weinheim, Germany, 2009; pp. 229–253. ISBN 9783527322060.

38. Combarros-Fuertes, P.; Estevinho, L.M.; Dias, L.G.; Castro, J.M.; Tomás-Barberán, F.A.; Tornadijo, M.E.; Fresno-Baro, J.M. Bioactive Components and Antioxidant and Antibacterial Activities of Different Varieties of Honey: A Screening Prior to Clinical Application. *J. Agric. Food Chem.* **2019**, *67*, 688–698. [CrossRef]
39. Krishnakumar, G.S.; Mahendiran, B.; Gopalakrishnan, S.; Muthusamy, S.; Malarkodi Elangovan, S. Honey Based Treatment Strategies for Infected Wounds and Burns: A Systematic Review of Recent Pre-Clinical Research. *Wound Med.* **2020**, *30*, 100188. [CrossRef]
40. Sukur, S.M.; Halim, A.S.; Singh, K.K.B. Evaluations of Bacterial Contaminated Full Thickness Burn Wound Healing in Sprague Dawley Rats Treated with Tualang Honey. *Indian J. Plast. Surg.* **2011**, *44*, 112–117. [CrossRef]
41. Natarajan, S.; Williamson, D.; Grey, J.; Harding, K.G.; Cooper, R.A. Healing of an MRSA-Colonized, Hydroxyurea-Induced Leg Ulcer with Honey. *J. Dermatol. Treat.* **2001**, *12*, 33–36. [CrossRef]
42. Gethin, G.; Cowman, S. Bacteriological Changes in Sloughy Venous Leg Ulcers Treated with Manuka Honey or Hydrogel: An RCT. *J. Wound Care* **2008**, *17*, 241–247. [CrossRef]
43. Johnson, D.W.; Van Eps, C.; Mudge, D.W.; Wiggins, K.J.; Armstrong, K.; Hawley, C.M.; Campbell, S.B.; Isbel, N.M.; Nimmo, G.R.; Gibbs, H. Randomized, Controlled Trial of Topical Exit-Site Application of Honey (Medihoney) versus Mupirocin for the Prevention of Catheter-Associated Infections in Hemodialysis Patients. *J. Am. Soc. Nephrol.* **2005**, *16*, 1456–1462. [CrossRef] [PubMed]
44. Sindi, A.; Chawn, M.V.B.; Hernandez, M.E.; Green, K.; Islam, M.K.; Locher, C.; Hammer, K. Anti-Biofilm Effects and Characterisation of the Hydrogen Peroxide Activity of a Range of Western Australian Honeys Compared to Manuka and Multifloral Honeys. *Sci. Rep.* **2019**, *9*, 1–7. [CrossRef]
45. Cooper, R.; Jenkins, L.; Rowlands, R. Inhibition of Biofilms through the Use of Manuka Honey. *Wounds* **2011**, *7*, 24–32.
46. Maddocks, S.E.; Lopez, M.S.; Rowlands, R.S.; Cooper, R.A. Manuka Honey Inhibits the Development of Streptococcus Pyogenes Biofilms and Causes Reduced Expression of Two Fibronectin Binding Proteins. *Microbiology* **2012**, *158*, 781–790. [CrossRef] [PubMed]
47. Merckoll, P.; Jonassen, T.Ø.; Vad, M.E.; Jeansson, S.L.; Melby, K.K. Bacteria, Biofilm and Honey: A Study of the Effects of Honey on "planktonic" and Biofilm-Embedded Chronic Wound Bacteria. *Scand. J. Infect. Dis.* **2009**, *41*, 341–347. [CrossRef] [PubMed]
48. Kot, B.; Sytykiewicz, H.; Sprawka, I.; Witeska, M. Effect of Manuka Honey on Biofilm-Associated Genes Expression during Methicillin-Resistant Staphylococcus Aureus Biofilm Formation. *Sci. Rep.* **2020**, *10*, 13552. [CrossRef]
49. Wasfi, R.; Elkhatib, W.F.; Khairalla, A.S. Effects of Selected Egyptian Honeys on the Cellular Ultrastructure and the Gene Expression Profile of Escherichia Coli. *PLoS ONE* **2016**, *11*, e0150984. [CrossRef]
50. Truchado, P.; López-Gálvez, F.; Gil, M.I.; Tomás-Barberán, F.A.; Allende, A. Quorum Sensing Inhibitory and Antimicrobial Activities of Honeys and the Relationship with Individual Phenolics. *Food Chem.* **2009**, *115*, 1337–1344. [CrossRef]
51. Jenkins, R.; Cooper, R. Improving Antibiotic Activity against Wound Pathogens with Manuka Honey In Vitro. *PLoS ONE* **2012**, *7*, e45600. [CrossRef] [PubMed]
52. Jenkins, R.E.; Cooper, R. Synergy between Oxacillin and Manuka Honey Sensitizes Methicillin-Resistant Staphylococcus Aureus to Oxacillin. *J. Antimicrob. Chemother.* **2012**, *67*, 1405–1407. [CrossRef]
53. Silva, B.; Biluca, F.C.; Gonzaga, L.V.; Fett, R.; Dalmarco, E.M.; Caon, T.; Costa, A.C.O. In Vitro Anti-Inflammatory Properties of Honey Flavonoids: A Review. *Food Res. Int.* **2021**, *141*, 110086. [CrossRef]
54. Yadav, A.; Verma, S.; Keshri, G.K.; Gupta, A. Combination of Medicinal Honey and 904 nm Superpulsed Laser-Mediated Photobiomodulation Promotes Healing and Impedes Inflammation, Pain in Full-Thickness Burn. *J. Photochem. Photobiol. B Biol.* **2018**, *186*, 152–159. [CrossRef] [PubMed]
55. Nooh, H.Z.; Nour-Eldien, N.M. The Dual Anti-Inflammatory and Antioxidant Activities of Natural Honey Promote Cell Proliferation and Neural Regeneration in a Rat Model of Colitis. *Acta Histochem.* **2016**, *118*, 588–595. [CrossRef] [PubMed]
56. Kassim, M.; Achoui, M.; Mansor, M.; Mohd, K. Fitoterapia The Inhibitory Effects of Gelam Honey and Its Extracts on Nitric Oxide and Prostaglandin E 2 in in Fl Ammatory Tissues. *Fitoterapia* **2010**, *81*, 1196–1201. [CrossRef]
57. Kassim, M.; Achoui, M.; Mustafa, M.R.; Mohd, M.A.; Yusoff, K.M. Ellagic Acid, Phenolic Acids, and Flavonoids in Malaysian Honey Extracts Demonstrate in Vitro Anti-Inflammatory Activity. *Nutr. Res.* **2010**, *30*, 650–659. [CrossRef]
58. Leong, A.G.; Herst, P.M.; Harper, J.L. Indigenous New Zealand Honeys Exhibit Multiple Anti-Inflammatory Activities. *Innate Immun.* **2011**, *18*, 459–466. [CrossRef]
59. Al-Mamary, M.; Al-Meeri, A.; Al-Habori, M. Antioxidant Activities and Total Phenolics of Different Types of Honey. *Nutr. Res.* **2002**, *22*, 1041–1047. [CrossRef]
60. Gheldof, N.; Engeseth, N.J. Antioxidant Capacity of Honeys from Various Floral Sources Based on the Determination of Oxygen Radical Absorbance Capacity and Inhibition of in Vitro Lipoprotein Oxidation in Human Serum Samples. *J. Agric. Food Chem.* **2002**, *50*, 3050–3055. [CrossRef]
61. Brudzynski, K.; Miotto, D. The Relationship between the Content of Maillard Reaction-like Products and Bioactivity of Canadian Honeys. *Food Chem.* **2011**, *124*, 869–874. [CrossRef]
62. Brudzynski, K.; Miotto, D. The Recognition of High Molecular Weight Melanoidins as the Main Components Responsible for Radical-Scavenging Capacity of Unheated and Heat-Treated Canadian Honeys. *Food Chem.* **2011**, *125*, 570–575. [CrossRef]
63. Pyrzynska, K.; Biesaga, M. Analysis of Phenolic Acids and Flavonoids in Honey. *TrAC Trends Anal. Chem.* **2009**, *28*, 893–902. [CrossRef]

64. Majtan, J. Honey: An Immunomodulator in Wound Healing. *Wound Repair Regen.* **2014**, *22*, 187–192. [CrossRef]
65. Alvarez-suarez, J.M.; Giampieri, F.; Battino, M. Honey as a Source of Dietary Antioxidants: Structures, Bioavailability and Evidence of Protective Effects against Human Chronic Diseases. *Curr. Med. Chem.* **2013**, *20*, 621–638. [CrossRef]
66. Subrahmanyam, M. A Prospective Randomised Clinical and Histological Study of Superficial Burn Wound Healing with Honey and Silver Sulfadiazine. *Burns* **1998**, *24*, 157–161. [CrossRef]
67. Kumar, P.; Kumar, S.; Udupa, E.P.; Kumar, U.; Rao, P.; Honnegowda, T. Role of Angiogenesis and Angiogenic Factors in Acute and Chronic Wound Healing. *Plast. Aesthetic Res.* **2015**, *2*, 243. [CrossRef]
68. Rossiter, K.; Cooper, A.J.; Voegeli, D.; Lwaleed, B.A. Honey Promotes Angiogeneic Activity in the Rat Aortic Ring Assay. *J. Wound Care* **2010**, *19*, 440–446. [CrossRef]
69. Chaudhary, A.; Bag, S.; Banerjee, P.; Chatterjee, J. Wound Healing Efficacy of Jamun Honey in Diabetic Mice Model through Reepithelialization, Collagen Deposition and Angiogenesis. *J. Tradit. Complement. Med.* **2020**, *10*, 529–543. [CrossRef]
70. Eteraf-oskouei, T.; Najafi, M. Traditional and Modern Uses of Natural Honey in Human Diseases: A Review. *Iran. J. Basic Med Sci.* **2013**, *16*, 731–742.
71. Stewart, J.A.; McGrane, O.L.; Wedmore, I.S. Wound Care in the Wilderness: Is There Evidence for Honey? *Wilderness Environ. Med.* **2014**, *25*, 103–110. [CrossRef]
72. Gethin, G.T.; Cowman, S.; Conroy, R.M. The Impact of Manuka Honey Dressings on the Surface PH of Chronic Wounds. *Proc. Int. Wound J.* **2008**, *5*, 185–194. [CrossRef]
73. Eteraf-Oskouei, T.; Najafi, M.; Gharehbagheri, A. Natural Honey: A New and Potent Anti-Angiogenic Agent in the Air-Pouch Model of Inflammation. *Drug Res.* **2013**, *64*, 530–536. [CrossRef] [PubMed]
74. Barui, A.; Mandal, N.; Majumder, S.; Das, R.K.; Sengupta, S.; Banerjee, P.; Ray, A.K.; Roychaudhuri, C.; Chatterjee, J. Assessment of Molecular Events during in Vitro Re-Epithelialization under Honey-Alginate Matrix Ambience. *Mater. Sci. Eng. C* **2013**, *33*, 3418–3425. [CrossRef] [PubMed]
75. Tonks, A.; Dudley, E.; Porter, N.G.; Parton, J.; Brazier, J.; Smith, E.L.; Tonks, A. A 5.8-kDa Component of Manuka Honey Stimulates Immune Cells via TLR4. *J. Leukoc. Biol.* **2007**, *82*, 1147–1155. [CrossRef] [PubMed]
76. Al-Waili, N.S.; Haq, A. Effect of Honey on Antibody Production against Thymus-Dependent and Thymus-Independent Antigens in Primary and Secondary Immune Responses. *J. Med. Food* **2004**, *7*, 491–494. [CrossRef]
77. Wink, D.A.; Hines, H.B.; Cheng, R.Y.S.; Switzer, C.H.; Flores-Santana, W.; Vitek, M.P.; Ridnour, L.A.; Colton, C.A. Nitric Oxide and Redox Mechanisms in the Immune Response. *J. Leukoc. Biol.* **2011**, *89*, 873–891. [CrossRef]
78. Al-Waili, N.S. Identification of Nitric Oxide Metabolites in Various Honeys: Effects of Intravenous Honey on Plasma and Urinary Nitric Oxide Metabolites Concentrations. *J. Med. Food* **2003**, *6*, 359–364. [CrossRef]
79. Azmi, M.F.; Abd Ghafar, N.; Che Hamzah, J.; Chua, K.H.; Ng, S.L. The Role of Gelam Honey in Accelerating Reepithelialization of Ex Vivo Corneal Abrasion Model. *J. Food Biochem.* **2021**, *45*, e13645. [CrossRef]
80. Angioi, R.; Morrin, A.; White, B. The Rediscovery of Honey for Skin Repair: Recent Advances in Mechanisms for Honey-Mediated Wound Healing and Scaffolded Application Techniques. *Appl. Sci.* **2021**, *11*, 5192. [CrossRef]
81. Majtan, J.; Kumar, P.; Majtan, T.; Walls, A.F.; Klaudiny, J. Effect of Honey and Its Major Royal Jelly Protein 1 on Cytokine and MMP-9 MRNA Transcripts in Human Keratinocytes. *Exp. Dermatol.* **2010**, *19*, e73–e79. [CrossRef]
82. Molan, P.C. Re-Introducing Honey in the Management of Wounds and Ulcers—Theory and Practice. *Ostomy/Wound Manag.* **2002**, *48*, 28–40.
83. Oluwatosin, O.M.; Olabanji, J.K.; Oluwatosin, O.A.; Tijani, L.A.; Onyechi, H.U. A Comparison of Topical Honey and Phenytoin in the Treatment of Chronic Leg Ulcers. *Afr. J. Med. Med Sci.* **2000**, *29*, 31–34. [PubMed]
84. Kwakman, P.H.S.; Zaat, S.A.J. Antibacterial Components of Honey. *IUBMB Life* **2012**, *64*, 48–55. [CrossRef]
85. Horniackova, M.; Bucekova, M.; Valachova, I.; Majtan, J. Effect of Gamma Radiation on the Antibacterial and Antibiofilm Activity of Honeydew Honey. *Eur. Food Res. Technol.* **2017**, *243*, 81–88. [CrossRef]
86. Molan, P.C.; Allen, K.L. The Effect of Gamma-Irradiation on the Antibacterial Activity of Honey. *J. Pharm. Pharmacol.* **1996**, *48*, 1206–1209. [CrossRef]
87. Gunduz, A.; Turedi, S.; Oksuz, H. The Honey, the Poison, the Weapon. *Wilderness Environ. Med.* **2011**, *22*, 182–184. [CrossRef]
88. Gunduz, A.; Turedi, S.; Russell, R.M.; Ayaz, F.A. Clinical Review of Grayanotoxin/Mad Honey Poisoning Past and Present. *Clin. Toxicol.* **2008**, *46*, 437–442. [CrossRef]
89. Scepankova, H.; Saraiva, J.A.; Estevinho, L.M. Honey health benefits and uses in medicine. In *Bee Products—Chemical and Biological Properties*; Springer International Publishing: Cham, Switzerland, 2017; pp. 83–96.
90. Hossain, L.; Lim, L.Y.; Hammer, K.; Hettiarachchi, D.; Locher, C. Honey-Based Medicinal Formulations: A Critical Review. *Appl. Sci.* **2021**, *11*, 5159. [CrossRef]
91. Bateman, S.; Graham, T. The Use of Medihoney^{TM} Antibacterial Wound Gel on Surgical Wounds Post-CABG. *Wounds* **2007**, *3*, 76–83.
92. Rezvani Ghomi, E.; Khalili, S.; Nouri Khorasani, S.; Esmaeely Neisiany, R.; Ramakrishna, S. Wound Dressings: Current Advances and Future Directions. *J. Appl. Polym. Sci.* **2019**, *136*, 1–12. [CrossRef]
93. Rajput, M.; Mandal, M.; Anura, A.; Mukhopadhyay, A.; Subramanian, B.; Paul, R.R.; Chatterjee, J. Honey Loaded Silk Fibroin 3D Porous Scaffold Facilitates Homeostatic Full-Thickness Wound Healing. *Materialia* **2020**, *12*, 100703. [CrossRef]

94. Datta, S.; Sarkar, R.; Vyas, V.; Bhutoria, S.; Barui, A.; Roy Chowdhury, A.; Datta, P. Alginate-Honey Bioinks with Improved Cell Responses for Applications as Bioprinted Tissue Engineered Constructs. *J. Mater. Res.* **2018**, *33*, 2029–2039. [CrossRef]
95. Schuhladen, K.; Mukoo, P.; Liverani, L.; Nescakova, Z.; Boccaccini, A.R. Manuka Honey and Bioactive Glass Impart Methylcellulose Foams with Antibacterial Effects for Wound-Healing Applications. *Biomed. Mater.* **2020**, *15*, 065002. [CrossRef]
96. Negut, I.; Dorcioman, G.; Grumezescu, V. Scaffolds for Wound Healing Applications. *Polymers* **2020**, *12*, 2010. [CrossRef]
97. Öhnstedt, E.; Lofton Tomenius, H.; Vågesjö, E.; Phillipson, M. The Discovery and Development of Topical Medicines for Wound Healing. *Expert Opin. Drug Discov.* **2019**, *14*, 485–497. [CrossRef]
98. Minden-Birkenmaier, B.A.; Bowlin, G.L. Honey-Based Templates in Wound Healing and Tissue Engineering. *Bioengineering* **2018**, *5*, 46. [CrossRef]
99. El-Kased, R.F.; Amer, R.I.; Attia, D.; Elmazar, M.M. Honey-Based Hydrogel: In Vitro and Comparative in Vivo Evaluation for Burn Wound Healing. *Sci. Rep.* **2017**, *7*, 9692. [CrossRef]
100. Mohd Zohdi, R.; Abu Bakar Zakaria, Z.; Yusof, N.; Mohamed Mustapha, N.; Abdullah, M.N.H. Gelam (Melaleuca Spp.) Honey-Based Hydrogel as Burn Wound Dressing. *Evid.-Based Complement. Altern. Med.* **2012**, *2012*, 843025. [CrossRef]
101. Noori, S.; Kokabi, M.; Hassan, Z.M. Poly(Vinyl Alcohol)/Chitosan/Honey/Clay Responsive Nanocomposite Hydrogel Wound Dressing. *J. Appl. Polym. Sci.* **2018**, *135*, 46311. [CrossRef]
102. Tavakoli, J.; Tang, Y. Honey/PVA Hybrid Wound Dressings with Controlled Release of Antibiotics: Structural, Physico-Mechanical and in-Vitro Biomedical Studies. *Mater. Sci. Eng. C* **2017**, *77*, 318–325. [CrossRef]
103. Afshari, M.J.; Sheikh, N.; Afarideh, H. PVA/CM-Chitosan/Honey Hydrogels Prepared by Using the Combined Technique of Irradiation Followed by Freeze-Thawing. *Radiat. Phys. Chem.* **2015**, *113*, 28–35. [CrossRef]
104. Wang, T.; Zhu, X.K.; Xue, X.T.; Wu, D.Y. Hydrogel Sheets of Chitosan, Honey and Gelatin as Burn Wound Dressings. *Carbohydr. Polym.* **2012**, *88*, 75–83. [CrossRef]
105. Zhang, Q.; Lin, Z.; Zhang, W.; Huang, T.; Jiang, J.; Ren, Y.; Zhang, R.; Li, W.; Zhang, X.; Tu, Q. Fabrication of Green Poly(Vinyl Alcohol) Nanofibers Using Natural Deep Eutectic Solvent for Fast-Dissolving Drug Delivery. *RSC Adv.* **2020**, *11*, 1012–1021. [CrossRef]
106. Croitoru, A.M.; Ficai, D.; Ficai, A.; Mihailescu, N.; Andronescu, E.; Turculet, C.F. Nanostructured Fibers Containing Natural or Synthetic Bioactive Compounds in Wound Dressing Applications. *Materials* **2020**, *13*, 2407. [CrossRef]
107. Abou Zekry, S.S.; Abdellatif, A.; Azzazy, H.M.E. Fabrication of Pomegranate/Honey Nanofibers for Use as Antibacterial Wound Dressings. *Wound Med.* **2020**, *28*, 100181. [CrossRef]
108. Sarkar, R.; Ghosh, A.; Barui, A.; Datta, P. Repositing Honey Incorporated Electrospun Nanofiber Membranes to Provide Anti-Oxidant, Anti-Bacterial and Anti-Inflammatory Microenvironment for Wound Regeneration. *J. Mater. Sci. Mater. Med.* **2018**, *29*, 31. [CrossRef]
109. Sarhan, W.A.; Azzazy, H.M.E. High Concentration Honey Chitosan Electrospun Nanofibers: Biocompatibility and Antibacterial Effects. *Carbohydr. Polym.* **2015**, *122*, 135–143. [CrossRef]
110. Nazeri, S.; Ardakani, E.M.; Babavalian, H.; Latifi, A.M. Evaluation of Effectiveness of Honey-Based Alginate Hydrogel on Wound Healing in Rat Model. *J. Appl. Biotechnol. Rep.* **2015**, *2*, 293–297.
111. Shamloo, A.; Aghababaie, Z.; Afjoul, H.; Jami, M.; Bidgoli, M.R.; Vossoughi, M.; Ramazani, A.; Kamyabhesari, K. Fabrication and Evaluation of Chitosan/Gelatin/PVA Hydrogel Incorporating Honey for Wound Healing Applications: An in Vitro, in Vivo Study. *Int. J. Pharm.* **2021**, *592*, 120068. [CrossRef]
112. Giusto, G.; Vercelli, C.; Comino, F.; Caramello, V.; Tursi, M.; Gandini, M. A New, Easy-to-Make Pectin-Honey Hydrogel Enhances Wound Healing in Rats. *BMC Complement. Altern. Med.* **2017**, *17*, 266. [CrossRef]
113. Sarhan, W.A.; Azzazy, H.M.E.; El-Sherbiny, I.M. The Effect of Increasing Honey Concentration on the Properties of the Honey/Polyvinyl Alcohol/Chitosan Nanofibers. *Mater. Sci. Eng. C* **2016**, *67*, 276–284. [CrossRef] [PubMed]
114. Sarhan, W.A.; Azzazy, H.M. Apitherapeutics and Phage-Loaded Nanofibers as Wound Dressings with Enhanced Wound Healing and Antibacterial Activity. *Nanomedicine* **2017**, *12*, 2055–2067. [CrossRef] [PubMed]
115. Amin, M.A.; Abdel-Raheem, I.T. Accelerated wound healing and anti-inflammatory effects of physically cross linked polyvinyl alcohol-chitosan hydrogel containing honey bee venom in diabetic rats. *Arch. Pharm. Res.* **2014**, *37*, 1016–1031. [CrossRef] [PubMed]
116. Maleki, H.; Gharehaghaji, A.A.; Dijkstra, P.J. A Novel Honey-Based Nanofibrous Scaffold for Wound Dressing Application. *J. Appl. Polym. Sci.* **2013**, *127*, 4086–4092. [CrossRef]
117. Khan, M.Q.; Lee, H.; Khatri, Z.; Kharaghani, D.; Khatri, M.; Ishikawa, T.; Im, S.S.; Kim, I.S. Fabrication and Characterization of Nanofibers of Honey/Poly(1,4-Cyclohexane Dimethylene Isosorbide Trephthalate) by Electrospinning. *Mater. Sci. Eng. C* **2017**, *81*, 247–251. [CrossRef]
118. Hixon, K.R.; Lu, T.; Mcbride-gagyi, S.H.; Janowiak, B.E.; Sell, S.A. A Comparison of Tissue Engineering Scaffolds Incorporated with Manuka Honey of Varying UMF. *BioMed Res. Int.* **2017**, *2017*, 4843065. [CrossRef] [PubMed]
119. Yang, X.; Fan, L.; Ma, L.; Wang, Y.; Lin, S.; Yu, F.; Pan, X.; Luo, G.; Zhang, D.; Wang, H. Green Electrospun Manuka Honey/Silk Fibroin Fibrous Matrices as Potential Wound Dressing. *Mater. Des.* **2017**, *119*, 76–84. [CrossRef]
120. Minden-Birkenmaier, B.A., Neuhalfen, R.M.; Janowiak, B.E.; Sell, S.A. Preliminary Investigation and Characterization of Electrospun Polycaprolactone and Manuka Honey Scaffolds for Dermal Repair. *J. Eng. Fibers Fabr.* **2015**, *10*, 155892501501000. [CrossRef]

molecules MDPI

Article

Anthocyanin Recovery from Grape by-Products by Combining Ohmic Heating with Food-Grade Solvents: Phenolic Composition, Antioxidant, and Antimicrobial Properties

Marta Coelho [1,2], Sara Silva [1], Eduardo Costa [1], Ricardo N. Pereira [2], António Sebastião Rodrigues [3], José António Teixeira [2] and Manuela Pintado [1,*]

1 CBQF-Centro de Biotecnologia e Química Fina-Laboratório Associado, Escola Superior de Biotecnologia, Universidade Católica Portuguesa, Rua Arquiteto Lobão Vital 172, 4200-374 Porto, Portugal; mccoelho@ucp.pt (M.C.); snsilva@ucp.pt (S.S.); emcosta@ucp.pt (E.C.)
2 CEB-Centre of Biological Engineering, University of Minho, 4710-057 Braga, Portugal; rpereira@deb.uminho.pt (R.N.P.); jateixeira@deb.uminho.pt (J.A.T.)
3 Centre for Toxicogenomics and Human Health, Genetics, Oncology and Human Toxicology, NOVA Medical School/Faculdade de Ciências Médicas, Universidade Nova de Lisboa, 1169-056 Lisbon, Portugal; sebastiao.rodrigues@nms.unl.pt
* Correspondence: mpintado@ucp.pt

Citation: Coelho, M.; Silva, S.; Costa, E.; Pereira, R.N.; Rodrigues, A.S.; Teixeira, J.A.; Pintado, M. Anthocyanin Recovery from Grape by-Products by Combining Ohmic Heating with Food-Grade Solvents: Phenolic Composition, Antioxidant, and Antimicrobial Properties. *Molecules* **2021**, *26*, 3838. https://doi.org/10.3390/molecules26133838

Academic Editor: Stefano Dall'Acqua

Received: 1 June 2021
Accepted: 20 June 2021
Published: 24 June 2021

Abstract: Usually, wine-making by-products are discarded, presenting a significant environmental impact. However, they can be used as a source of bioactive compounds. Moreover, consumers' increasing demand for naturally nutritious and healthy products requires new formulations and food product improvement, together with sustainable, environmentally friendly extraction methods. Thus, this work aimed to compare ohmic heating (OH) with conventional methodology (CONV), using food-grade solvents, mainly water, compared to standard methanol extraction of anthocyanins. No significant differences were found between the CONV and OH for total phenolic compounds, which were 2.84 ± 0.037 and 3.28 ± 0.46 mg/g DW gallic acid equivalent, respectively. The same tendency was found for antioxidant capacity, where CONV and OH presented values of 2.02 ± 0.007 g/100 g and 2.34 ± 0.066 g/100 g ascorbic acid equivalent, respectively. The major anthocyanins identified were malvidin-3-*O*-acetylglucoside, delphinidin-3-*O*-glucoside, petunidine-3-*O*-glucoside, cyanidin-3-*O*-glucoside, and peonidine-3-*O*-glucoside. These extracts displayed antimicrobial potential against microorganisms such as *Yersinia enterocolitica*, *Pseudomonas aeruginosa*, *Salmonella enteritidis*, methicillin-sensitive *Staphylococcus aureus*, a methicillin-resistant *Staph. aureus* (MRSA), and *Bacillus cereus*. In conclusion, OH provides similar recovery yields with reduced treatment times, less energy consumption, and no need for organic solvents (green extraction routes). Thus, OH combined with water and citric acid allows a safe anthocyanin extraction from grape by-products, thus avoiding the use of toxic solvents such as methanol, and with high biological potential, including antimicrobial and antioxidant activity.

Keywords: grape by-products; ohmic heating; conventional methods; biological properties; phenolic compounds; anthocyanins; antimicrobial; antioxidant activity

1. Introduction

The wine-making process produces a large number of by-products that have a significant environmental impact. This process generates a high amount of solid organic waste, namely stalks, pomace (including skins, seeds, grape pulp l) and lees, which may be disposed of or beneficial use [1,2]. Nevertheless, these by-products can also be used as a source of bioactive compounds, such as dietary fiber, grape seed oil, hydrocolloids, and phenolic compounds, which might be applied by the agri-food and feed industries promoting economic value. Its reuse follows the actual circular economy concept imposed by the European Union. According to it, strategies for smart, sustainable, and inclusive growth must be adopted, promoting environmental protection [3–5].

Directive 2008/98/EC of the European Parliament and of the Council of 19 November 2008 established a legal framework for treating waste in the European Union, emphasizing the importance of proper waste management, recovery, and recycling techniques to reduce the environmental and human health impact [6]. Value-added compounds could be isolated from the by-products to be used either as natural ingredients for the formulations of functional foods or as additives. These by-products have drawn the attention of scientists and the food industry. Traditionally, grape pomace has been used to produce wine beverages, nutrient colorants, and grape oil. More recently, research has concentrated on creating different value-added products, e.g., bioactive compounds, primarily phenols, healing of tartaric acid, and the production of flours [5,7].

These wine industry by-products are frequently undervalued but represent a possible source of bioactive compounds, such as polyphenols, that could be applied in many industries. Phenolic substances from grapes, including anthocyanins, are reduced at the skin and seeds, more precisely, the portion that remain as pomace after the processing of grapes [2,7,8].

Anthocyanins belong to the flavonoid class and represent the most significant set of water-soluble plant pigments [2,9]. Anthocyanin's color depends on the solvent's pH; the red-color anthocyanins are stable at lower pH (3.5–4) [7,9,10].

As they are localized in black grape peels, anthocyanins are usually extracted (30 to 40%) through wine-making operations. However, previous research has shown that the anthocyanin content of a given cultivar is not necessarily positively linked with anthocyanin concentration in the resulting wine. The lack of association was due to the partial preservation by cell-wall polymers of these anthocyanins in the skin cells [2,11,12].

Various methods have been developed to extract bioactive compounds, and their effects on the preparation and functionality of extracts from agro-industrial wastes have been evaluated. These commonly applied methods use harmful and toxic compounds, restricting the use of grape by-product extracts [2,7,13]. Nevertheless, given the need for sustainability, the political agenda is fostering the development of "clean label" processes toward the reduction of environmental impacts and a strong bioeconomy. In addition, the conversion of wine-making by-products into added-value products could be possible through the development of environment-friendly technologies. New technologies, such as microwave-, ultrasound-, and ohmic heating (OH)-assisted extraction, have been used to improve the recovery of bioactive compounds from food samples [2,8,14]. These methods have attracted significant attention from the scientific community. In particular, OH is a thermal process where an alternating electric flow is forced to pass through the food materials [2]. The thermal effect needed to assist the extraction process is rendered internally due to the passage of electric flow through the materials (Joule effect). The general purpose of OH technology is to improve food, cosmetics, and pharmaceutical products that are safe and beneficial to human health [2,15,16]. Despite these advantages, the OH application can be impaired by products' physical and chemical properties, such as poor electrical conductivity due to higher fat or sugar contents in their composition [16].

El Darra and colleagues performed an assisted extraction by pulsed OH from red grape pomace. The authors explored the pulsed OH effects on cell membrane damage to increased polyphenol recovery. They also studied the effects of the electrical field strength, temperature, and the proportion of ethanol/water used. Pre-treatment caused cell membrane permeabilization. In addition, pulsed OH, which was used as a pre-treatment, increased the recovery kinetics of TPC. Other researchers reinforce high yields of recovery of phytochemicals obtained by OH from black rice bran. They suggest OH as a promising technology to extract anthocyanins with a future application in the production of natural colorants [17].

Pereira et al. [18] explore how ohmic heating (OH) influences phytochemical components recovery from colored potatoes (*Solanum tuberosum* L. var. *Vitelotte*) using moderate electric fields. Their results reveal that low-energy electrical fields and thermal effects may be integrated and adjusted into a single phase of treatment by recovering anthocyanins and phenolic chemicals from vegetable tissues, which delivers a high recovery rate with lower treatment duration, decreased energy consumption, and no organic solvents (green extraction). More recently, the authors [13] have also shown that OH has the potential to be used as an efficient and environmentally friendly technology toward sustainable food processes; it has been shown that OH can be used as a pre-treatment for enhanced aqueous extraction of anthocyanins from grape skins, particularly when high-temperature short-time (HTST) treatments are applied. There still is scarce information regarding the effects of combining OH pre-treatments with food-grade solvents to enhance the aqueous extraction of phenolic compounds, namely anthocyanins, comparing with conventional methods based on solvent extraction. Accordingly, this study aimed to evaluate the effectiveness of OH pre-treatments in the aqueous extraction of anthocyanins from red grape pomace by-products using food-grade extraction solvents and compared it with traditional solvent extraction methodologies.

2. Results

2.1. Total Phenolic Compounds and Antioxidant Capacity

Concerning total phenolic compounds (TPC) (Figure 1), a better recovery yield was obtained with MeOH acidified in all extraction methods ohmic (OH), negative control (CN), and positive control (CP). Comparing the extraction methods, OH presented higher values than CP for all solvents tested. In addition, higher amounts of TPC were obtained when combined OH with MeOH acidified. Significant differences between extraction methods were only verified for MeOH extraction ($p < 0.001$). In addition, values of 423 ± 0.2 mg/100 g DW gallic acid equivalent and 360 ± 0.8 mg /100 gallic acid equivalent were observed for MeOH and citric acid, respectively, with OH extraction. These results are corroborated by results from antioxidant analysis measured by ABTS, where citric acid showed the highest values of antioxidant capacity (AA) when compared with other solvents (Figure 2).

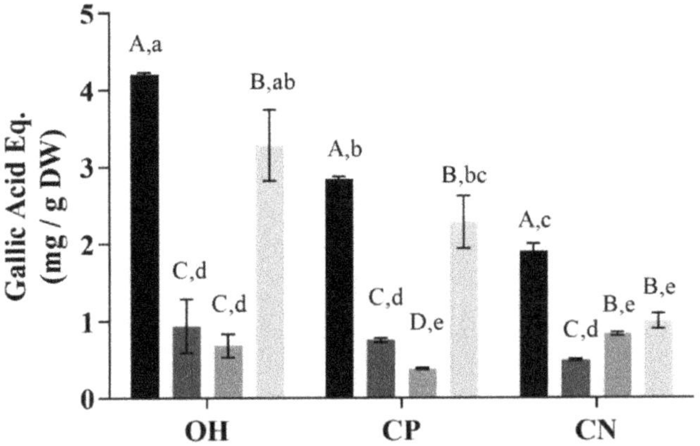

Figure 1. Total phenolic content of extracts performed with methanolic and aqueous solutions from grape wine-making by-products (gallic acid equivalent mg/g DW). OH—ohmic heating; CP—positive control; CN—negative control. Extraction solvents. ▪ MeOH with 1% HCl, ▪ H_2O, ▪ lactic acid, ▪ citric acid. Different capital letters in the same extraction method indicate a statistically significant difference ($p < 0.01$). The small letters represent statistical differences between solvents and extraction methods ($p < 0.05$).

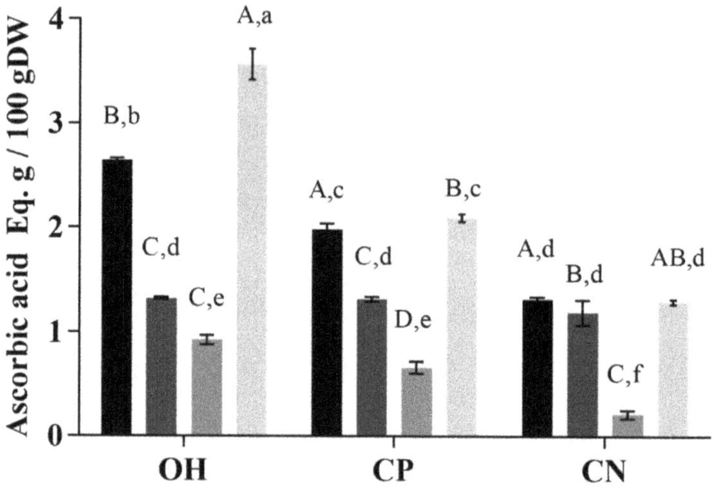

Figure 2. Antioxidant activity by ABTS test of extracts from grape by-products (g ascorbic acid equivalent/100 g DW). OH—ohmic heating; CP—positive control; CN—negative control. Extraction solvents. ▪ MeOH with 1% HCl, ▪ H_2O, ▪ lactic acid, ▪ citric acid. Different capital letters in the same extraction method indicate a statistically significant difference ($p < 0.01$). The small letters represent statistical differences between solvents and the extraction method ($p < 0.05$).

AA was also analyzed by the ORAC method. According to the results of AA with the ORAC method presented in Figure 3, this method resulted in different patterns compared with ABTS. Two-way ANOVA showed that there is a significant interaction between solvent and treatment, $p < 0.0001$. Methanol CP extract showed ORAC quantities of 0.489 ± 0.0443 µmol/g Trolox equivalents, which were analogous to those in citric extracts with OH (0.351 ± 0.022 micromol/g Trolox equivalents).

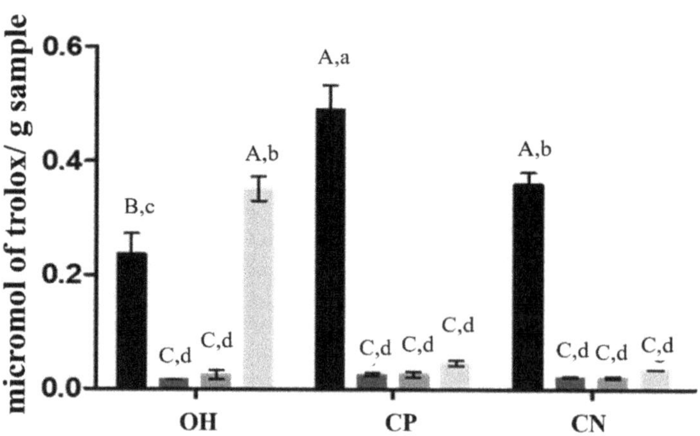

Figure 3. Antioxidant capacity by the ORAC method. Results are expressed in micromol of Trolox equivalent per gram of sample. OH—ohmic heating; CP—positive control; CN—negative control; extraction solvents. ▪ MeOH with 1% HCl, ▪ H_2O, ▪ lactic acid, ▪ citric acid. Different capital letters in the same extraction method indicate a statistically significant difference ($p < 0.01$). The small letters represent statistical differences between solvents and the extraction method ($p < 0.05$).

2.2. Total and Individual Anthocyanins Content

Results showed an increase of anthocyanin recovery (Figure 4) in OH samples when compared with the CN method for all solvents extraction ($p < 0.05$), while when compared with CP, similar results were obtained. This increase could be explained by non-thermal effects on plant cell permeabilization, which were probably due to electrical disturbances in the membranes of cells or by electroporation impacts [13,15,19]. However, it is essential to highlight that compared with CN, OH treatment resulted in a reduced thermal load. This may have contributed to less degradation of extracted anthocyanins and justify the presented results. Regarding individual compounds, the malvidin-3-*O*-glucoside is the main anthocyanin present for all extracts (Table 1 and Figure 5).

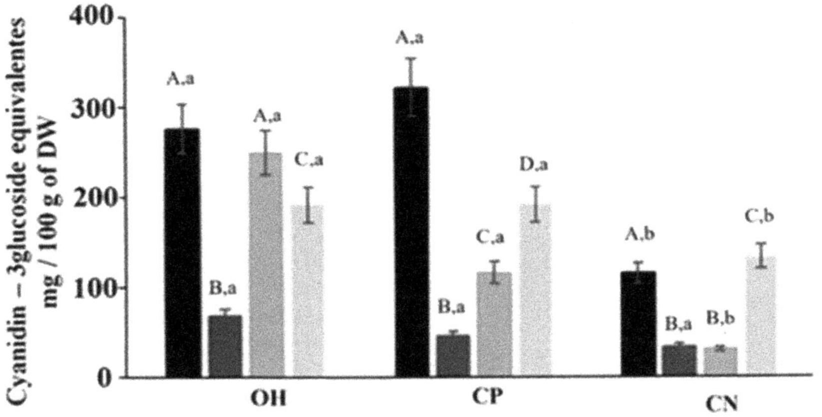

Figure 4. Total anthocyanins for all extraction process of grape by-products. OH—ohmic heating; CP—positive control; CN—negative control. extraction solvents. ▪ MeOH with 1% HCl, ▪ H_2O, ▪ lactic acid, ▪ citric acid. Different capital letters in the same extraction method indicate a statistically significant difference ($p < 0.05$). Different small letters represent a statistical difference between methods of extraction ($p < 0.05$).

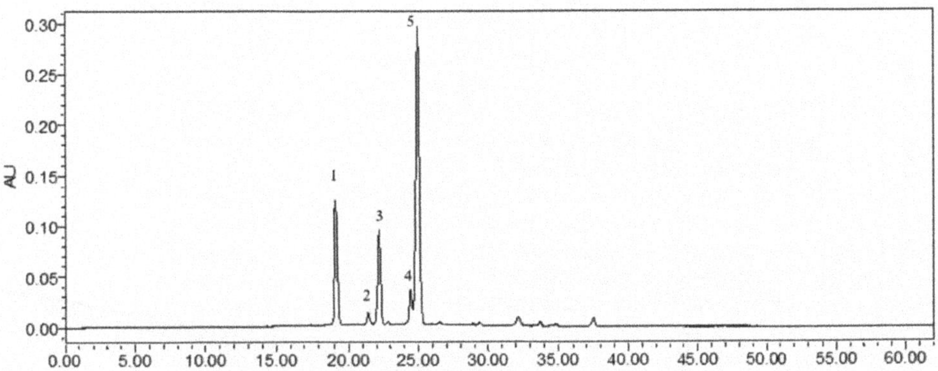

Figure 5. Chromatogram example of anthocyanins profile detected in samples by HPLC-DAD. (1) delphinidin-3-*O*-glucoside; (2) cyanidin-3-*O*-glucoside; (3) petunidin-3-*O*-glucoside; (4) peonidin-3-*O*-glucoside; and (5) malvidin-3-*O*-glucoside.

showed that the highest concentration of 1 mg /mL of water extracts tested did not inhibit the cellular metabolism (negative values of metabolism inhibition), thus not showing cytotoxicity for these cells (Figure 6). Only citric acid presented an inhibition above 10% for the CN, which indicates some inhibition of cell viability, but also with no relevance, since values are lower than 30% cell metabolism inhibition. All extracts obtained by OH and some from the CN extracts presented negative values, suggesting that they promote cell growth, mainly the extracts produced by OH with citric acid.

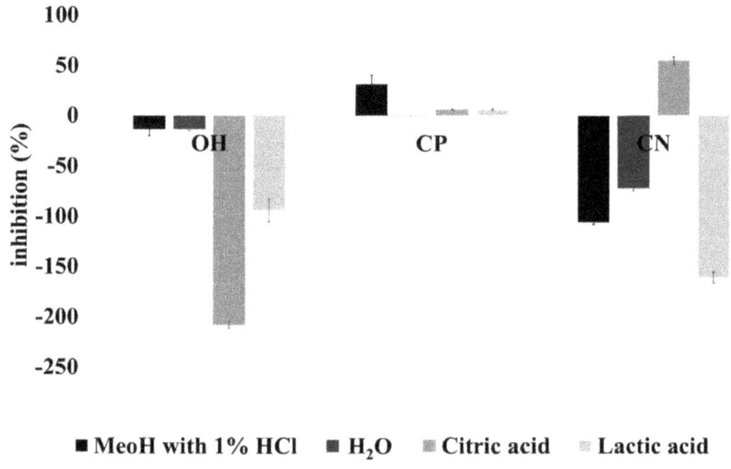

Figure 6. Cytotoxicity analysed by the XTT method. The line represents the non-toxic extracts. OH—ohmic heating; CP—positive control; CN—negative control. The extraction solvents. ▪ MeOH with 1% HCl, ▪ H_2O, ▪ lactic acid, ▪ citric acid.

3. Discussion

3.1. Total Phenolic Compounds and Antioxidant Capacity

The results showed that all treatment and solvents influenced the total polyphenols extraction. Significant differences were found between OH and the CN. A decrease in TPC was observed in CN when compared with OH and the CP method. Although OH and CN treatments use high temperatures in the OH, a maximum of 2 s was required for the temperature to reach 100 °C, while in the CT method, 20 min were necessary to reach the same temperature.

Both heating processes and solvent used affected the TPC. It can cause leaching, and consequently, the TPC can decrease, or it could affect the rupture of the cell wall and, as a result, the release of cell-bound polyphenols. Several studies reported the effect of heating processes on polyphenols content [13,23–26]. A study carried out by Pereira et al. [13] showed that it is possible to obtain higher extraction yields of TPC and anthocyanins with aqueous extracts with an OH pre-treatment. The fast heating prevents the increase of compounds degradation. Additionally, the polarity of the solution used for polyphenols extraction affects its cell availability, and it also changes the recovery of the TPC. In moderate settings (23 °C), pre-treatments with limited permeabilization effects and no organic solvent have only encouraged the diffusion of small molecular weight hydrophilic components. This explains the differences in the TPC levels across pre-treated OH and CP ($p > 0.05$) samples. In general, the flavonoids in which anthocyanins are contained are low molecular weight molecules and hence easily extracted due to further OH-induced permeabilization.

A significant impact of the treatment was observed when compared to the impact of the solvent by the application of the two-way ANOVA and Tukey's tests. The treatment accounts for 14.94% of the total variance, with significance $p < 0.001$, while the solvent account for 66.64% of the total variance. The same was observed for AA, according to the ABTS method (Figure 2). Regarding AA, there were differences between results obtained by ABTS and ORAC methods, which could be explained by the sensitivity of methods and the compounds recovered during the extraction processes [27,28]. The ORAC evaluates the AA and determines the antioxidant status in biological systems, while ABTS measures the reduction of the specific force. In addition, different solvents and extraction methods could recover different bioactive compounds, which could justify the differences in values [29].

This result suggested that for TPC and AA, the solvent has more impact than the treatment used.

Mostly, according to several studies, the inhibition of polyphenol oxidases and acid hydrolysis could occur; as a result, the TPC increases, and consequently, the AA also increases [18,30,31]. The differences found in TPC and AA when citric acid ($C_6H_8O_7$) or lactic acid ($C_3H_6O_3$) were used could be explained by its chemical structure and its pH solution, 2.55 and 2.98, respectively. Studies have shown that the nature of the acid employed, along with its pka, influences the selectivity of the extracting medium toward phenolic compounds [32]. Furthermore, the chemical properties of anthocyanins make them susceptible to reactive oxygen molecules.

3.2. *Total and Individual Anthocyanin Content*

Significant differences in total anthocyanin in OH extracts were observed compared to the CP method for food-grade solvents, indicating the greater anthocyanin recovery capacity of red pigments with OH. Anthocyanins are water-soluble glycosides of polyhydroxy and polymethoxy derivatives of 2-phenyl-benzopyryliurn. Furthermore, the pH solution influences the anthocyanin's colour due to its ionic structure. They present a flavylium form in an acidic condition, being highly soluble in water, and they are also more stable [10,33]. In addition, some studies have shown a disintegration of cell wall caused by electric fields during OH, which could increase the compound's availability and, consequently, their extraction capacity [15,17,34].

Regarding CN, the time to reach 100 °C is significant. In this work, higher levels of anthocyanin degradation were observed in CT samples (exposed to high temperatures for an extended time). These results showed that higher temperatures could promote the degradation of anthocyanins compounds. According to results reported with the literature, the anthocyanins degradation levels could reach 55% with higher temperature usage [35]. Furthermore, some studies also indicate that the heating process and the electric field could potentiate synergies affecting enzymatic activities, such as the polyphenol oxidase activity, which indirectly degrades the monomeric anthocyanins during enzymatic browning [36,37]. Eventually, the electric effects can activate or inactivate enzymes, since there are studies that report effects on the enzymes [38–41]. However, further studies would be needed to confirm whether the same is true in this case. In addition, the electric fields can promote a selective extraction or even a change in the structure, similar to what happens with proteins [42–45]. However, it is not known whether the same is true for anthocyanins, and further studies are needed to understand at a fundamental level the interactions between electric fields and anthocyanins molecules.

Regarding individual anthocyanins content, Pereira et al. [13] used water to recover phenolic compounds, and they obtained higher rate yields using OH as pre-treatment to anthocyanins extraction. They only compared OH extraction with the heating process without comparing CONV methods (organic solvents). The authors extracted anthocyanins with OH, which avoids chemical solvents, by using water and reduced treatment times. The main anthocyanins found in this study were malvidin-3-*O*-glucoside, cyanidin-3-*O*-glucoside, and delphinidin-3-*O*-glucoside, which is in agreement with our results. In addition, they present higher cyanidin-3-*O*-glucoside values than delphinidin-3-*O*-glucoside and petunidin-3-*O*-glucoside, while we present lesser cyanidin values compared with delphinidin. The differences found between anthocyanins content could be explained by the extraction method and solvents used. Rackic et al. [46] showed that an alkaline pH increases the cyanidin contents in the extracts, while the opposite happens for pH values ranging from 2.0 and 4.0.

3.3. Antimicrobial Properties

Polyphenolic compounds, including anthocyanins, have antimicrobial activity against a wide variety of microorganisms, particularly in inhibiting the development of food-borne pathogens. Anthocyanins demonstrate antimicrobial action through various mechanisms, e.g., causing cell damage by damaging the cell wall, membrane, and intercellular matrix [10]. The amounts of phenolic compounds showing antibacterial action correspond to those previously reported. Polyphenols from Touriga Nacional and Preto Martinho wine by-products were isolated by Silva et al. [47], and they showed strong antibacterial action against different pathogens. These are similar to the findings obtained for the non-conventional extraction OH with water acidified with citric acid, presenting significant differences compared with CN ($p < 0.05$). The CP method displayed similar results regarding antibacterial activity. These microorganisms are usually linked with food as indicators of pathogenic microorganisms. The extraction process of bioactive compounds affects the antimicrobial activity of the recovered compounds [48,49].

Additional investigations with *Staph. aureus* showed that doses as low as 1.6 g/100 g of total phenolics might have a large inhibitory impact on the development of MRSA and MSSA biofilm, but chlorogenic acid was the primary component in this case rather than anthocyanin [50]. The possibility that the OH method could be a selective extraction method for certain compounds, as mentioned above, may potentiate this antimicrobial effect when compared to CN. According to Table 1, the main anthocyanin present in this extract is the malvidin 3- *O*-glucoside. The antibacterial activity of anthocyanin-containing extracts may be caused by the diverse processes and synergy effects of distinct extract phytochemicals such as anthocyanin, phenolic acids, and their chemical combinations [10].

In addition, CN uses a longer thermal effect (100 °C, 20 min). Therefore, it can degrade some more sensitive compounds, including anthocyanins, which have this antimicrobial effect, as described in the literature [51,52]. Other studies have shown that Gram-negative bacteria but not Gram-positive bacteria are inhibited in anthocyanin-rich extracts. This may be related to the distinct cell wall structures between Gram-negative and Gram-positive bacteria in which the outer membrane of Gram-negatives functions as a preventative barrier for hydrophobic compounds but not on hydrophilic compounds [10]. Côté et al. [53] showed the antibacterial activity of cranberry extract in vancomycin-resistant *Enterococcus faecium*, *P. aeruginosa*, *Staph aureus*, and *E. coli*. The antibacterial action of cranberries extracts is not related to their low pH, but it is likely to be attributable to bioactive elements, such anthocyanin and flavonols, in pH-adjusted cranberry extracts.

The results show the potential use of OH extracts against both Gram-negative and Gram-positive bacteria.

3.4. Cytotoxicity

The XTT method is an excellent technique for measuring cell viability. Only methanol extracts presented inhibition of cell viability in the case of the CP samples higher than 30%, demonstrating a cytotoxic effect. The results are following the literature, which reports the cytotoxicity of methanolic extracts [54]. In addition, no evidence of cytotoxicity was found with water grape extracts [55].

4. Materials and Methods

4.1. Chemicals

The 2, 20-azo-bis-(2-methylpropionamidine)-dihydrochloride (AAPH), fluorescein, 2, 2-azinobis-3-ethylbenzothiazoline-6-sulphonic acid (ABTS diammonium salt), potassium sorbate, sodium carbonate, ethylenediaminetetraacetic acid (EDTA), sodium sulfite, and sodium lauryl sulfate were purchased from Sigma-Aldrich (Sintra, Portugal). Methanol, acetonitrile, and hydrochloric acid were purchased from Fischer Scientific Portugal. Folin–

Ciocalteu's reagent, potassium persulfate, citric acid, and lactic acid were purchased from Merck (Algés, Portugal). Standards of ascorbic acid, Trolox, and gallic acid were purchased from Sigma-Aldrich (Sintra, Portugal), while delphinidin-3-*O*-glucoside, cyanidin3-*O*-glucoside, petunidin-3-*O*-glucoside, peonidin-3-*O*-glucoside, and malvidin-3-*O*-glucoside were purchased from Extrasynthese (Lyon, France).

4.2. Samples

The red grape pomace, obtained from a wine manufacture using Vinhão cultivar, was used for the study. Grape pomace includes a mixture of pulp, skins, and seeds, which were separated randomly in aliquots of 50 g and dried in an oven at 50 °C ± 3 °C. After samples were milled in a cuisine robot (Bimby Vorwerk, Wuppertal, Germany, TM5), the powder was sieved manually at 150 µm. The powder was used for the following phenolic extractions.

4.3. Extraction Procedures

One of the most critical factors that affects the bioactive compounds recovery yields is the solvents used in the extraction method. Solvents differ in polarity and comprise methanol, hexane, acetone, chloroform, and diethyl ether [56]. According to the literature, the traditional solvent for anthocyanins recovery is acidified methanolic solution [57,58].

4.3.1. Pre-Treatments

The grape by-products are non-conductive samples; thus, 2.5 g of grape by-products were placed in 5 mL sodium chloride (NaCl) 0.1 M solution to increase the conductivity to 4.6 mS/cm at room temperature. Three methodologies were performed with each solvent described before—ohmic heating (OH), which reaches 100 °C in 13 s; a control negative (CN) reaches 100 °C in 20 min; and the control positive (CP) used at room temperature (Figure 7)—to understand if the effect came from the ohmic heating (OH) and not from the temperature and solvent during the extraction process. After, all samples were cooled in ice to stop the reactions and proceed with the solvents' extractions.

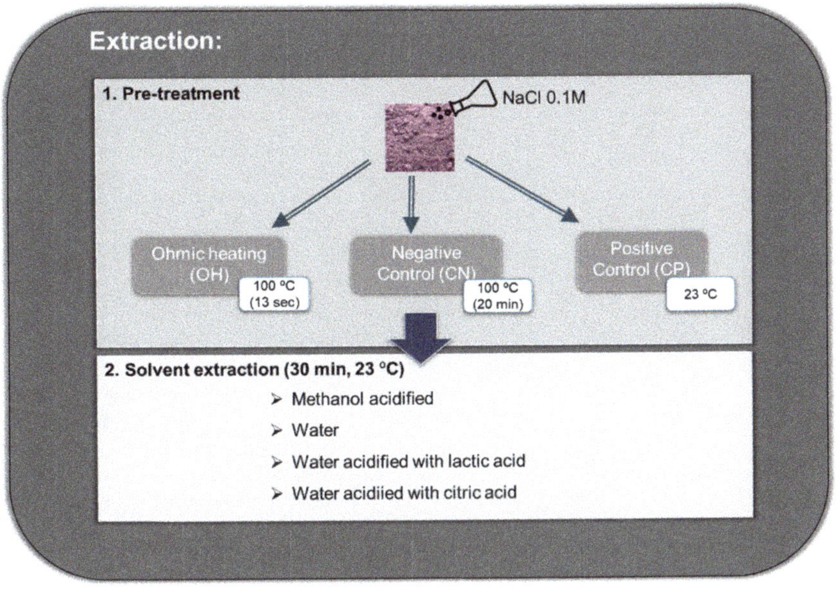

Figure 7. Scheme of experimental analysis.

OH Pre-Treatment

The OH was carried out as a pre-treatment before solvent extraction. A high frequency of 25 kHz and an electric field of ≈30 V/cm was applied. Within these conditions, the temperature reached 100 °C in 13 s, after which the system was turned off; then, samples were placed on ice to stop the residual temperature effect and promote a fast decrease toward room temperature (23 °C).

Control Negative

Control of temperature was also performed to observe the temperature effects. It consisted of placing 2.5 g of grape by-products (placed in the same sodium chloride solution described before) in a bath, and when samples reached 100 °C (≈20 min), they were placed on ice to stop the reaction and also promote a fast decrease toward room temperature (23 °C).

4.3.2. Solvent Extraction

Water, acidified water (lactic and citric acid 1%), and methanol/water solution (80:19:1 *v*/*v*) acidified with hydrochloric acid usually are used as a conventional method (CONV) for anthocyanins extractions, as described by Silva et al. [8]. Both lactic and citric acid are food-grade acids usually used in the food industry.

Extraction solvents (25.0 mL) were added to pre-treated samples (OH and CN) and directly to 2.5 g of grape by-products pre-treated with NaCl and CP, and the material suspensions were kept under gentle stirring at room temperature (23 °C) for 30 min, allowing the bioactive compound recovery. Water solution and water acidified with two food-grade acids (lactic, citric) were used to decrease the pH (pH 3) to favor extraction and stabilize the color of anthocyanins.

Then, the extract was centrifuged at 4000× *g*, 4 °C for 10 min, and the supernatant was filtered through a 0.45 mm cellulose acetate filter (Orange Scientific, Braine-l'Alleud, Belgium), and the pellets were stored at −80 °C. This procedure was used for total activities measurement.

4.4. Total Antioxidant Capacity and Phenolic Content

4.4.1. Total Antioxidant Activity

After the extraction's procedure, the extracts were evaluated in terms of antioxidant activity to understand the effects of treatments and solvents extraction.

AA was performed using the ABTS method, according to [14]. The sample was added to a colored solution of 2,2′-azinobis-(3-ethylbenzothiazoline-6-sulfonic acid radical cation) ($ABTS^{\bullet+}$), with an optical density (OD) measured at 734 nm and adjusted to 0.700 ± 0.020 in a spectrophotometric microplate reader (Sunrise Tecan, Grödig, Austria). After 6 min of reaction, the final OD was read, and the results were given in ascorbic acid equivalent. The standard was ascorbic acid (0–500 mg/L), and the regression equation for ascorbic acid and samples was calculated (R^2 = 0.999).

The Oxygen Radical Absorbance Capacity (ORAC-FL) was also measured to evaluate the AA, according to Ubeda et al. [45]. Briefly, 20 µL of OH, CN, and CP extracts were mixed with 120 µL of fluorescein (70 nM) in black, untreated 96-well microplates (Nunc, Roskilde, Denmark) and incubated at 40 °C for 10 min. Then, 60 µL of 2,2′-azo-bis-(2-methylpropionamidine)-dihydrochloride (APPH) solution (12 mM, final concentration), Sigma-Aldrich, AAPH solution (60 µL; 12 mM, final concentration) was rapidly added using a multichannel pipet. The microplate was immediately placed in the multidetector plate reader (Synergy H1, Burlington, VT, USA), and the fluorescence was recorded at intervals of 1 min for 140 min. The excitation wavelength was set at 485 nm, and the emission wavelength was set at 528 nm [45]. The microplate was automatically shaken before each reading. The area under the curve (AUC) was calculated for each sample by integrating the relative fluorescence curve. Trolox (10–80 µM) was used as the standard, and regression equations for Trolox and samples were calculated. The ORAC-FL values were calculated by the ratio of sample slope to Trolox slope obtained in the same assay.

Final ORAC-FL values were expressed as micromol of Trolox equivalent (TE) per mg of sample.

4.4.2. Total Phenolic Content

The TPC of extracts was evaluated through the Folin–Ciocalteu spectrophotometric method as described by Ferreira-Santos et al. [59]. A mixture of the sample (5 μL), Folin–Ciocalteu reagent (15 μL), sodium carbonate at 75 g/L (60 μL, Sigma-Aldrich) and distilled water (200 μL) were performed, and the solutions were mixed. Afterwards, samples were heated at 60 °C for 5 min, and the OD was read at 734 nm using a spectrophotometric microplate reader (Sunrise Tecan, Grödig, Austria). The gallic acid standard (0–500 mg/L) was used to measure TPC, and regression equations for gallic acid and samples were calculated (R^2 0.997) and expressed as a milligram of gallic acid equivalent per dry weight material (mg GAE/100 g). The analyses were performed in triplicate, and a standard deviation was calculated.

4.5. Total Anthocyanins

Total anthocyanins (TA) were assessed using a spectrophotometric analysis, as Pereira et al. [25] described. The results are expressed in equivalent cyanidine-3-glucoside equivalents (Cy-3-GE) and compared to a range of standards prepared to start from a stock ethanolic solution of cyanidin-3-glucoside. The analysis was performed in triplicate.

4.6. High-Performance Liquid Chromatography (HPLC) Analysis

Polyphenol profiles (quantitative and qualitative) were assessed according to Coelho et al. [14]. The analysis was conducted on HPLC-DAD (Waters Series 600. Mildford MA. USA). A Symmetry® C18 column, 250 × 4.6 mm i.d. 5 μm particle size, and 125 Å pore size with a guard column (waters) was used, and solvents elution consisted of solvent A—acetonitrile (100%) with 0.2% TFA; Solvent B—acetonitrile/water (5:95 *v*/*v*) (Merck pure grade and pure water) with 0.2% TFA (Sigma-Aldrich, Darmstadt, Germany). A linear gradient at a flow rate of 1 mL/min was applied: 0–20 min (100%B); 30–60 min (60% B); and 10 min (100% B). Samples were analyzed in triplicate. Calibration curves were obtained at a detection wavelength of 520 nm. Standards solutions over the concentration range from 0.10 to 100.00 mg/L were prepared for the identification and quantification of the following compounds: delphinidin-3-*O*-glucoside; cyanidin-3-*O*-glucoside; petunidin-3-*O*-glucoside; peonidin-3-*O*-glucoside; and malvidin-3-*O*-glucoside expressed as μg per mL of dry weight (DW) biomass of grape. All calibration curves were linear over the concentration ranges tested with correlation coefficients of 0.999.

4.7. Antimicrobial Analysis

4.7.1. Microorganisms

A few pathogenic microorganisms were utilized in this study. Clinical bacterial isolates from urine were kindly given by CHTMAD—Hospital Center of Trás-os-Montes e Alto Douro (through PhD Maria José Alves). The isolated strains comprised *E. coli* (*E. coli* CI resistant to ampicillin, nalidixic acid, norflaxin, and ciproflaxin), a *P. aeruginosa* (*Ps. aeruginosa* CI intermediately resistant to cefotaxime), a methicillin-resistant *Staph. aureus* ((MRSA) resistant to oxacillin, ciprofloxacin, and levofloxacin), and a methicillin-sensitive *Staph aureus* (MSSA) [60]. Additionally, there were three references (R) strains of food isolate from ESB collection: *S.* Enteritidis ATCC 13076, *Y. enterocolitica* NCTC 10406, and *L. monocytogenes*. An inoculum of each bacteria was prepared at a density equivalent to 0.5 on the McFarland scale (~1.5×10^8 CFU mL^{-1}). Next, serial decimal dilutions were performed in saline solution, obtaining suspensions with about 1.5×10^6 CFU mL^{-1}.

4.7.2. Plate Test

The different extracts at (1 $mg.mL^{-1}$) were used to evaluate their effects on antimicrobial properties. The assays were performed after 48 and 72 h of sample preparation. To perform these assays, 100 μL of inoculum was transferred to Petri dishes containing the Nutrient Agar medium by the spread plate method, and each plate was incubated at 37 °C for 48 h. As a positive control, nutrient agar plates containing bacterial suspension and saline solution were used as well as plaques containing the bacterial suspension, and the solvents were used to perform the extracts (methanol, water, water acidified with citric and lactic acid). All tests were performed in triplicate according to the method described by Ramos et al. [61].

4.8. Cytotoxicity

The colorimetric method using the XTT was performed to assess cell viability as a function of redox potential, according to Jiang et al. [62]. In the presence of metabolic activity, the water-soluble XTT is converted to a water-soluble, orange-colored formazan product. Shortly, 100 μL cell suspension aliquots were seeded in a 96 micro-plating well (1 μL to 105 cell/mL) (Nucleon Delta Surface, Thermo Scientific, Roskilde, Denmark). Then, after seeding, the cultivated media was carefully changed and incubated in the dark by the various test solutions. After 24 h, cell viability was tested as follows: a 10 mmol/L of PMS solution was prepared in phosphate-buffered saline (PBS, 0.01 M; pH 7.4), and a 1 mg/mL XTT solution was prepared using the appropriate culture media, previously heated to 37 °C. OH, CN, and CP extracts were used at concentrations of 1.0 mg/mL. Both solutions were sterilized using sterile membrane filters of 0.22 μm (Millipore, Billerica, MA, USA) and combined (2.5 μL PMS per mL XTT solution just before application). In each well, aliquots (25 μL) of mixture were added, and cells were incubated for about 2 h in the dark. The optical density was measured with a microplate reader at 485 nm (FluoSTAR, OPTIMA, BMG Labtech, Ortenberg, Germany). The findings were shown as the percentage of inhibition of cell metabolism. Plain culture media was used as a negative control. All assays were performed in quintuplicate.

4.9. Statistical Analysis

All experiments were performed in triplicate, and the results were expressed as the mean ± standard deviation. The SPSS v. 19 (Chicago IL, USA) software was used to evaluate the statistical differences within different treatments determined by ANOVA, using the Shapiro–Wilk for variance homogeneity and Tukey's test for multiple comparisons. Differences were considered significant at a 5% confidence level ($p < 0.05$).

5. Conclusions

The present study assessed the recovery of anthocyanins based on thermal and solvent treatments of grape by-products by OH, CP, and CT extraction methods. The present study assessed for the first time the recovery and characterization of anthocyanins based on thermal treatments of grape by-products by OH combined with acidified food-grade solvents. OH with water acidified with citric acid allowed higher extraction yields of total polyphenols content when compared with other methods. Furthermore, it is possible to obtain extracts with higher AA in the case of OH with water acidified with citric acid than obtained with MeOH. Total anthocyanins recovery was higher with OH and citric acid application. This treatment yielded similar results when compared with the CP method. The main anthocyanins recovery was of malvidin-3-*O*-glucoside, delphinidin-3-*O*-glucoside, and petunidin-3-*O*-glucoside. No cytotoxicity was found for OH extracts obtained with citric acid at 1 mg/mL. On the other hand, for the CP method with MeOH at the same concentrations, there was an inhibition of cell viability of 80%. Additionally, OH with citric acid at 1 mg/mL exhibited antimicrobial properties against pathogens, namely *P. aeruginosa*, *Y. enterocolitica*, *S. enteritidis*, MSSA, MRSA, and *B. cereus*.

OH combined with food-grade solvent (water and citric acid) allows the recovery of stable anthocyanins, which is in line with the European Union Directive 2009/32/EC. These results demonstrate a relevant opportunity to valorize red grape by-products in a circular economy context.

Author Contributions: Conceptualization, M.C. and M.P.; Methodology, M.C., S.S., E.C., R.N.P., A.S.R. and J.A.T.; Formal Analysis: M.C.; Investigation, M.C. and M.P.; Writing—Original Draft Preparation, M.C.; Writing—Review and Editing, M.C., M.P., R.N.P., A.S.R. and J.A.T. All authors have read and agreed to the published version of the manuscript.

Funding: This research was funded by "MultiBiorefinery: Estratégias multiuso para a valorização de uma gama alargada de subprodutos agroflorestais e das pescas: Um passo em frente na criação de uma biorrefinaria" financiado pelo Programa Operacional Competitividade e Internacionalização (POCI-01-0145-FEDER-016403) e pelo Programa Operacional Regional de Lisboa(LISBOA-01-0145-FEDER-016403), na sua componente FEDER e pela Fundação para a Ciência e Tecnologia, I.P. na componente nacional (SAICTPAC/0040/2015).

Institutional Review Board Statement: Not applicable.

Informed Consent Statement: Not applicable.

Data Availability Statement: The data presented in this study are available on request from the corresponding author.

Acknowledgments: The authors would like to thank "MultiBiorefinery: Estratégias multiuso para a valorização de uma gama alargada de subprodutos agroflorestais e das pescas: Um passo em frente na criação de uma biorrefinaria" financiado pelo Programa Operacional Competitividade e Internacionalização (POCI-01-0145-FEDER-016403) e pelo Programa Operacional Regional de Lisboa (LISBOA-01-0145-FEDER-016403), na sua componente FEDER e pela Fundação para a Ciência e Tecnologia, I.P. na componente nacional (SAICTPAC/0040/2015). The authors would also like to thank the scientific collaboration under the FCT project UID/Multi/50016/2019, UID-BIM-00009-2020, and UID/BIO/04469/2020. The author Marta Coelho would like to acknowledge FCT for your PhD grant with the reference [grant number SFRH/BD/111884/2015].

Conflicts of Interest: The authors declare no conflict of interest.

References

1. Lucarini, M.; Durazzo, A.; Kiefer, J.; Santini, A.; Lombardi-Boccia, G.; Souto, E.B.; Romani, A.; Lampe, A.; Nicoli, S.F.; Gabrielli, P.; et al. Grape seeds: Chromatographic profile of fatty acids and phenolic compounds and qualitative analysis by FTIR-ATR spectroscopy. *Foods* **2020**, *9*, 10. [CrossRef] [PubMed]
2. Coelho, M.C.; Pereira, R.N.; Rodrigues, A.S.; Teixeira, J.A.; Pintado, M.E. The use of emergent technologies to extract added value compounds from grape by-products. *Trends Food Sci. Technol.* **2020**, *106*, 182–197. [CrossRef]
3. Escribano-Bailón, M.T.; Rivas-Gonzalo, J.C.; García-Estévez, I. Wine Color Evolution and Stability. In *Red Wine Technology*; Morata, A., Ed.; Charlotte Cockle: Oxford, UK, 2019; pp. 195–205.
4. Mendes, J.A.S.; Prozil, S.O.; Evtuguin, D.V.; Lopes, L.P.C. Towards comprehensive utilization of winemaking residues: Characterization of grape skins from red grape pomaces of variety Touriga Nacional. *Ind. Crops Prod.* **2013**, *43*, 25–32. [CrossRef]
5. Chowdhary, P.; Gupta, A.; Gnansounou, E.; Pandey, A.; Chaturvedi, P. Current trends and possibilities for exploitation of Grape pomace as a potential source for value addition. *Environ. Pollut.* **2021**, *278*, 116796. [CrossRef]
6. European Parliament and Council. Directive 2008/98/EC of the European Parliament and of the Council of 19 November 2008 on waste and repealing certain Directives. *Off. J. Eur. Union* **2008**, *34*, 99–126.
7. Hogervorst, J.C.; Miljić, U.; Puškaš, V. 5—Extraction of Bioactive Compounds from Grape Processing By-Products A2—Galanakis, Charis M. In *Handbook of Grape Processing By-Products: Sustainable Solutions*; Galanakis, C.M.B.T., Ed.; Academic Press: Cambridge, MA, USA, 2017; pp. 105–135. ISBN 978-0-12-809870-7.
8. Maroun, R.G.; Rajha, H.N.; Vorobiev, E.; Louka, N. Emerging Technologies for the Recovery of Valuable Compounds from Grape Processing By-Products. In *Handbook of Grape Processing By-Products: Sustainable Solutions*; Galanakis, C.M.B.T., Ed.; Academic Press: Cambridge, MA, USA, 2017; ISBN 9780128098714.
9. Silva, S.; Costa, E.M.; Coelho, M.C.; Morais, R.M.; Pintado, M.E. Variation of anthocyanins and other major phenolic compounds throughout the ripening of four Portuguese blueberry (*Vaccinium corymbosum* L.) cultivars. *Nat. Prod. Res.* **2017**, *31*, 93–98. [CrossRef]
10. Khoo, H.E.; Azlan, A.; Tang, S.T.; Lim, S.M. Anthocyanidins and anthocyanins: Colored pigments as food, pharmaceutical ingredients, and the potential health benefits. *Food Nutr. Res.* **2017**, *61*, 1361779. [CrossRef]
11. El Gengaihi, S.; Ella, F.; Emad, M.; Shalaby, E.; Doha, H. Food Processing & Technology Antioxidant Activity of Phenolic Compounds from Different Grape Wastes. *J. Food Process. Technol.* **2014**, *5*, 1–5. [CrossRef]
12. Hanušovský, O.; Gálik, B.; Bíro, D.; Šimko, M.; Juráček, M.; Rolinec, M.; Zábranský, L.; Philipp, C.; Puntigam, R.; Slama, J.A.; et al. The nutritional potential of grape by-products from the area of Slovakia and Austria. *Emirates J. Food Agric.* **2020**. [CrossRef]
13. Pereira, R.N.; Coelho, M.I.; Genisheva, Z.; Fernandes, J.M.; Vicente, A.A.; Pintado, M.E.; Teixeira, eJ.A. Using Ohmic Heating effect on grape skins as a pretreatment for anthocyanins extraction. *Food Bioprod. Process.* **2020**, *124*, 320–328. [CrossRef]

14. Barba, F.J.; Zhu, Z.; Koubaa, M.; Sant'Ana, A.S.; Orlien, V. Green alternative methods for the extraction of antioxidant bioactive compounds from winery wastes and by-products: A review. *Trends Food Sci. Technol.* **2016**, *49*, 96–109. [CrossRef]
15. Coelho, M.; Pereira, R.; Rodrigues, A.S.; Teixeira, J.A.; Pintado, M.E. Extraction of tomato by-products' bioactive compounds using ohmic technology. *Food Bioprod. Process.* **2019**, *117*, 329–339. [CrossRef]
16. Kumar, T. A Review on Ohmic Heating Technology: Principle, Applications and Scope. *Int. J. Agric. Environ. Biotechnol.* **2018**, *11*. [CrossRef]
17. Loypimai, P.; Moongngarm, A.; Chottanom, P.; Moontree, T. Ohmic heating-assisted extraction of anthocyanins from black rice bran to prepare a natural food colourant. *Innov. Food Sci. Emerg. Technol.* **2015**, *27*, 102–110. [CrossRef]
18. Pereira, R.N.; Rodrigues, R.M.; Genisheva, Z.; Oliveira, H.; de Freitas, V.; Teixeira, J.A.; Vicente, A.A. Effects of ohmic heating on extraction of food-grade phytochemicals from colored potato. *LWT Food Sci. Technol.* **2016**, *74*, 493–503. [CrossRef]
19. Rocha, C.M.R.; Genisheva, Z.; Ferreira-Santos, P.; Rodrigues, R.; Vicente, A.A.; Teixeira, J.A.; Pereira, R.N. Electric field-based technologies for valorization of bioresources. *Bioresour. Technol.* **2018**, *254*, 325–339. [CrossRef] [PubMed]
20. García-Lomillo, J.; González-SanJosé, M.L.; Del Pino-García, R.; Rivero-Pérez, M.D.; Muñiz-Rodríguez, P. Antioxidant and antimicrobial properties of wine byproducts and their potential uses in the food industry. *J. Agric. Food Chem.* **2014**, *62*, 12595–12602. [CrossRef] [PubMed]
21. Luchian, C.E.; Cotea, V.V.; Vlase, L.; Toiu, A.M.; Colibaba, L.C.; Răschip, I.E.; Nadăş, G.; Gheldiu, A.M.; Tuchiluş, C.; Rotaru, L. Antioxidant and Antimicrobial Effects of Grape Pomace Extracts. Available online: https://www.bio-conferences.org/articles/bioconf/full_html/2019/04/bioconf-oiv2019_04006/bioconf-oiv2019_04006.html (accessed on 16 September 2020).
22. Oliveira, D.A.; Salvador, A.A.; Smânia, A.; Smânia, E.F.A.; Maraschin, M.; Ferreira, S.R.S. Antimicrobial activity and composition profile of grape (*Vitis vinifera*) pomace extracts obtained by supercritical fluids. *J. Biotechnol.* **2013**, *164*, 423–432. [CrossRef]
23. Achir, N.; Dhuique-Mayer, C.; Hadjal, T.; Madani, K.; Pain, J.P.; Dornier, M. Pasteurization of citrus juices with ohmic heating to preserve the carotenoid profile. *Innov. Food Sci. Emerg. Technol.* **2016**, *33*, 397–404. [CrossRef]
24. Marra, F.; Zell, M.; Lyng, J.G.; Morgan, D.J.; Cronin, D.A. Analysis of heat transfer during ohmic processing of a solid food. *J. Food Eng.* **2009**, *91*, 56–63. [CrossRef]
25. Oliveira, C.M.; Barros, A.S.; Silva Ferreira, A.C.; Silva, A.M.S. Influence of the temperature and oxygen exposure in red Port wine: A kinetic approach. *Food Res. Int.* **2015**, *75*, 337–347. [CrossRef]
26. Rocha, C.; Coelho, M.; Lima, R.C.; Campos, F.M.; Pintado, M.; Cunha, L.M. Increasing phenolic and aromatic compounds extraction and maximizing liking of lemon verbena (*Aloysia triphylla*) infusions through the optimization of steeping temperature and time. *Food Sci. Technol. Int.* **2019**, *25*, 701–710. [CrossRef]
27. Abreu, J.; Quintino, I.; Pascoal, G.; Postingher, B.; Cadena, R.; Teodoro, A. Antioxidant capacity, phenolic compound content and sensory properties of cookies produced from organic grape peel (*Vitis labrusca*) flour. *Int. J. Food Sci. Technol.* **2019**, *54*, 1215–1224. [CrossRef]
28. Coscueta, E.R.; Campos, D.A.; Osório, H.; Nerli, B.B.; Pintado, M. Enzymatic soy protein hydrolysis: A tool for biofunctional food ingredient production. *Food Chem. X* **2019**, *1*, 100006. [CrossRef]
29. Coelho, M.C.; Ribeiro, T.B.; Oliveira, C.; Batista, P.; Castro, P.; Monforte, A.R.; Rodrigues, A.S.; Teixeira, J.; Pintado, M. In Vitro Gastrointestinal Digestion Impact on the Bioaccessibility and Antioxidant Capacity of Bioactive Compounds from Tomato Flours Obtained after Conventional and Ohmic Heating Extraction. *Foods* **2021**, *10*, 554. [CrossRef]
30. Fallis, A. Phytochemical Mehods—A guide to Modern Techniques of Plant Analysis. *J. Chem. Inf. Model.* **2013**, *53*, 1689–1699. [CrossRef]
31. Oliveira, A.; Coelho, M.; Alexandre, E.M.C.; Almeida, D.P.F.; Pintado, M. Long-Term Frozen Storage and Pasteurization Effects on Strawberry Polyphenols Content. *Food Bioprocess Technol.* **2015**, *8*. [CrossRef]
32. Tzima, K.; Kallithraka, S.; Kotseridis, Y.; Makris, D.P. A Comparative Evaluation of Aqueous Natural Organic Acid Media for the Efficient Recovery of Flavonoids from Red Grape (*Vitis vinifera*) Pomace. *Waste Biomass Valorization* **2015**, *6*, 391–400. [CrossRef]
33. Lee, K.G.; Shibamoto, T. Determination of antioxidant potential of volatile extracts isolated from various herbs and spices. *J. Agric. Food Chem.* **2002**, *50*, 4947–4952. [CrossRef] [PubMed]
34. Moongngarm, A.; Loypimai, P.; Fitriati, A.; Moontree, T. Ohmic heating assisted extraction improves the concentrations of phytochemicals in rice bran oil and unsaponifiable matter. *Int. Food Res. J.* **2019**, *26*, 1389–1396.
35. Queiroz, F.; Oliveira, C.; Pinho, O.; Ferreira, I.M.P.L.V. Degradation of anthocyanins and anthocyanidins in blueberry jams/stuffed fish. *J. Agric. Food Chem.* **2009**, *57*, 10712–10717. [CrossRef]
36. Minatel, I.O.; Borges, C.V.; Ferreira, M.I.; Gomez, H.A.G.; Chen, C.-Y.O.; Lima, G.P.P. Phenolic Compounds: Functional Properties, Impact of Processing and Bioavailability. In *Phenolic Compounds—Biological Activity*; Soto-Hernández, M., Ed.; IntechOpen: London, UK, 2017.
37. Turfan, Ö.; Türkyilmaz, M.; Yemi, O.; Özkan, M. Anthocyanin and colour changes during processing of pomegranate (*Punica granatum* L.; Cv. Hicaznar) juice from sacs and whole fruit. *Food Chem.* **2011**, *129*, 1644–1651. [CrossRef]
38. Brochier, B.; Mercali, G.D.; Marczak, L.D.F. Effect of ohmic heating parameters on peroxidase inactivation, phenolic compounds degradation and color changes of sugarcane juice. *Food Bioprod. Process.* **2018**, *111*, 62–71. [CrossRef]
39. Funcia, E.S.; Gut, J.A.W.; Sastry, S.K. Effect of Electric Field on Pectinesterase Inactivation during Orange Juice Pasteurization by Ohmic Heating. *Food Bioprocess Technol.* **2020**, *13*, 1206–1214. [CrossRef]

40. Leite, T.S.; Samaranayake, C.P.; Sastry, S.K.; Cristianini, M. Polyphenol oxidase inactivation in viscous fluids by ohmic heating and conventional thermal processing. *J. Food Process Eng.* **2019**. [CrossRef]
41. Kubo, M.T.; Siguemoto, É.S.; Funcia, E.S.; Augusto, P.E.; Curet, S.; Boillereaux, L.; Sastry, S.K.; Gut, J.A. Non-thermal effects of microwave and ohmic processing on microbial and enzyme inactivation: A critical review. *Curr. Opin. Food Sci.* **2020**, *35*, 36–48. [CrossRef]
42. Pereira, R.N.; Teixeira, J.A.; Vicente, A.A.; Cappato, L.P.; da Silva Ferreira, M.V.; da Silva Rocha, R.; da Cruz, A.G. Ohmic heating for the dairy industry: A potential technology to develop probiotic dairy foods in association with modifications of whey protein structure. *Curr. Opin. Food Sci.* **2018**, *22*, 95–101. [CrossRef]
43. Ferreira-Santos, P.; Nunes, R.; De Biasio, F.; Spigno, G.; Gorgoglione, D.; Teixeira, J.A.; Rocha, C.M.R. Influence of thermal and electrical effects of ohmic heating on C-phycocyanin properties and biocompounds recovery from Spirulina platensis. *LWT* **2020**, *128*, 109491. [CrossRef]
44. Pereira, R.N.; Rodrigues, R.M.; Machado, L.; Ferreira, S.; Costa, J.; Villa, C.; Barreiros, M.P.; Mafra, I.; Teixeira, J.A.; Vicente, A.A. Influence of ohmic heating on the structural and immunoreactive properties of soybean proteins. *LWT* **2021**, *148*, 111710. [CrossRef]
45. Yoon, S.W.; Lee, C.Y.J.; Kim, K.M.; Lee, C.H. Leakage of cellular materials from Saccharomyces cerevisiae by ohmic heating. *J. Microbiol. Biotechnol.* **2002**, *12*, 183–188.
46. Rakic, V.; Skrt, M.; Miljkovic, M.; Kostic, D.; Sokolovic, D.; Poklar-Ulrih, N. Effects of pH on the stability of cyanidin and cyanidin 3-O-β-glucopyranoside in aqueous solution. *Hem. Ind.* **2015**, *69*, 511–522. [CrossRef]
47. Silva, V.; Igrejas, G.; Falco, V.; Santos, T.P.; Torres, C.; Oliveira, A.M.P.; Pereira, J.E.; Amaral, J.S.; Poeta, P. Chemical composition, antioxidant and antimicrobial activity of phenolic compounds extracted from wine industry by-products. *Food Control* **2018**, *92*, 516–522. [CrossRef]
48. Pourhashemi, A.; Deka, S.C.; Haghi, A.K. *Research Methods and Applications in Chemical and Biological Engineering*, 1st ed.; Pourhashemi, A., Deka, S.C., Haghi, A.K., Eds.; Apple Academic Press: Boca Raton, FL, USA, 2019.
49. Valle, D.L.; Cabrera, E.C.; Puzon, J.J.M.; Rivera, W.L. Antimicrobial activities of methanol, ethanol and supercritical CO_2 extracts of philippine Piper betle L. on clinical isolates of Gram positive and Gram negative bacteria with transferable multiple drug resistance. *PLoS ONE* **2016**, *11*, e0146349. [CrossRef]
50. Silva, S.; Costa, E.M.; Mendes, M.; Morais, R.M.; Calhau, C.; Pintado, M.M. Antimicrobial, antiadhesive and antibiofilm activity of an ethanolic, anthocyanin-rich blueberry extract purified by solid phase extraction. *J. Appl. Microbiol.* **2016**, *121*, 693–703. [CrossRef] [PubMed]
51. Ghada, B.; Pereira, E.; Pinela, J.; Prieto, M.A.; Pereira, C.; Calhelha, R.C.; Stojkovic, D.; Sokóvic, M.; Zaghdoudi, K.; Barros, L.; et al. Recovery of anthocyanins from passion fruit epicarp for food colorants: Extraction process optimization and evaluation of bioactive properties. *Molecules* **2020**, *25*, 3203. [CrossRef] [PubMed]
52. Xia, E.-Q.; Deng, G.-F.; Guo, Y.-J.; Li, H.-B. Biological Activities of Polyphenols from Grapes. *Int. J. Mol. Sci.* **2010**, *11*, 622–646. [CrossRef] [PubMed]
53. Côté, J.; Caillet, S.; Doyon, G.; Dussault, D.; Sylvain, J.F.; Lacroix, M. Antimicrobial effect of cranberry juice and extracts. *Food Control* **2011**, *22*, 1413–1418. [CrossRef]
54. Bozkurt, E.; Atmaca, H.; Kisim, A.; Uzunoglu, S.; Uslu, R.; Karaca, B. Effects of Thymus serpyllum extract on cell proliferation, apoptosis and epigenetic events in human breast cancer Cells. *Nutr. Cancer* **2012**, *64*, 1245–1250. [CrossRef] [PubMed]
55. Costa, J.R.; Amorim, M.; Vilas-Boas, A.; Tonon, R.V.; Cabral, L.M.C.; Pastrana, L.; Pintado, M. Impact of: In vitro gastrointestinal digestion on the chemical composition, bioactive properties, and cytotoxicity of *Vitis vinifera* L. cv. Syrah grape pomace extract. *Food Funct.* **2019**, *10*, 1856–1869. [CrossRef]
56. Grumezescu, A.M.; Holban, A.M. Preface for Volume 4: Ingredients Extraction by Physicochemical Methods in Food. In *Ingredients Extraction by Physicochemical Methods in Food*; Grumezescu, A., Holban, A.M., Eds.; Academic Press: Cambridge, MA, USA, 2017; pp. xxi–xxiv. ISBN 9780128112021.
57. Rodriguez-Saona, L.E.; Wrolstad, R.E. Extraction, Isolation, and Purification of Anthocyanins. In *Handbook of Food Analytical Chemistry*; John Wiley & Sons, Inc.: Hoboken, NJ, USA, 2005; Volume 2, pp. 7–17. ISBN 9780471709084.
58. Hong, H.T.; Netzel, M.E.; O'Hare, T.J. A dataset for anthocyanin analysis in purple-pericarp sweetcorn kernels by LC-DAD-MS. *Data Brief* **2020**, *30*, 105495. [CrossRef]
59. Ferreira-Santos, P.; Genisheva, Z.; Pereira, R.N.; Teixeira, J.A.; Rocha, C.M.R. Moderate Electric Fields as a Potential Tool for Sustainable Recovery of Phenolic Compounds from Pinus pinaster Bark. *ACS Sustain. Chem. Eng.* **2019**, *7*, 8816–8826. [CrossRef]
60. Alves, M.J.; Ferreira, I.C.F.R.; Martins, A.; Pintado, M. Antimicrobial activity of wild mushroom extracts against clinical isolates resistant to different antibiotics. *J. Appl. Microbiol.* **2012**, *113*, 466–475. [CrossRef] [PubMed]
61. Ramos, O.L.; Silva, S.I.; Soares, J.C.; Fernandes, J.C.; Poças, M.F.; Pintado, M.E.; Malcata, F.X. Features and performance of edible films, obtained from whey protein isolate formulated with antimicrobial compounds. *Food Res. Int.* **2012**, *45*, 351–361. [CrossRef]
62. Jiang, W.; Akagi, T.; Suzuki, H.; Takimoto, A.; Nagai, H. A new diatom growth inhibition assay using the XTT colorimetric method. *Comp. Biochem. Physiol. Part C Toxicol. Pharmacol.* **2016**, *185–186*, 13–19. [CrossRef]

molecules

Article

Valonea Tannin: Tyrosinase Inhibition Activity, Structural Elucidation and Insights into the Inhibition Mechanism

Jiaman Liu [1], Yuqing Liu [1], Xiaofeng He [1], Bo Teng [1,2,*] and Jacqui M. McRae [3]

[1] College of Science, Shantou University, Shantou 515063, China; 19jmliu1@stu.edu.cn (J.L.); 15yqliu2@stu.edu.cn (Y.L.); 19xfhe@stu.edu.cn (X.H.)
[2] Guangdong Provincial Key Laboratory of Marine Biotechnology, Shantou University, Shantou 515063, China
[3] School of Chemical Engineering and Advanced Materials, The University of Adelaide, Adelaide, SA 5005, Australia; jacqui.mcrae@adelaide.edu.au
* Correspondence: bteng@stu.edu.cn

Abstract: Valonea tannin is a natural product readily extracted from acorn shells that has been suggested to have potential skin whitening properties. This study investigated the tyrosinase inhibition activity of extracted valonea tannin and the associated structure–function activity. Nuclear magnetic resonance spectroscopy and molecular weight analysis with gel permeation chromatography revealed that valonea tannin could be characterized as a hydrolysable tannin with galloyl, hexahydroxydiphenoyl and open formed-glucose moieties and an average molecular weight of 3042 ± 15 Da. Tyrosinase inhibition assays demonstrated that valonea tannin was 334 times more effective than gallic acid and 3.4 times more effective than tannic acid, which may relate to the larger molecular size. Kinetic studies of the inhibition reactions indicated that valonea tannin provided tyrosinase inhibition through mixed competitive–uncompetitive way. Stern–Volmer fitted fluorescence quenching analysis, isothermal titration calorimetry analysis and in silico molecule docking showed valonea tannin non-selectively bound to the surface of tyrosinase via hydrogen bonds and hydrophobic interactions. Inductively coupled plasma-optical emission spectroscopy and free radical scavenging assays indicated the valonea tannin had copper ion chelating and antioxidant ability, which may also contribute to inhibition activity. These results demonstrated the structure–function activity of valonea tannin as a highly effective natural tyrosinase inhibitor that may have commercial application in dermatological medicines or cosmetic products.

Keywords: valonea tannin; hydrolysable tannin; tyrosinase; enzyme binding; enzyme inhibition; inhibition mechanism

Citation: Liu, J.; Liu, Y.; He, X.; Teng, B.; McRae, J.M. Valonea Tannin: Tyrosinase Inhibition Activity, Structural Elucidation and Insights Into the Inhibition Mechanism. *Molecules* **2021**, *26*, 2747. https://doi.org/10.3390/molecules26092747

Academic Editors: Manuela Pintado, Ezequiel Coscueta and Maria Emilia Brassesco

Received: 14 April 2021
Accepted: 5 May 2021
Published: 7 May 2021

Publisher's Note: MDPI stays neutral with regard to jurisdictional claims in published maps and institutional affiliations.

1. Introduction

Valonea tannin is a hydrolysable tannin [1] with many uses, including as wood adhesives [2], metal chelating depressants in the mining industry [3], leather tanning or potentially as a replacement for plastics [4]. Valonea tannin is abundantly present in the acorn cups of Valonia oak (*Quercus ithaburensis* subsp. *macrolepis* (Kotschy, Hedge and Yaltirik), a common tree throughout Eurasia, and can be readily extracted using hot water [5]. These properties make it a potentially useful resource for many applications that might provide greater economic value.

Skin whitening agents are in large demand around the world with almost 15% of the human population investing in these products and the worldwide market reaching U.S. $23 billion in 2020 [6]. These agents are commonly used to treat a range of dermatological problems caused by the over-production of melanin pigment (known as disordered melanogenesis), such as melanoma, melasma, solar lentigo, freckles, pigmented acne scars and age spots [7,8]. Tyrosinase is a reaction rate-limiting enzyme of melanin synthesis and is also a target enzyme for the treatment of pigmentation related disorders [6,9,10]. Chemicals with tyrosinase inhibition abilities, such as hydroquinone, arbutin, kojic acid, corticosteroids, azelaic acid and hydroxyanisole, are widely applied in medicine and cosmetics as functional ingredients [6,9,10]. However, application of these chemicals is often associated with drawbacks and side effects, including contact dermatitis, irritation, burning and hypochromia [11,12]. Thereby, finding new tyrosinase inhibitors, with effective performance but without safety issues, has become a great concern for both the medical and cosmetic industries.

Tannins are quantitatively abundant plant secondary metabolites that have protein binding capabilities. They are classified as either condensed or hydrolysable tannins depending on their molecular structure. Hydrolysable tannins are composed of glucose, galloyl, hexahydroxydiphenoyl (HHDP) and other derivatives subdivided into ellagitannin and gallotannin structures [1]. In contrast, condensed tannins (also known as proanthocyanidins) are composed of flavan-3-ol subunits which are linked by covalent bonds [13]. Condensed tannins have tyrosinase inhibition abilities which are comparable to commercial inhibitors [14,15]. This tyrosinase inhibition activity may be due to structural similarities between condensed tannin subunits and substrates (tyrosine and L-DOPA) (Figure 1) [15]. The mechanism of action may also be because condensed tannins behave as competitors in both L-DOPA and DOPA quinone formation processes [16].

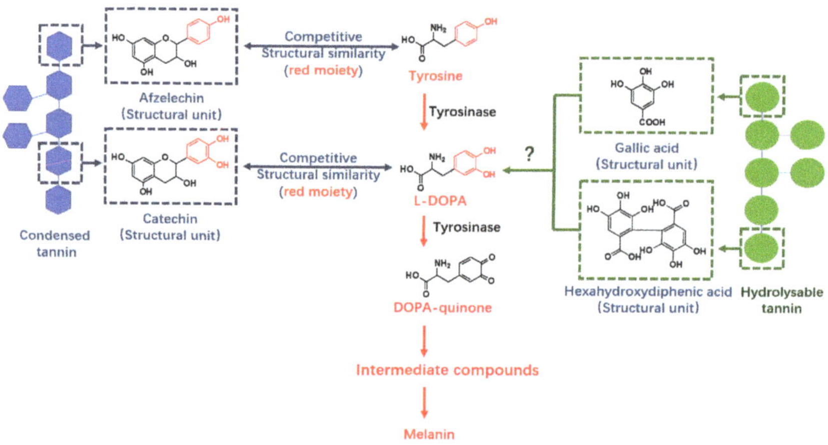

Figure 1. Condensed tannin has structural similarities with tyrosine and L-DOPA and thereby provides tyrosinase inhibition ability, but the structure of hydrolysable tannin differs from condensed tannin and its tyrosinase inhibition mechanism remains unknown.

Hydrolysable tannins may also have tyrosinase inhibition activity due to their protein binding capabilities. Previous studies have shown that the structural subunits of hydrolysable tannins, ellagic acid and gallic acid, have significant tyrosinase inhibition abilities [17]; however, the mechanism of action remains unknown. It is therefore likely that valonea tannin, a hydrolysable tannin, may exhibit tyrosinase inhibition activity, providing a novel, abundant and natural source of these important medical and cosmetic agents.

In order to provide more theoretical information about the tyrosinase inhibition ability from hydrolysable tannin, also to extend the potential application of hydrolysable tannins. In the current study, the structures of valonea tannins extracted from acorn cups were elucidated and the tyrosinase inhibition activity compared to gallic acid and tannic acid. The inhibition mechanism was explored on the aspects of: (1) inhibition kinetic analysis; (2) tyrosinase binding ability; (3) antioxidant activity; and (4) copper ion (coenzyme) chelating ability, using fluorescence quenching accompanied with Stern–Volmer fitting analysis, isothermal titration calorimetry (ITC) analysis, as well as in silico molecular docking (Autodock Vina).

2. Results and Discussion

2.1. Structure Elucidation of Valonea Tannins

Tannin is a mixture of oligomers and polymers with similar structures and physical properties [1]. Composition of structural moieties and molecular weights were considered as crucial structural characteristics because these characters have a substantial impact on the functions of tannins [18]. The valonea tannin was purified with Sephadex LH-20 because it was proven previously that this method can eliminate the simple phenolics, proteins and polysaccharides efficiently, therefore it commonly used for tannin purification [19]. Then, valonea tannin structure was elucidated using ^{13}C NMR (Figure S1) to obtain the composition of the structural moieties. Chemical shifts were compared with those of standard tannic and gallic acids as well as those of similar moieties from previous reports [20] to give the proposed structure of valonea tannin based on the previous 1D and 2D-NMR research of ellagitannins and isolated flavonoid oligomers; resonances from 170 to 55 ppm were assigned to the carbons on carbonyl, phenyl and alkyl moieties, respectively, and shown on Table 1.

Table 1. ^{13}C NMR chemical shifts assignment for the Valonea tannin [1].

Chemical Shift (ppm)	Assignment	Note
	Valonea tannin	
164.41	C7″	Carbonyl C=O, HHDP
163.98	C7	Carbonyl C=O, galloyl
158.50	C7′	Carbonyl C=O, HHDP
144.67	C3	Phenolic –OH, galloyl
143.65	C5, C3′, C5′, C3″, C5″	Phenolic –OH, galloyl and HHDP
135.92	C4, C4′, C4″	Phenolic –OH, galloyl and HHDP
123.88	C1, C1′, C1″	Phenolic carbons, galloyl and HHDP
114.89	C2′	Phenolic C-C bridges, HHDP
113.63	C2″	Phenolic C-C bridges, HHDP
109.61	C2, C6	Phenolic carbons, Galloyl
70.04	C-5‴	Open glucose form
62.71	C-3‴, C-4‴, C-6‴	Open glucose form
57.40	C-2‴	Open glucose form
	Gallic acid	
169.1	C7	Carbonyl C=O
144.9	C3, C5	Phenolic –OH
138.2	C4	Phenolic –OH
120.5	C1	Phenolic carbons
109.0	C2, C6	Phenolic carbons
	Tannic acid	
169.7	C7	Carbonyl C=O
165.4	C7′	Carbonyl C=O
144.9	C3, C5, C3′, C5′	Phenolic –OH
139.2	C4	Phenolic –OH
138.0	C4′	Phenolic –OH
121.3	C1, C1′	Phenolic carbons
109.1	C2, C6, C2′, C6′	Phenolic carbons
92.5	C1″	Closed glucoside
71.9	C2″	Closed glucoside
67.1	C4″	Closed glucoside
66.2	C6″	Closed glucoside
63.0	C5″	Closed glucoside

[1] Spectra were applied in Figure S1, the numbering of carbons was referred to Figure 2.

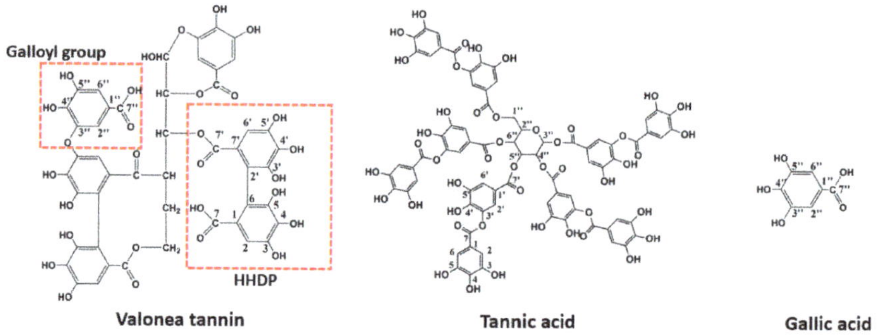

Figure 2. Example of the elucidated structure of valonea tannin, ^{13}C NMR spectra indicate that tannic acid and gallic acid were dominant moieties in valonea tannin.

Resonances from 100 to 170 ppm were used to identify the composition of hydrolysable tannins [20]. The chemical shifts appeared at 164.41, 158.50, 114.89 and 113.63 ppm were attributed to the carbonyl C=O and phenolic C-C bridges on hexahydroxydiphenoyl (HHDP) moieties [20]. These shifts can be considered as direct evidence to prove the existence of HHDP in valonea tannins. It also implied that valonea tannin belonged to the typical ellagitannin in the hydrolysable tannin family. This finding is in agreement with the report from Özgünay et al., who also found the HHDP moieties based on precursor and fragments ions in MALDI TOF MS spectrum of the valonea tannin [21].

Resonances at 163.98, 144.67 and 109.61 ppm were attributed to the carbonyl C=O, phenolic –OH and phenolic carbon on galloyl moieties [20]. Similarly, these shifts can be also found in gallic acid (169.1, 144.9 and 109.0 ppm) and tannic acid (169.7, 144.9 and 109.1 ppm).

The resonances from 70 to 54 ppm, were characteristic resonances from the open glucose forms [22], while the closed glucose form (95 to 80 ppm) was not detected in the current study. Other resonances were attributed to the phenolic carbon linked with or without hydroxyl groups and were attributed to both HHDP and galloyl moieties [20].

These chemical shifts not only confirmed the existence of galloyl and HHDP moieties in valonea tannins, but also indicated the structural characteristics of the valonea tannins were significantly different to condensed tannins (composed of flavanol subunits) [23]. Based on the information obtained from ^{13}C NMR analysis, the typical structure of the valonea tannin was deduced and shown on Figure 2, accompanied with the structure of tannic acid and gallic acid.

Tannin molecular weight is directly related to the biochemical properties and is a crucial parameter for protein interaction and enzyme inhibition [15]. Determining the average molecular weight of a tannin can also assist in elucidating the average size of the tannins [15,18]. The molecular weight distribution of the isolated valonea tannin was analysed using gel permeation chromatography (GPC) (Figure 3).

Based on the GPC results, 10%, 50% and 90% elution mass of the valonea tannins were 817 ± 12, 3042 ± 15 and 17,469 ± 257 g/mol, respectively. This indicated that 80% of the tannin molecules had a mass ranging from 817 to 17,469 g/mol. Furthermore, 50% valonea tannin molecules eluted earlier than tannic acid and 90% of the valonea tannin showed higher molecular weight than gallic acid. This indicated the average molecular weight of valonea tannin should be higher than the other two compounds. Based on the results from Kennedy's report, 50% elution mass of tannin is in good agreement with the average molecular weight determined by other method [19]. Therefore, the 50% elution mass (3042 ± 15 g/mol) of the valonea tannin was taken for the following analysis.

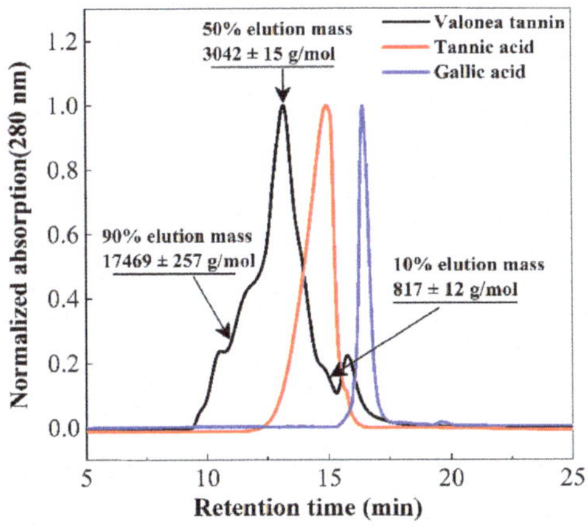

Figure 3. Gel permeation chromatogram (280 nm) showed the elution profile of valonea tannin, tannic acid and gallic acid, average molecular mass of valonea tannin is shown as the 50% elution mass.

2.2. Tyrosinase Inhibition Activity of the Valonea Tannin

Valonea tannin was assessed for tyrosinase inhibition to determine if this natural product may have value for commercial application as a whitening agent in cosmetic products or dermatologic medicines. Tannic acid and gallic acid were also assessed for tyrosinase inhibition for comparison as these compounds were: (1) all composed of pyrogallol moieties (Figure 2) that played critical roles in the enzyme inhibition processes [9]; (2) have molecular weights that differ to each other and are lower than valonea tannins (Figure 3); and (3) accepted as tyrosinase inhibitors with better performances than the commercial tyrosinase inhibitors [9].

The tyrosinase inhibition activities were assessed across a range of concentrations for valonea tannins as well as tannic and gallic acids. Inhibition was measured as an increase in absorbance over time (Figure S2). The IC_{50} values were calculated as the inhibitor concentration inducing a 50% reduction in maximum absorbance and are presented in Table 2.

Table 2. Tyrosinase inhibition ability of the valonea tannin, tannic acid and gallic acid.

Sample	Valonea Tannin	Tannic Acid	Gallic Acid	Hydroquinone
IC_{50} (mM) [1]	1.15 ± 0.37	4.00 ± 0.10	389.56 ± 4.77	1809.38 ± 5.74
IC_{50} (g/L)	3.50 ± 0.11	6.80 ± 0.17	66.23 ± 0.81	199.21 ± 0.63
Fold [2]	-	3.4	339	

[1] IC_{50} was the half maximal inhibitory concentration, data were expressed as mean of 3 replicates ± standard deviation; Kruskal–Wallis test showed $p < 0.05$, indicated significant differences between IC_{50} obtained from different samples. [2] The fold was calculated through: Fold = IC_{50} of tannic acid or gallic acid/IC_{50} of valonea tannin.

The IC_{50} values of the involved inhibitors followed a trend: valonea tannin < tannic acid < gallic acid. This suggested that tyrosinase inhibition capability was, positively related to molecular weight, which is also in agreement with a conclusion obtained by condensed tannins with different molecular weights [15].

Gallic acid has previously been shown to have excellent tyrosinase inhibition ability with an IC_{50} value 100-fold lower than kojic acid [24]. In the current study, the IC_{50} value of valonea tannin was 339-fold lower than gallic acid (Table 2), which indicated that the valonea tannin may has better tyrosinase inhibition ability than the commercial inhibitors. To explore why the valonea tannin showed such an outstanding tyrosinase inhibition ability, the inhibition mechanisms were studied and presented in the following sections.

2.3. The Kinetic Study of Tyrosinase Inhibition Mechanism

To answer why valonea tannin has inhibition ability, kinetic characteristics of the inhibition reactions were studied to better understand the mechanism of action of valonea tannin's tyrosinase inhibition. The kinetic studies were applied and the results are provided in Figure 4.

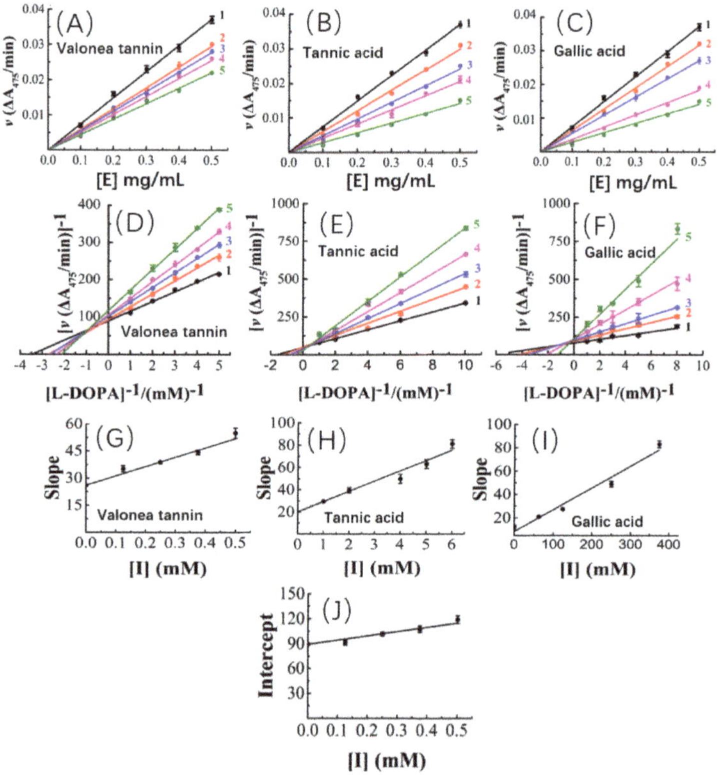

Figure 4. Tyrosinase concentration-reaction rate plots showed the valonea tannin (**A**), tannic acid (**B**) and gallic acid (**C**) reversibly inhibited tyrosinase activities (1 to 5, gradually increase the concentration of inhibitors); The Lineweaver–Burk plots of valonea tannin (**D**), tannic acid (**E**) and gallic acid (**F**); the plot of slope (or intercept) versus inhibitor concentrations of valonea tannin (**G**), tannic acid (**H**) and gallic acid (**I**); the plot of intercept versus valonea tannin concentration for determining inhibition constants K_{IS} (**J**).

The tyrosinase concentration-reaction rate plots showed that, with a constant valonea tannin concentration, the initial velocities linearly fitted with tyrosinase concentrations and the fitted lines all passed through origin of the axis (Figure 4A). Therefore, a deduction can be made as: despite varying valonea tannin concentrations, the reactions will continue proceeding until tyrosinase is eliminated completely from the reaction systems ([E] = 0). Based on the characteristics mentioned above, referencing the previous reports [16,25], the valonea tannin inhibition reaction was classified as a reversible type. As for tannic acid or gallic acid, their tyrosinase concentration-reaction rate plots also showed significant characteristics of the reversible type as well (Figure 4B,C), in agreement with the results obtained by previous research [26].

Based on the adjusted Michaelis–Menten equation [27], the Lineweaver–Burk plots of the reactions were obtained by analysing velocity under different valonea tannin concentrations (Figure 4D). It was evident that the inverse of the velocity (1/v) were all linearly fitted with the inverse of the substrate concentration (R^2 and *p*-values are shown in Table S4) while meanwhile it intersected in the second quadrant of the plot. Based on Waldrop's finding, these graphical characters indicated the maximum reaction speed (Vmax = 1/intersect on y-axis) and Michaelis's constant (K_M) of the reactions were all affected with varying inhibitor concentrations [27]. Thus, the inhibition provided by valonea tannins belonged to a competitive–noncompetitive mixed type. In other words, the valonea tannin presents inhibition through either entering the active centre or binding on the surface (non-active centre) of tyrosinase. This is similar to the inhibition mechanism provided by condensed tannins [16]. In contrast to valonea tannins, the Lineweaver–Burk plots of the tannic acid and gallic acid showed different characteristics (Figure 4E,F). The fitted lines were intersected on y-axis, which indicated these compounds inhibit tyrosinase competitively, which is in agreement with a previous report [28].

Slopes and y-axis intercepts of the fitted Lineweaver–Burk lines were taken for secondary fittings, where they all were linearly fitted with inhibitor concentrations (Figure 4G–I). The slopes and y-axis intercepts of the new lines were calculated to obtain the inhibition constants of inhibitor-tyrosinase (K_I) and inhibitor–tyrosinase–substrate complexes (K_{IS}) (Figure 4J), respectively [27]. K_I is the dissociation constant between inhibitor and enzyme, and K_{IS} is the dissociation constant between inhibitor and the enzyme–substrate complex. The smaller the K_I or K_{IS} values provided by inhibitor, the better competition abilities to substrate were indicated [27].

In the current research, the K_I and K_{IS} values of the involved inhibitors followed a trend as: valonea tannin (K_I) < valonea tannin (K_{IS}) < tannic acid (K_I) < gallic acid (K_I) (Table 3). Indicated in comparison with tannic acid and gallic acid, the valonea tannin showed better competition ability than substrate. Furthermore, the higher the molecular weight of tested inhibitors, the better the competition ability that was shown. This trend is not only in agreement with the results observed from IC_{50}, but also similar to the conclusions that were obtained from condensed tannins [29]. The results obtained from the kinetic study are summarized on Table 3.

Table 3. Effects of Valonea tannin on tyrosinase activities [1].

	Valonea Tannin	Tannic Acid	Gallic Acid
Inhibition mechanism	Reversible	Reversible	Reversible
Inhibition type	Competitive-noncompetitive mixed	Competitive	Competitive
Inhibition constants (mM) [2]	K_{IS} = 1.68 ± 0.18 K_I = 0.51 ± 0.05	K_I = 1.97 ± 0.32	K_I = 56.58 ± 3.65

[1]. Data are expressed as mean of 3 replicates ± standard deviation, the model fitting results, including R^2 and *p* values, are shown in Table S4; Kruskal–Wallis test showed $p < 0.05$, indicated significant differences between K_I and K_{IS} obtained from different samples.
[2]. K_I was the inhibition constant, K_{IS} was the inhibition constant in enzyme–substrate complex, K_{IS} was attribute to the valonea tannin–tyrosianse-L-DOPA complex.

To find more information to explain why valonea tannin showed a better competition ability to substrate, the mechanism of valonea tannin–tyrosinase interactions were investigated using fluorescence and isothermal titration calorimetry (ITC) as described below.

2.4. Fluorescence and Thermodynamic Characters of the Valonea Tannin-Tyrosinase Interaction

The cause of the observed improved competition ability of valonea tannin was explored using fluorescence quenching (Figure 5) and isothermal titration calorimetry (ITC) (Figure 6).

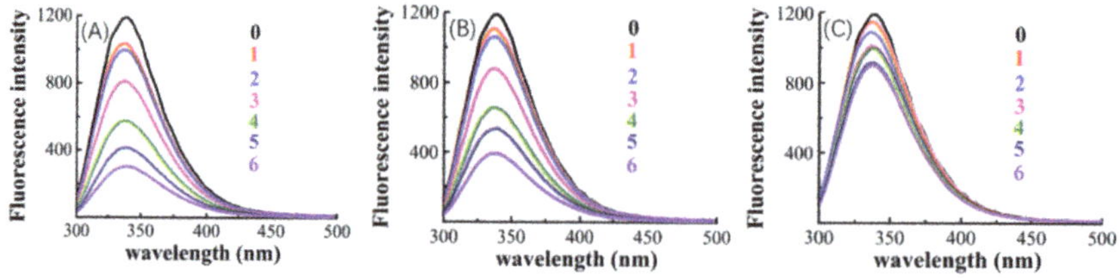

Figure 5. Fluorescence emission spectra of mushroom tyrosinase in solutions combined with valonea tannin (**A**), tannic acid (**B**) and gallic acid (**C**) with a range of concentrations (0 = 0 μM control; 1–6 = 5, 10, 25, 50, 75 and 100 μM, respectively).

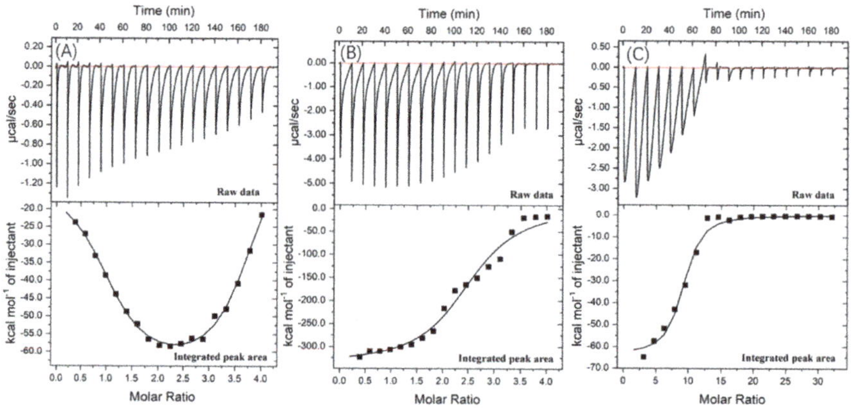

Figure 6. Interaction of (**A**) valonea tannin, (**B**) tannic acid and (**C**) gallic acid with tyrosinase studied by ITC at 25 °C showed the thermogram and binding isotherm, the titration curve from valonea tannin was fitted using two sets of sites binding model while tannic acid and gallic acid were fitted by one set of sites.

The tryptophan inside tyrosinase is the fluorophore that has florescence emitting with a 280 nm excitation wavelength [30]. The tyrosinase–inhibitor binding process may lead to a fluorescence quenching induced by the structural changes of the tyrosinase around tryptophan, consequently, fluorescence quenching was used to provide primary information about inhibitor–tyrosinase interaction [30].

Figure 5 indicated that, the tyrosinase showed a florescence emission at 340 nm, and fluorescence quenching appeared after adding inhibitors (valonea tannin, tannic acid or gallic acid, respectively). Greater concentrations of inhibitors also increased the observed quenching, which implies that the quenching may be induced by the interaction between inhibitors and tyrosinase.

The quenching phenomenon can be classified into dynamic and static types, while the former one does not involve chemical interaction between fluorophore and quencher, only the static quenching refers to binding interactions [29]. Based on the study provided by Van De Weert and Stella [29], the Stern–Volmer quenching rate constant (K_q) and apparent binding constant (K_a) were consequently obtained by linear regressions of the Stem–Volmer equation and double log Stem–Volmer equation (Figure S3) and shown in Table 4.

The fluorescence analysis results showed that K_q presented by valonea tannin–tyrosinase complex was significantly higher than the maximum scatter collision quenching rate constant for dynamic quenching (2×10^{10} L/mol/s) [29], meanwhile their maximum emission wavelengths were observed as constant at 340 nm (Table 4). This phenomenon implied a non-covalent bond was formed between valonea tannin and tyrosinase [29], which is in agreement with the results obtained through hydrolysable tannin–protein interactions [31]. As for the tannic acid and gallic acid, the tyrosinase showed similar quenching characters. This further suggests that these interactions were non-covalent as previously reported for interactions between tyrosinase and tannic acid [32]. The quenching efficiency as well as the K_a of the involved inhibitors followed: valonea tannin > tannic acid > gallic acid. These results preliminary indicated the valonea tannin maybe more strongly bound to tyrosinase than the other two inhibitors [29].

Table 4. Information of valonea tannin–tyrosinase binding process obtained by fluorescence spectra [1].

	Valonea Tannin	Tannic Acid	Gallic Acid
Maximum emitting wavelength (nm)	340	340	340
Quenching efficiency (%) [2]	74.3 ± 0.2	66.5 ± 0.4	24.2 ± 0.4
K_q ($\times 10^{13}$)	2.85 ± 0.00	1.94 ± 0.00	0.28 ± 0.00
Quenching type	Static	Static	Static
Linkage type	Non-covalent	Non-covalent	Non-covalent
K_a (L/mol) × 10^5	2.23 ± 0.00	1.32 ± 0.00	0.73 ± 0.00

[1] Data are expressed as mean of 3 replicates ± standard deviation, the model fitting results, including R^2 and p values, are shown in Table S4; K_q and K_a was Stern–Volmer quenching rate constant and apparent binding constant, respectively; Kruskal–Wallis test showed $p < 0.05$ for quenching efficiency, K_q and K_a, respectively, indicated significant differences between different samples. [2] Calculated as: Quenching efficiency = 100 × emission at 340 nm (with 10 μM inhibitor)/emission at 340 nm (without inhibitor).

Further information about the valonea tannin–tyrosinase binding process was further investigated using ITC analysis. The tyrosinase is a typical globular protein which has complicated stereochemical structures, and tannin is a natural product with polydispersity, therefore tannin–protein interactions are believed to be complex [33]. ITC enables measurement of the binding strength of tannin–protein interactions by quantifying the thermodynamic changes of complex reactions [34]. The ITC curves of valonea tannin, tannic acid and gallic acid titrated into tyrosinase are shown in Figure 6. These negative peaks, shown on the raw data, were attributed to the negative enthalpy from tannin–tyrosinase interaction. In other words, the interaction was exothermic [33]. These titration signals were typical of protein–ligand interactions, resulting from the combination of hydrogen bonding and hydrophobic interactions as has been previously seen in the initial section of ITC curves from earlier reports [34].

The rapid reduction of exothermic peaks ($\Delta H < 0$) upon titration indicated the valonea tannins showed a strong affinity to tyrosinase. With increasing valonea tannin addition, the number of available binding sites decreased, which resulted in the associated enthalpy changes decreasing. The titration curve was found to better fitted with the "two sets of binding sites" model, which implied the two kinds of binding sites were occupied by valonea tannins. On the contrary, the titration curves obtained from tannic acid and gallic acid showed better fitted with the "one set of binding sites" model. The inhibition mechanism results classified valonea tannin as a competitive–noncompetitive mixed type, while classified gallic acid and tannic acid were competitive types (Table 3). This also indicated the valonea tannin bound on both active and non-active sites of the tyrosinase, while gallic acid and tannic acid bound on the active sites further confirming the proposed inhibition mechanism.

Enthalpy change (ΔH) was calculated from the area under each ITC curve, and entropy changes ($T\Delta S$), stoichiometry (molar ratio of inhibitor to site, expressed as "n" [34]) and equilibrium binding constants (K) were calculated from the graphical characters of the curve using the Origin 7.0 software package. Results are shown in Table 5. The negative ΔH values of valonea tannin–tyrosinase interaction indicating that interactions on these two sites were exothermic. The negative ΔH and $T\Delta S$, observed from inhibitor–tyrosinase complexes, demonstrate both hydrophobic interactions and hydrogen bonding were essential for the binding [34].

3. Materials and Methods

3.1. Extraction and Purification of Valonea Tannins

Acorn caps (valonea, 600 g) were collected from Valonia Oak trees (*Quercus macrolepis*) in Wuming (108°16′27.34″ E, 23°8′36.66″ N, South China) in 2019. Valonea were washed with distilled water, air-dried at 25 °C and ground into a fine powder using a plant tissue pulveriser (800A, Jinfeng Machinery Factory, Yongkang, China). Tannins were extracted from the powder (100 g) using 70% acetone: water solution (2.0 L) in a temperature-controlled shake incubator (THZ-C-1, Taicang experimental equipment factory, Hebei, China) with rotation speed at 30 rpm, 25 °C for 8 h. After extraction, the valonea residue was removed via filter paper filtration, and acetone removed by rotary evaporation under reduced pressure (35 °C) to obtain a crude aqueous extract (140 mL). Lipids were removed from the aqueous extract by liquid–liquid separation via the addition of dichloromethane (1:1, *v*:*v*). Finally, the residual organic solvent was removed from the aqueous extract by rotary evaporation (35 °C) and the extract lyophilized to obtain an 80 g crude extract of powdered valonea tannins.

Crude valonea extracts (1 g in 10 mL 50% MeOH solution) was purified with Sephadex LH-20 (Cytiva, Switzerland), as previously described [18]. Briefly, the Sephadex LH-20 column (φ = 3.5 cm, bed volume 260 mL) was equilibrated with 50% MeOH (1% *v*/*v* formic acid) prior to loading the extract. Sugars and simple phenolics were removed using 50% MeOH solution (5 L, flow rate 10 mL/min) and purified tannin eluted with 70% acetone (1.5 L), dried via rotary evaporation (35 °C) followed by lyophilization to obtain the purified valonea tannin powder (240 mg). Purity of the tannin extract was confirmed using MALDI-TOF MS analysis, which demonstrated the absence of polysaccharides, proteins and lipids (Figure S5). The tannin purity was also tested by a methyl cellulose precipitation assay [39] and the results showed tannin concentration was 81.85 ± 0.93% (standard curve is shown in Figure S6).

3.2. Characterization of Valonea Tannin Structure

Characterization of the purified tannin structure was achieved using Nuclear Magnetic Resonance (NMR) spectroscopy and the average molecular mass measured using gel permeation chromatography (GPC). For the NMR analysis, purified valonea tannin powder (30 mg in 750 μL 1:1, *v*/*v*, CD_4O:D_2O) ^{13}C NMR spectra were obtained using an Ascend 400 MHz NMR spectrometer (Bruker, Switzerland) [20], 1.36 s acquisition time, 20.80 μs dwell time, sweep width 24,038 Hz, frequency 100.60 MHz, relaxation delay 2 s, receiver gain 203. Power level for pulse was set as 66 W and 90° high power pulse was set at 10.57 μs. The ^{13}C NMR spectra were also obtained for standard samples of gallic acid (Aladdin Biochemical Technology Co., LTD, Shanghai, China) and tannic acid (Macklin Biochemical Technology Co., LTD, Shanghai, China), and chemical shifts were assigned (Figure S1). GPC was used to obtain the average molecular mass of the purified tannin as previously described [18,19]. Briefly, the tannin sample (20 μL in mobile phase) was injected into 2 x PLgel GPC columns in tandem (300 × 7.5 mm, 5 μm, 100 Å followed by 10^4 Å) on an HPLC (Agilent 1100, Palo Alto, USA) and protected by a guard column of the same material (50 × 7.5 mm, 5 μm) using a mobile phase of dimethylformamide solution (1% acetic acid, 5% water and 0.15 M lithium chloride) at 1 mL/min, column temperature 50 °C, with detection at 280 nm. The average molecular mass of valonea tannin was obtained by comparing elution time with a standard curve made by isolated grape skin tannins with molecular weight 2035 g/mol, 5016 g/mol, 7993 g/mol and 17,674 g/mol, respectively. These grape tannins were prepared in accordance with our previous report [18]. The molecular weights of the grape tannins were determined with phloroglucinolysis in accordance with Kennedy's report [19].

3.3. Tyrosinase Inhibition Activity Assessment and Inhibition Type Analysis

Tyrosinase inhibition activity was evaluated as previously described [17] with L-DOPA as the substrate and the enzyme activity calculated by monitoring DOPA-quinone formation. Tyrosinase (EC 1.14.18.1, from mushroom) and L-DOPA (both from Aladdin Biochemical Technology Co., LTD, Shanghai, China) were dissolved in the sodium phosphate buffer (PBS, 50 mM sodium phosphate in distil water, pH = 6.8, same as below), respectively, to obtain L-DOPA solution (0.5 mM) and tyrosinase solution (0.4 mg/mL). Valonea tannin was prepared in water to a range of concentrations (0 mM, 0.2 mM, 0.4 mM, 0.6 mM, 0.8 mM, 1.0 mM, 1.3 mM and 1.5 mM). For the inhibition assays, valonea tannin solutions (50 µL) at each concentration were mixed with L-DOPA solution (1 mL) and heated in a water bath (30 °C) for 10 min. Tyrosinase solution (50 µL) was then added to each solution, mixed and an aliquot (300 µL) immediately transferred into a 96-well plate. Absorbances were measured at 475 nm at the beginning (0 min) and end (10 min) of the reactions using a microplate reader (HTX, Bior Tek Synergy, Vermont, USA). During the analysis, the temperature was kept constant at 30 °C by using a temperature controller equipped on the microplate reader. Reaction solution without tannin was used as negative control. The inhibition rate was calculated in accordance with Equation (1).

$$\text{Inhibition rate (\%)} = [(A_2 - A_1) - (B_2 - B_1)]/(A_2 - A_1) \times 100\% \quad (1)$$

where A_1 and A_2 represented the absorbance of the control at the beginning and end of the reaction; B_1 and B_2 represented the absorbance of the sample at beginning and end of the reaction, respectively.

The extent of inhibition was calculated using the obtained inhibition rate-inhibitor concentration plots (Figure S2) and expressed as the half maximal inhibitory concentration (IC_{50}).

Inhibition type was analysed based on the kinetic characteristics of the DOPA-quinone formation processes, which is concomitant with valonea tannins. The reversible–irreversible assessment was provided as previously described [16] with modifications. The initial velocities of DOPA-quinone formations were tested in reaction solutions with a constant L-DOPA concentration (0.5 mM), and different valonea tannin (0 mM, 0.125 mM, 0.25 mM, 0.375 mM and 0.5 mM) and tyrosinase concentrations (0.1 mg/mL, 0.2 mg/mL, 0.3 mg/mL, 0.4 mg/mL and 0.5 mg/mL). The absorbances of reaction solutions were recorded at 475 nm and the initial velocities were taken and expressed as ΔA/min.

The competitive–noncompetitive assessment was performed in accordance with a previous report [25]. Briefly, initial velocities (recorded at 475 nm) of reactions were taken from solutions prepared using a constant tyrosinase concentration (0.1 mg/mL), and different valonea tannin (0 mM, 0.125 mM, 0.25 mM, 0.375 mM and 0.5 mM) and L-DOPA concentrations (1 mM, 0.5 mM, 0.33 mM, 0.25 mM and 0.20 mM). The Michaelis–Menten equation was then adjusted to fit Lineweaver–Burk plot as Equation (2) [27]:

$$(1/V) = (K_M/V_{max}) \times (1/[S]) + (1/V_{max}) \quad (2)$$

where [S] was the substrate concentration, V was the corresponding initial velocity. K_M and V_{max} were the Michaelis's constant and maximum reaction speed, respectively, which were obtained from the vertical and horizontal intercept from the Lineweaver–Burk plot.

The inhibition constant (K_I) and inhibition constant in enzyme−substrate complex (K_{IS}) were obtained by secondary plot of inhibitor concentration with slope and intercept of the Lineweaver–Burk plot as Equations (3) and (4) [27]:

$$\text{Slope} = K_M/V_{max}\ (1 + [I]/K_I) \quad (3)$$

$$\text{Intercept} = 1/V_{max}\ (1 + [I]/K_{IS}) \quad (4)$$

The tyrosinase inhibition activity of tannic acid and gallic acid were also measured along with the inhibition type analysis for comparison with valonea tannin.

3.4. Fluorescence Quenching Analysis of Tyrosinase in the Presence of Valonea Tannins

Fluorescence quenching was used to measure the change in tyrosinase activity in the presence of valonea tannins. The analysis was conducted in accordance with a published protocol [30], and the method was adjusted as follows: The fluorescence intensities were recorded by a fluorescence spectrometer (Hitachi F-7000, Kyoto, Japan) equipped with a xenon lamp source. A 280 nm excitation wavelength, 240 nm/min scan speed, 5 nm emission slit width and 2.5 nm excitation slit width were chosen for the analysis. In this assay, valonea tannins with different concentrations (0, 5, 10, 25, 50, 75 and 100 μM) were mixed with tyrosinase solution at a fixed concentration (0.4 mg/mL). The prepared samples were then transferred for fluorescence analysis, while intensities were recorded from 300 to 500 nm. In the current study, emissions at 340 nm were chosen for quenching related calculations since the maximum fluorescence appeared at this wavelength. Quenching was calculated using the Stern–Volmer [29] Equation (5):

$$F_0/F = 1 + K_{sv}\,[Q] = 1 + K_q\tau_0\,[Q] \quad (5)$$

where F and F_0 were the fluorescence intensities (at 340 nm) with or without valonea tannin. [Q] was the tannin concentration and K_{sv} is the Stern–Volmer quenching constant. The K_{sv} was the slope calculated from the linear regression plot of F_0/F against [Q]. K_q was the biomolecular quenching rate constant. τ_0 was the average lifetime of the fluorophore in the absence of the tannin ($\tau_0 = 10^{-8}$ s) [29].

As for the static quenching, the F_0, F and [Q] were taken for apparent binding constant calculation based on double log Stern–Volmer [29] Equation (6):

$$\text{Log}\,[(F_0 - F)/F] = \log K_a + n\log\,[Q] \quad (6)$$

where K_a was the apparent binding constant that was obtained by the slope of log $[(F_0 - F)/F]$ versus log [Q] plots. Double log plotting needed restrict requisite otherwise false n values could be obtained [29]. Therefore, in the current study, the n value was not taken into consideration. The fluorescence quenching of tyrosinase was analysed with gallic acid and tannic acid for comparison, the Stern–Volmer plot (F_0/F against [Q]) and double log Stern–Volmer plot (log $[(F_0 - F)/F]$ versus log [Q]) are provided in Supplementary Information S3.

3.5. Isothermal Titration Calorimetry Analysis

Isothermal Titration Calorimetry (ITC) analysis was conducted as previously described [34] but modified as follows: Tyrosinase solution (50 μM PBS buffer 50 mM, pH = 6.8) added to the calorimeter cell (200 μL) on an ITC instrument (ITC 200, MicroCal, Northampton, MA, USA) instrument and equilibrated at 25 °C for 30 min under a rotation speed of 1000 r/min. To calculate the thermal character during binding, valonea tannin solution (1 mM in PBS buffer 50 mM, pH = 6.8) was injected into the sample cell at 25 °C. The injection volume, number of injections and the spacing time between injections were set as 2 μL, 18 and 600 s, respectively. During analysis, the valonea tannin solution was injected into the reference cell which contained a buffer solution; the thermal behaviour was recorded and used as a negative control.

Peak interpretation, stoichiometry (n), the binding constant (K), change in enthalpy (ΔH) and change in entropy (ΔS) were calculated on Origin 7.0 software package (Northampton, MA, USA), while changes in Gibbs free energy were calculated based on a previously report [40]. A "two set of identical sites" model was applied for valonea tannins. The tannic acid–tyrosinase reaction and gallic acid-tyrosinase reaction were also analysed as described above, and the data were fitted with the "one set of identical sites". All standard deviations shown in the ITC results were based on the accuracy of the curve fit to the data and obtained by Origin software package.

3.6. In Silico Molecule Docking

In silico molecule docking models were performed using AutoDock Vinasoftware (DeLano Scientific LLC, Palo Alto, CA, USA) to provide an understanding of the mechanism of tyrosinase inhibition by valonea tannins. The crystallographic structure of the tyrosinase–tropolone complex from *Agaricus bisporus* (PDB: 2Y9X, obtained by X-ray diffusion) was obtained from RCSB Protein Databank. Then, tropolone and water molecules on tyrosinase were removed, followed by a polar hydrogen atoms addition, a missing atoms correction and a Gasteiger charges assignment [41]. Gallic acid, HHDP and tannic acid were constructed with Chem Draw 17.0 (Cambridge, UK) and taken as ligands. Structures of the ligands were subsequentially geometry optimized with an MM2 force field to minimize the energy and obtain the preferential conformations [41].

In accordance with Heitz's report [41], the blind docking simulations were performed within a grid box which was centred on the geometric central position of the tropolone on tyrosinase complex (x, y, z: −7.392, −24.898, −39.626). The grid box was set to have a 30 × 30 × 30 size with grid spacing of 0.1 nm. The energy range was set at 10, while exhaustiveness was 20. Other operator weights for crossover, mutation and elitism were default parameters. A semi-flexible docking was chosen for all processes, while the sophisticated gradient optimization method in its local optimization was employed to search for the preferential conformations, and the empirical scoring function was used for docking score calculation. The predicted binding energies were collected from 20 ligand–receptor complexes with lowest energies, from which the complex with the lowest binding energy was preserved and pose of the ligand, hydrogen bond and related amino acid residuals were analysed with PyMOL 2.2 (Schrödinger Inc, New York, NY, USA).

3.7. Antioxidant Activity Analysis

The antioxidant activity of the valonea tannin was evaluated using scavenging assays with 3-ethylbenzthiazolin-6-sulfonic acid ($ABTS^{\cdot+}$) as well as 2,2-diphenyl-1-picrylhydrazyl ($DPPH^{\cdot}$). The $DPPH^{\cdot}$ scavenging ability was analysed in accordance with Brand-Williams's report [42]. $DPPH^{\cdot}$ solutions (1.5 mL, 25 mg/Lin methanol) were mixed with 50µL valonea tannin solutions at different concentrations (0 mM, 0.02 mM, 0.03 mM, 0.04 mM, 0.05 mM and 0.06 mM in methanol), and placed in the dark for 30 min. After the reaction time, 300 µL of the mixed solution was transferred into a 96 well microplate reader, and the absorbance measured at 517 nm and corrected using methanol. This method was used in place of the valonea tannins as a negative control.

The $ABTS^{\cdot+}$ scavenging ability was measured as previously described [43]. Briefly, 7 mM ABTS and 2.45 mM potassium persulfate were mixed and placed in the dark for 16 h (at 25 °C) to form a stable oxidation state of $ABTS^{\cdot+}$ radical cations. $ABTS^{\cdot+}$ solution was then diluted with 80% ethanol to reach an absorbance of 0.700 ± 0.05 at wavelength 734 nm. Valonea tannin solutions (0.1 mL) with different concentrations (0 mM, 0.02 mM, 0.03 mM, 0.04 mM, 0.05 mM, 0.06 mM and 0.07 mM, in 80% ethanol) were mixed with 3.9 mL $ABTS^{\cdot+}$ solution. After 6 min reaction (at 25 °C), 300 µL of the mixed solution was transferred into a 96 well microplate and the absorbance at 734 nm was recorded, corrected with methanol. Methanol (0.1 mL) was used to replace the valonea tannins as a negative control.

The $DPPH^{\cdot}$ and $ABTS^{\cdot+}$ free radical scavenging rates were calculated by using Equation (7):

$$\text{Free radical elimination rate (\%)} = [(A_1 - A_2)/A_1] \times 100 \quad (7)$$

where A_1 was the absorbance of control, and A_2 was the absorbance of sample.

The IC_{50} was obtained using the linear regression of a valonea tannin concentration to free radical elimination rate plot. Following the method described above, IC_{50} values were also obtained for gallic acid, tannic acid and ascorbic acid for comparison.

3.8. Inductively Coupled Plasma-Optical Emission Spectroscopy (ICP-OES)

The Cu^{2+}-binding capacity of valonea tannin was measured using ICP-OES as previously described [44]. Valonea tannin solution (1 mg/mL, in distilled water) and copper sulphate solution (0.5 M, in distilled water) were mixed (1:1, v:v), incubated for 3 h (37 °C) and centrifuged for 30 min (4390 g). The supernatant was subjected to an ICP-OES analysis (Agilent 5110, Palo Alto, CA, USA) and the remnant Cu^{2+} concentration measured. A 12.0 L/min plasma flow rate, 1.5 L/min auxiliary flow rate and 0.70 L/min nebulizer flow rates were provided during the analysis, while sample uptake and instrument stabilization delay were 15 s. Gallic acid and tannic acid were also used to react with copper sulphate and remnant Cu^{2+} were quantified following above method for comparison.

3.9. Statistical Analysis

Non-parametric test (Kruskal–Wallis) was performed to determine differences between the experimental samples (triplicated) using Minitab 18 (Minitab Inc., State College, PA, USA). The model fitting results are all applied in Supplementary Information S4.

4. Conclusions

Elucidation of the isolated valonea tannin revealed the tannin was composed of gallic acid, HHDP and glucose with an average molecular weight of 3042 ± 15 Da. Tyrosinase inhibition ability provided by valonea tannin was 334 times greater than that of gallic acid and 3.4 times higher than tannic acid. Mechanism studies suggested that the observed enzyme inhibition was driven by a combination of hydrogen bonding and hydrophobic interactions which involved non-selective binding to the surface of tyrosinase, resulting in a competitive and non-competitive mixed inhibition mechanism. Antioxidant activity and copper ion chelating ability of valonea tannin also contributed to tyrosinase inhibition. These results suggest that extracted valonea tannins may be of value as a whitening agent in cosmetics or dermatological medicines as an effective replacement for commercial tyrosinase inhibitors.

Supplementary Materials: The following are available online, Figure S1. C NMR spectrum of valonea tannin, tannic acid and gallic acid; Figure S2. The tyrosinase inhibition activity evaluated by conducting the catalysis reaction with different inhibitor concentrations. Figure S3. Stern–Volmer plot (A, F_0/F against [Q]) and double log Stern–Volmer plot (B, log $[(F_0 - F)/F]$ versus log [Q]) obtained from calculating valonea tannin, tannic acid and gallic acid induced fluorescence quenching of the tyrosinase. Table S4. The model fitting results from liner regression of the kinetic and fluorescence quenching analysis. Figure S5. The MALDI TOF MS spectra of the purified valonea tannin. Figure S6. The standard curve for methyl cellulose precipitation assay.

Author Contributions: Conceptualization, B.T.; methodology, B.T. and J.L.; formal analysis, J.L.; investigation, Y.L.; resources, X.H.; writing—original draft preparation, J.L. and B.T.; writing—review and editing, J.M.M.; visualization, X.H.; supervision, B.T.; project administration, B.T.; funding acquisition, B.T. All authors have read and agreed to the published version of the manuscript.

Funding: This research is funded by the Natural Science Foundation of Guangdong Province, China (2020A1515011357), the Department of Education of Guangdong Province, China (2018KQNCX081) and the Research and Development Start-up Foundation (NTF18031).

Institutional Review Board Statement: Not applicable.

Informed Consent Statement: Not applicable.

Data Availability Statement: The data presented in this study are available on reasonable request from the corresponding author.

Acknowledgments: Stella Kassara and Keren Bindon from the Australian Wine Research Institute are thanked for their assistance in providing standard samples for GPC analysis. Yizhao Ren from Northwest Agriculture and Forest University is thanked for his help and support during GPC analysis. Thanks also to Jinwei Zhang from Sichuan University for support in the valonea sample collection.

Conflicts of Interest: The authors declare no conflict of interest.

Sample Availability: Samples of the compounds are not available from the authors.

References

1. Romani, A.; Ieri, F.; Turchetti, B.; Mulinacci, N.; Vincieri, F.; Buzzini, P. Analysis of condensed and hydrolysable tannins from commercial plant extracts. *J. Pharm. Biomed. Anal.* **2006**, *41*, 415–420. [CrossRef] [PubMed]
2. Li, C.; Wang, W.; Mu, Y.; Zhang, J.; Zhang, S.; Li, J.; Zhang, W. Structural Properties and Copolycondensation Mechanism of Valonea Tannin-Modified Phenol-formaldehyde Resin. *J. Polym. Environ.* **2018**, *26*, 1297–1309. [CrossRef]
3. Ren, Z.; Yu, F.; Gao, H.; Chen, Z.; Peng, Y.; Liu, L. Selective Separation of Fluorite, Barite and Calcite with Valonea Extract and Sodium Fluosilicate as Depressants. *Minerals* **2017**, *7*, 24. [CrossRef]
4. Wang, H.; Wang, L. Developing a bio-based packaging film from soya by-products incorporated with valonea tannin. *J. Clean. Prod.* **2017**, *143*, 624–633. [CrossRef]
5. Onem, E.; Gulumser, G.; Akay, S.; Yesil-Celiktas, O. Optimization of tannin isolation from acorn and application in leather processing. *Ind. Crops Prod.* **2014**, *53*, 16–22. [CrossRef]
6. Pillaiyar, T.; Manickam, M.; Namasivayam, V. Skin whitening agents: Medicinal chemistry perspective of tyrosinase inhibitors. *J. Enzym. Inhib. Med. Chem.* **2017**, *32*, 403–425. [CrossRef] [PubMed]
7. Iozumi, K.; Hoganson, G.E.; Pennella, R.; Everett, M.A.; Fuller, B.B. Role of Tyrosinase as the Determinant of Pigmentation in Cultured Human Melanocytes. *J. Investig. Dermatol.* **1993**, *100*, 806–811. [CrossRef] [PubMed]
8. Lianza, M.; Mandrone, M.; Chiocchio, I.; Tomasi, P.; Marincich, L.; Poli, F. Screening of ninety herbal products of commercial interest as potential ingredients for phytocosmetics. *J. Enzym. Inhib. Med. Chem.* **2020**, *35*, 1287–1291. [CrossRef] [PubMed]
9. Khan, M.T.H. Molecular design of tyrosinase inhibitors: A critical review of promising novel inhibitors from synthetic origins. *Pure Appl. Chem.* **2007**, *79*, 2277–2295. [CrossRef]
10. Zolghadri, S.; Bahrami, A.; Khan, M.T.H.; Munoz-Munoz, J.; Garcia-Molina, F.; Garcia-Canovas, F.; Saboury, A.A. A comprehensive review on tyrosinase inhibitors. *J. Enzym. Inhib. Med. Chem.* **2019**, *34*, 279–309. [CrossRef]
11. Weiss, R.M.; Del Fabbro, E.; Kolisang, P. Cosmetic ochronosis caused by bleaching creams containing 2% hydroquinone. *South Afr. Med J.* **1990**, *77*, 373.
12. A Ibrahim, Z.; Gheida, S.F.; El Maghraby, G.M.; E Farag, Z. Evaluation of the efficacy and safety of combinations of hydroquinone, glycolic acid, and hyaluronic acid in the treatment of melasma. *J. Cosmet. Dermatol.* **2015**, *14*, 113–123. [CrossRef] [PubMed]
13. Schofield, P.; Mbugua, D.; Pell, A. Analysis of condensed tannins: A review. *Anim. Feed. Sci. Technol.* **2001**, *91*, 21–40. [CrossRef]
14. Sarikurkcu, C.; Kocak, M.S.; Tepe, B.; Uren, M.C. An alternative antioxidative and enzyme inhibitory agent from Turkey: Robinia pseudoacacia L. *Ind. Crops Prod.* **2015**, *78*, 110–115. [CrossRef]
15. Chai, W.-M.; Shi, Y.; Feng, H.-L.; Xu, L.; Xiang, Z.-H.; Gao, Y.-S.; Chen, Q.-X. Structure Characterization and Anti-tyrosinase Mechanism of Polymeric Proanthocyanidins Fractionated from Kiwifruit Pericarp. *J. Agric. Food Chem.* **2014**, *62*, 6382–6389. [CrossRef] [PubMed]
16. Chai, W.-M.; Wang, R.; Wei, M.-K.; Zou, Z.-R.; Deng, R.-G.; Liu, W.-S.; Peng, Y.-Y. Proanthocyanidins Extracted from Rhododendron pulchrum Leaves as Source of Tyrosinase Inhibitors: Structure, Activity, and Mechanism. *PLoS ONE* **2015**, *10*, e0145483. [CrossRef]
17. Manosroi, A.; Jantrawut, P.; Akazawa, H.; Akihisa, T.; Manosroi, J. Biological activities of phenolic compounds isolated from galls of Terminalia chebula Retz. (Combretaceae). *Nat. Prod. Res.* **2010**, *24*, 1915–1926. [CrossRef] [PubMed]
18. Teng, B.; Hayasaka, Y.; Smith, P.A.; Bindon, K.A. Effect of Grape Seed and Skin Tannin Molecular Mass and Composition on the Rate of Reaction with Anthocyanin and Subsequent Formation of Polymeric Pigments in the Presence of Acetaldehyde. *J. Agric. Food Chem.* **2019**, *67*, 8938–8949. [CrossRef]
19. A Kennedy, J.; Taylor, A.W. Analysis of proanthocyanidins by high-performance gel permeation chromatography. *J. Chromatogr. A* **2003**, *995*, 99–107. [CrossRef]
20. Xu, M.; Zha, Z.-J.; Qin, X.-L.; Zhang, X.-L.; Yang, C.-R.; Zhang, Y.-J. Phenolic Antioxidants from the Whole Plant of Phyllanthus urinaria. *Chem. Biodivers.* **2007**, *4*, 2246–2252. [CrossRef]
21. Ozgunay, H.; Sari, O.; Tozan, M. Molecular investigation of valonea tannin. *J. Am. Leather Chem. Assoc.* **2007**, *102*, 154–157.
22. Abdalla, S.; Pizzi, A.; Bahabri, F.; Ganash, A. Analysis of Valonia Oak (Quercus aegylops) Acorn Tannin and Wood Adhesives Application. *BioResources* **2015**, *10*, 7165–7177. [CrossRef]
23. Chung, K.-T.; Wong, T.Y.; Wei, C.-I.; Huang, Y.-W.; Lin, Y. Tannins and Human Health: A Review. *Crit. Rev. Food Sci. Nutr.* **1998**, *38*, 421–464. [CrossRef]
24. Chang, T.-S. An Updated Review of Tyrosinase Inhibitors. *Int. J. Mol. Sci.* **2009**, *10*, 2440–2475. [CrossRef] [PubMed]
25. Waldrop, G.L. A qualitative approach to enzyme inhibition. *Biochem. Mol. Biol. Educ.* **2009**, *37*, 11–15. [CrossRef]
26. Si, C.-L.; Wu, L.; Shen, T.; Huang, X.-F.; Du, Z.-G.; Ren, X.-D.; Luo, X.-G.; Hu, W.-C. Recovery of Low-molecular Weight Galloyltannins from Agricultural Residue of Juglans sigillata Dode Seed Husks and their Tyrosinase Inhibitory Effect. *BioResources* **2014**, *9*, 2226–2236. [CrossRef]
27. Whiteley, C.G. Enzyme kinetics: Partial and complete uncompetitive inhibition. *Biochem. Educ.* **2000**, *28*, 144–147. [CrossRef]
28. Kubo, I.; Kinst-Hori, I.; Nihei, K.-I.; Soria, F.; Takasaki, M.; Calderón, J.S.; Céspedes, C.L. Tyrosinase Inhibitors from Galls of Rhus javanica Leaves and Their Effects on Insects. *Zeitschrift für Naturforschung C* **2003**, *58*, 719–725. [CrossRef]
29. Van de Weert, M.; Stella, L. Fluorescence quenching and ligand binding: A critical discussion of a popular methodology. *J. Mol. Struct.* **2011**, *998*, 144–150. [CrossRef]

30. Chai, W.-M.; Wei, M.-K.; Wang, R.; Deng, R.-G.; Zou, Z.-R.; Peng, Y.-Y. Avocado Proanthocyanidins as a Source of Tyrosinase Inhibitors: Structure Characterization, Inhibitory Activity, and Mechanism. *J. Agric. Food Chem.* **2015**, *63*, 7381–7387. [CrossRef] [PubMed]
31. Deaville, E.R.; Green, R.J.; Mueller-Harvey, I.; Willoughby, A.I.; Frazier, R.A. Hydrolyzable Tannin Structures Influence Relative Globular and Random Coil Protein Binding Strengths. *J. Agric. Food Chem.* **2007**, *55*, 4554–4561. [CrossRef]
32. Patil, S.; Sistla, S.; Jadhav, J. Interaction of small molecules with human tyrosinase: A surface plasmon resonance and molecular docking study. *Int. J. Biol. Macromol.* **2016**, *92*, 1123–1129. [CrossRef] [PubMed]
33. McRae, J.M.; Kennedy, J.A. Wine and Grape Tannin Interactions with Salivary Proteins and Their Impact on Astringency: A Review of Current Research. *Molecules* **2011**, *16*, 2348–2364. [CrossRef]
34. McRae, J.M.; Falconer, R.J.; Kennedy, J.A. Thermodynamics of Grape and Wine Tannin Interaction with Polyproline: Impiications for Red Wine Astringency. *J. Agric. Food Chem.* **2010**, *58*, 12510–12518. [CrossRef]
35. Haslam, E. Natural Polyphenols (Vegetable Tannins) as Drugs: Possible Modes of Action. *J. Nat. Prod.* **1996**, *59*, 205–215. [CrossRef]
36. Noh, H.; Lee, S.J.; Jo, H.-J.; Choi, H.W.; Hong, S.; Kong, K.-H. Histidine residues at the copper-binding site in human tyrosinase are essential for its catalytic activities. *J. Enzym. Inhib. Med. Chem.* **2020**, *35*, 726–732. [CrossRef] [PubMed]
37. Ramsden, C.A.; Riley, P.A. Mechanistic studies of tyrosinase suicide inactivation. *Arkivoc* **2010**, *2010*, 260–274. [CrossRef]
38. Ros, J.R.; Rodríguez-López, J.N.; García-Cánovas, F. Effect of l-ascorbic acid on the monophenolase activity of tyrosinase. *Biochem. J.* **1993**, *295*, 309–312. [CrossRef]
39. Teng, B.; Petrie, P.; Smith, P.; Bindon, K. Comparison of water addition and early-harvest strategies to decrease alcohol concentration in Vitis vinifera cv. Shiraz wine: Impact on wine phenolics, tannin composition and colour properties. *Aust. J. Grape Wine Res.* **2020**, *26*, 158–171. [CrossRef]
40. Jelesarov, I.; Bosshard, H.R. Isothermal titration calorimetry and differential scanning calorimetry as complementary tools to investigate the energetics of biomolecular recognition. *J. Mol. Recognit.* **1999**, *12*, 3–18. [CrossRef]
41. Heitz, M.P.; Rupp, J.W. Determining mushroom tyrosinase inhibition by imidazolium ionic liquids: A spectroscopic and molecular docking study. *Int. J. Biol. Macromol.* **2018**, *107*, 1971–1981. [CrossRef] [PubMed]
42. Brand-Williams, W.; Cuvelier, M.E.; Berset, C. Use of a free radical method to evaluate antioxidant activity. *LWT Food Sci. Technol.* **1995**, *28*, 25–30. [CrossRef]
43. Re, R.; Pellegrini, N.; Proteggente, A.; Pannala, A.; Yang, M.; Rice-Evans, C. Antioxidant activity applying an improved ABTS radical cation decolorization assay. *Free Radic. Biol. Med.* **1999**, *26*, 1231–1237. [CrossRef]
44. Zeng, X.; Du, Z.; Ding, X.; Jiang, W. Characterization of the direct interaction between apple condensed tannins and cholesterol in vitro. *Food Chem.* **2020**, *309*, 125762. [CrossRef] [PubMed]

molecules MDPI

Article

Phytochemical Profiling of *Lavandula coronopifolia* Poir. Aerial Parts Extract and Its Larvicidal, Antibacterial, and Antibiofilm Activity against *Pseudomonas aeruginosa*

Mahmoud Emam [1,2], Doaa R. Abdel-Haleem [3], Maha M. Salem [2], Lina Jamil M. Abdel-Hafez [4], Rasha R. Abdel Latif [2], Shaimaa Mahmoud Farag [3], Mansour Sobeh [5,*] and Mohamed A. El Raey [2,*]

1 College of Pharmaceutical Science & Collaborative Innovation Center of Yangtze River Delta Region Green Pharmaceuticals, Zhejiang University of Technology, Hangzhou 310014, China; me.hegazy@nrc.sci.eg
2 Phytochemistry and Plant Systematics Department, National Research Centre, Dokki, Giza 12622, Egypt; maha_abdelmejed@yahoo.com (M.M.S.); rasharefaat38@yahoo.com (R.R.A.L.)
3 Department of entomology, Faculty of Science, Ain Shams University, Abbasia, Cairo 11566, Egypt; doaaramadan@sci.asu.edu.eg (D.R.A.-H.); shaimaa.mahmoudfarag@sci.asu.edu.eg (S.M.F.)
4 Department of Microbiology and Immunology, Faculty of Pharmacy, October 6 University, 6th October City, Giza 12585, Egypt; linajamil@o6u.edu.eg
5 AgroBioSciences Research Division, Mohammed VI Polytechnic University, Lot 660–Hay MoulayRachid, 43150 Ben-Guerir, Morocco
* Correspondence: mansour.sobeh@um6p.ma (M.S.); ma.aziz@nrc.sci.eg (M.A.E.R.)

Citation: Emam, M.; Abdel-Haleem, D.R.; Salem, M.M.; Abdel-Hafez, L.J.M.; Latif, R.R.A.; Farag, S.M.; Sobeh, M.; El Raey, M.A. Phytochemical Profiling of *Lavandula coronopifolia* Poir. Aerial Parts Extract and Its Larvicidal, Antibacterial, and Antibiofilm Activity against *Pseudomonas aeruginosa*. *Molecules* **2021**, *26*, 1710. https://doi.org/10.3390/molecules26061710

Academic Editors: Manuela Pintado, Ezequiel Coscueta and María Emilia Brassesco

Received: 20 February 2021
Accepted: 15 March 2021
Published: 19 March 2021

Abstract: Infections associated with the emergence of multidrug resistance and mosquito-borne diseases have resulted in serious crises associated with high mortality and left behind a huge socioeconomic burden. The chemical investigation of *Lavandula coronopifolia* aerial parts extract using HPLC–MS/MS led to the tentative identification of 46 compounds belonging to phenolic acids, flavonoids and their glycosides, and biflavonoids. The extract displayed larvicidal activity against *Culex pipiens* larvae (LC_{50} = 29.08 µg/mL at 72 h). It significantly inhibited cytochrome P-450 monooxygenase (CYP450), acetylcholinesterase (AChE), and carboxylesterase (CarE) enzymes with the comparable pattern to the control group, which could explain the mode of larvae toxification. The extract also inhibited the biofilm formation of *Pseudomonas aeruginosa* by 17–38% at different Minimum Inhibitory Concentrations (MICs) (0.5–0.125 mg/mL) while the activity was doubled when combined with ciprofloxacin (ratio = 1:1 *v:v*). In conclusion, the wild plant, *L. coronopifolia*, can be considered a promising natural source against resistant bacteria and infectious carriers.

Keywords: *Lavandula coronopifolia*; *Culex pipiens*; larvicidal; antibiofilm formation; LC-MS/MS; molecular networking

1. Introduction

Mosquito vector-borne diseases are considered a global problem, which highlights the necessity for new prospects and cost-effective agents for vector control. Around 100 species of mosquitoes transmit viral and bacterial disorders such as malaria, lymphatic filariasis, dengue, and yellow fever, affecting several millions of people worldwide. In 2017, WHO recorded the highest mortality and morbidity due to mosquito-borne disorders that affect human health and economic society. Therefore, the development of novel mosquito repellents and antibacterial agents to overcome the microbial resistance threat is highly demanded. This could also help to avoid disrupting the ecological balance [1–3]. Plant secondary metabolites could furnish safe, efficacious, and multi-mechanistic candidates that might be useful as insecticidal and antibacterial agents.

The genus *Lavandula* (commonly known as lavender) comprises 45 species that are mainly distributed in subtropical and tropical regions [4,5]. Plants of the genus have been used in folk medicine since ancient times to treat pain, headache, migraine, and antiepileptic, antidiuretic, antirheumatic, and carminative agents. They become famous for their multiple uses in different pharmaceuticals, aroma, and food products [6,7]. The phytochemistry of the genus is centered on mono- and sesquiterpenoids, together with traces of alkaloids, and phenolic structures [7].

Lavandula coronopifolia Poir. (Arabic: Khozama) is a shrublike perennial, growing in the rocky environment and desert plains mainly distributed in subtropical and tropical regions [8]. The first attention to *L. coronopifolia* traces back to 1999, when El-Garf et al. isolated and identified numerous hydroxyl flavones such as hypolaetin, isoscutellarien, and luteolin from its dried aerial parts [4]. Then, several studies reported the presence of polyhydroxyoleanolic acids, polyhydroxyursolic acids and their glycosides, caffeic acid, rosmarinic acid, rutin, quercetin, and hesperidin [9,10].

L. coronopifolia showed a plethora of substantial biological activities. These include antioxidant [11], antimicrobial [12], α-glucosidase inhibitory [10], and hepatoprotective [13] activities. These activities were attributed to the presence of flavonoids, especially flavones and their glucuronides, in addition to triterpenes [9,10]. Moreover, *L. coronopifolia* essential oils possessed substantial antibacterial activity against the Gram-negative bacteria and methicillin-resistant *Staphylococcus aureus* bacteria [14].

In this work, we comprehensively characterized the phytoconstituents of the aerial parts of *L. coronopifolia* extract by HPLC–MS/MS and confirmed the existence of possible skeletons quantitatively using ^{1}H-NMR and molecular networking. We also evaluated the insecticidal activities against *Culex pipiens* larvae and the antibiofilm formation activity against isolates of *Pseudomonas aeruginosa*. We explored several biochemical parameters to investigate the mechanism of the insecticidal activities.

2. Material and Methods

2.1. Plant Material, Extraction, and Preliminary Qualitative Analysis

Lavandula coronopifolia Poir. was collected from the Western Desert of Egypt in March 2018. A voucher sample was placed at the international herbarium of the National Research Centre (CAIRC) (S.N: 1023). The plant aerial parts (250 g) were crushed into small pieces and then extracted by dipping into 70% MeOH:H_2O (*v:v*) at ambient temperature for one week. The solution was filtered (Whatman no. 1), then concentrated till dryness using Rotavapor® (Heizbad Hei-VAP, Heidolph, Germany), yielding 28.45 g, and stored at 4 °C for further experiments. The *L. coronopifolia* extract was screened for its phyto-constituents as flavonoids (Shinoda's test), phenolics ($FeCl_3$ test), ellagitannins ($NaNO_2$ assay), and gallotannins (KIO_3 test) [15,16]. The proton NMR (Jeol ECA-500 MHz, Japan) experiment using DMSO-*d6* was done for the total extract.

2.2. HPLC-MS/MS

HPLC-PDA-MS^n mass spectra was performed through a ThermoFinnigan (Thermo Electron Corporation, Austin, TX, USA) LC system coupled with a mass spectrometer (LCQ-Duo ion trap) having an ESI source (ThermoQuest, Thermo Scientific, Waltham, MA, USA) [15]. The injection process, flow rate, elution solvents, resolution, and negative MS operating parameters were described previously [17]. In brief, a Zorbax Eclipse XDB-C18, rapid resolution, 150 × 4.6 mm, 3.5 µm column was used (Agilent, Santa Clara, CA, USA). A gradient consisting of water and acetonitrile (ACN), each having 0.1% formic acid, was applied, and ACN was increased from 5% to 30% within 60 min and then to 90% within the next 30 min at a flow rate of 1 mL/min and a 1:1 split before the ESI source [17].

2.3. Molecular Networking Workflow Description

The mgf formatting mass file was uploaded to the online platform of GNPS (http://gnps.ucsd.edu) (accessed on 27 June 2020). Then the data were filtered as described previously and the visualization of molecular networking (MNW) workflow was carried out using Cytoscape 3.6.1 software [15,18].

2.4. Larvicidal Assay

2.4.1. Insects

A laboratory susceptible strain of *Culex pipiens* was obtained from the Research and Training Center on Vectors of Diseases (RTC), Ain Shams University. It was colonized in the entomology department insectary at 27 ± 2 °C, 75 ± 5% relative humidity (RH), and a 14 h/10 h light/dark photoperiod following standard procedures [19]. The larvae were reared in enamel dishes containing 2000 mL of distilled water. Newly hatched larvae were fed on Tetra-Min, Germany. Adults were reared in (30 × 30 × 30 cm) wooden cages and provided with 10% sucrose solution daily, as well as a pigeon for female blood feeding.

2.4.2. Bioassay

The larvicidal activity was evaluated against the third larval instar of *C. pipiens* under the same controlled laboratory conditions. The bioassay was assessed using the standard method described in [20]. The extract was dissolved in water to prepare the stock solution. Batches of 25 of 3rd instar larvae of *C. pipiens* were transferred by a plastic dropper to small disposable test cups and treated with different concentrations of the extract (10, 25, 50, 100, 150, and 200 μg/mL prepared in distilled water) and control with distilled water only, in a triplicate manner. Mortality was recorded after 24, 48, and 72 h post treatment.

2.4.3. Preparation of Samples for Biochemical Assay

The 3rd larval instar of *C. pipiens* was treated by LC_{50} values, and then the insects were prepared as described by Amin et al. [21]. The whole bodies of larvae were homogenized in distilled water (50 mg/1 mL). The homogenates were centrifuged at 8000 r.p.m. for 15 min at 4 °C. The supernatants were used for biochemical analyses. Acetylcholinesterase (AChE) and carboxylesterase assays were measured according to the method described by Simpson et al. [22], using acetylcholine bromide (AchBr) and methyl n butyrate (MeB) as substrates, respectively. Alpha esterase (α-esterase) activity was determined according to Van Asperen [23] using α-naphthyl acetate as substrate. Glutathione S-transferase (GST) catalyzes the conjugation of reduced glutathione (GSH) with 1-chloro 2,4-dinitrobenzene (CDNB) via the –SH group of glutathione. The conjugate S-(2,4-dinitro-phenyl)-L-glutathione could be detected as described by the method of Habig et al. [24]. Cytochrome P-450 monooxygenase activity was determined using *p*-nitroanisole o-demthylation according to the method of Hansen and Hodgson [25] with slight modifications.

2.5. Microbiological Assay

2.5.1. Sample Collection and Identification of Isolated Bacteria

Clinical isolates of *Pseudomonas aeruginosa* were collected from burn wounds, otitis media, and urine as previously described [26] and were kept for scientific research. The isolates were grown on Tryptic soya agar (TSA) (Difco™, Strasbourg, France) for 24 h at 37 °C, then one single colony of each isolate was inoculated into 2 mL Tryptic soya broth (TSB) (Difco™, France) with overnight incubation at 37 °C. The samples were cultivated on Cetrimide agar media, after which the isolated species were identified by morphology (pale yellow colonies on MacConkey and green exopigment on Cetrimide agar), Gram staining (Gram-negative bacilli), and biochemical reactions (oxidase-positive). The Microbact™ Gram-negative system was implemented in compliance with the manufacturer's protocol (Oxoid, Hampshire, UK). A standard strain of *P. aeruginosa* (ATCC 12924) was kindly provided by NAMRU as a frozen cultural broth containing 40% glycerol.

2.5.2. The Antimicrobial Susceptibility Testing

The antimicrobial susceptibility testing was done using cup agar diffusion method [27]. The bacterial cultures were adjusted to an optical density (OD) of 0.5 at 600 nm, TSA plates were covered with 100 μL of each bacterial isolate, and 5 mm pores were filled with 100 μL of extract dissolved in sterile water at 2.5 mg/mL. The plates were incubated for 24 h at 37 °C. The zone of inhibitions was measured in mm.

2.5.3. Minimum Inhibitory Concentration (MIC)

MIC is used as the gold standard method for detecting the sensitivity of the organisms to antimicrobial agents [28]. The final concentrations of the extract ranging from 2.5 to 0.0195 mg/mL were prepared and then added to test tubes containing 1 mL of sterile TSB media. The bacterial suspensions with OD 0.5 at 600 nm were diluted 1:100 (≈106 CFU/mL), and then 50 µL of inoculums were added to each tube. The tubes were incubated at 37 °C ± 2 °C for 24 h. A tube containing TSB broth without extract was taken as control. The MIC was defined as the lowest concentration of the tested extract that restricted the visible growth of tested strains compared to the blank [29].

2.5.4. Minimal Bactericidal Concentration (MBC)

The minimal bactericidal concentration (MBC) was determined by the Petri dish sowing method [30]. This procedure was dependent on the procedures for the determination of MIC. After the incubation period for the determination of the MIC, an aliquot of 0.1 mL was taken from each of the test tubes that were not showing growth, and then was inoculated into a TSA agar plate. The plates were then incubated at a temperature of 37 ± 2 °C for 24 h. After this period, the presence of bacterial colonies was observed in each plate. The MBC was defined as the lowest concentration of the plant extract that was able to prevent microbial growth in a culture medium (formation of bacterial colonies). The bioassays were performed in duplicate with three repetitions for each bacterial isolate.

2.5.5. Biofilm Formation Assay and Quantification

The biofilms were assayed as described in [31,32] using sterile 96-well microtiter plates, each well containing 180 µL TSB broth and 20 µL of bacterial suspension with OD 0.5 at 600 nm. After 24 h incubation at appropriate conditions, all the planktonic cells were removed, and the biofilms were gently washed twice with phosphate buffer saline (PBS) to remove any free-floating bacteria. The biofilm cells formed in each well were stained with 200 µL crystal violet (0.1% *w*/*v*) and incubated at room temperature (28 °C) for 10 min. The stain was removed and washed with distilled water for 30–60 s. After 5 min of air drying, the biofilms were solubilized by 200 µL of 98% ethanol, then the optical densities of stained adherent biofilms were measured at 620 nm using a microplate reader. The evaluation of biofilm production was categorized according to the criteria of Stepanović et al. as follows: OD ≤ ODc: not a biofilm producer (non-adherent); ODc < OD ≤ 2ODc: a weak biofilm producer (weakly adherent); 2ODc < OD ≤ 4ODc: a moderate biofilm producer (moderately adherent); 4ODc < OD: a strong biofilm producer (strongly adherent). ODc and OD were defined as the mean OD of the blank wells and wells with biofilm, respectively [33].

2.5.6. Biofilm Inhibition Assay

The ability of the extract to inhibit the biofilms of the clinical isolates of *P. aeruginosa* was evaluated according to Stepanović et al. [34] with some modifications. Microbial biofilms were developed in a round-bottom 96-well microtiter plate. Each clinical isolate was inoculated into each well of the 96-well microtiter plate. The extract was added to each well at 1/2, 1/4, and 1/8 MICs and incubated for 24 h at 37 °C. After the incubation period, non-adherent cells were detached by dipping each sample three times in sterile PBS. The samples were fixed for one hour, and the biofilms were stained with 0.1% solution of crystal violet in H_2O. After staining, the samples were washed with distilled H_2O (DW). The measurable biofilm production was achieved by adding 125 µL of 30% acetic acid to de-stain the samples. Afterwards, the OD at 620 nm was detected using the microplate reader. The percentage (%) of inhibition formula is as follows:

$$\%\text{Inhibition} = \frac{\text{Abs control} - \text{Abs sample}}{\text{Abs control}} \times 100$$

2.5.7. Combination of the Extract with Ciprofloxacin

The ability of ciprofloxacin/extract (1:1) to inhibit the biofilm of the strong biofilm isolate of *P. aeruginosa* was evaluated according to Stepanović et al. [34] with some modifications. Microbial biofilms were developed in a round-bottom 96-well microtiter plate. The clinical isolate C4 was inoculated into each well of the 96-well microtiter plate, and ciprofloxacin/extract (1:1) was added to each well at 1/2, 1/4, and 1/8 MICs and incubated for 24 h at 37 °C. After incubation (24 h), non-adherent cells were detached by dipping each sample three times in sterile PBS. The samples were fixed for 1 h, and the biofilms were stained with 0.1% solution of crystal violet in H_2O. After staining, the samples were washed with DW (distilled H_2O). The quantitative analysis of biofilm production was achieved by adding 125 µL of 30% acetic acid to de-stain the samples. Afterwards, the OD at 620 nm was measured using the microplate reader. The percentage (%) of inhibition formula is as follows:

$$\%\text{Inhibition} = \frac{\text{Abs control} - \text{Abs sample}}{\text{Abs control}} \times 100$$

2.6. *Statistical Analysis*

The biofilm formation inhibiting activities of different concentrations of the extract were compared by two-way ANOVA (Bonferroni post hoc tests). $p < 0.05$ was used to detect the significance of differences. The obtained larvicidal data were analyzed using a statistics package (LDP-line) for goodness of fit (chi square test) and to detect LC_{50} and LC_{95} values with corresponding 95% confidence limits (CL), slope, correlation coefficient and standard error. The results of biochemical determinations were investigated by one-way analysis of variance (ANOVA) using Costat statistical software (Cohort software, Berkeley). When the ANOVA statistics were significant ($p < 0.01$), the means were compared by the Duncan's multiple range test [35]. GraphPad Prism 5.0 software (GraphPad Prism Software Inc., San Diego, CA, USA) was used to draw most of the figures.

3. Results and Discussion

3.1. *Phytochemical Screening and LC-MS/MS Profile of* L. coronopifolia

The preliminary phytochemical screening of *L. coronopifolia* extract revealed the presence of dihydroxy phenolics and flavonoids and/or their glycosides, as well as the absence of ellagic and gallotannins moieties. The mass spectrometry analysis (full scan and product ion scan mode) provided structural information of 46 metabolites, including organic and phenolic acids, flavonoids and their glycosides, and bioflavonoids (Table 1 and Figure 1).

3.2. *Molecular Networking (MNW) of* L. coronopifolia *Aerial Parts' Metabolite Perception*

The symmetrical chemical entities were facilitated to be visualized through the molecular networking between the identical fragments (*m/z*) of definite metabolites. Through the network, each node was characterized with the parent mass $[M–H]^-$ and the contiguous arrows (edges) connected between the similar nodes. The network was built for the negative ionization (–ve) mode using the GNPS 2 platform (Figure 2). The (–ve) network involved 148 nodes, 71 self-looped (individual) nodes, and 92 connected components. The designed networks facilitated the visual examination of the different compound families and analogues and assisted in isomer differentiation.

In the negative network, five clusters, A, B, C, D, and E, were mentioned and annotated as apigenin derivatives, which were grouped as methylated flavones, *O*-glycosidic flavones, *C*-glycosidic flavones, and/or biflavones. In addition, the other self-looped nodes asterisked within the network were denoted as phenolic-*O*-glycosides, N-acetylamino acid, flavan-3-ol, phenolic acids, and biflavones.

Organic acids: Quinic acid was determined with an [M-H]$^-$ ion at *m/z* 191 fragmented to *m/z* 111, and 173, whereas malic acid was characterized with an [M-H]$^-$ ion at *m/z* 133 and fragmented to *m/z* 115, 89, and 71. **Phenolic acids:** These structures were tentatively identified as eucomic acid, syringic acid-4-*O*- hexoside, sinapic acid 3-*O*-glucoside, and dihydrosinapic acid hexoside. They showed negative molecular ions at *m/z* 239, 359, 375, 385, and 387, respectively. These structures were tentatively identified depending on the main fragments that are shown in Table 1. **Flavonoids and their glycosides:** Most of the aglycones were found to be apigenin derivatives and their *O*- and/or *C*-linkage of mono and/or diglycosides. The MS of *C*-glycosides was characterized through the main fragmentations by the loss of different masses of 60, 90, 120, and 240 Daltons [36]. A molecular ion [M-H]$^-$ at *m/z* 289 and yielding a main fragment at *m/z* 245 was identified as catechin. **Biflavonoids:** Several signals that gave parent ions [M–H]$^-$ of *m/z* 551, 555, 565, 579, 581, and 609 were fragmented into specific fragments that characterized bioflavonoid derivatives [37]. Their identification, retention times, molecular weights, and fragmentation pattern are shown in Table 1.

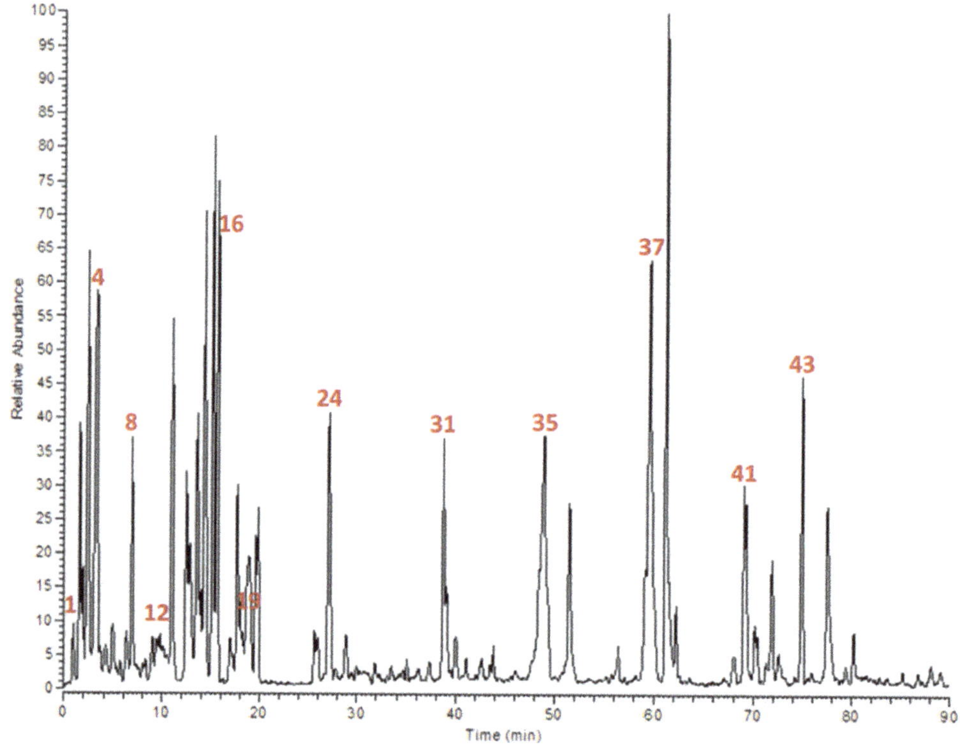

Figure 1. Base peak chromatogram of *L. coronopifolia* Poir. aerial parts' extract.

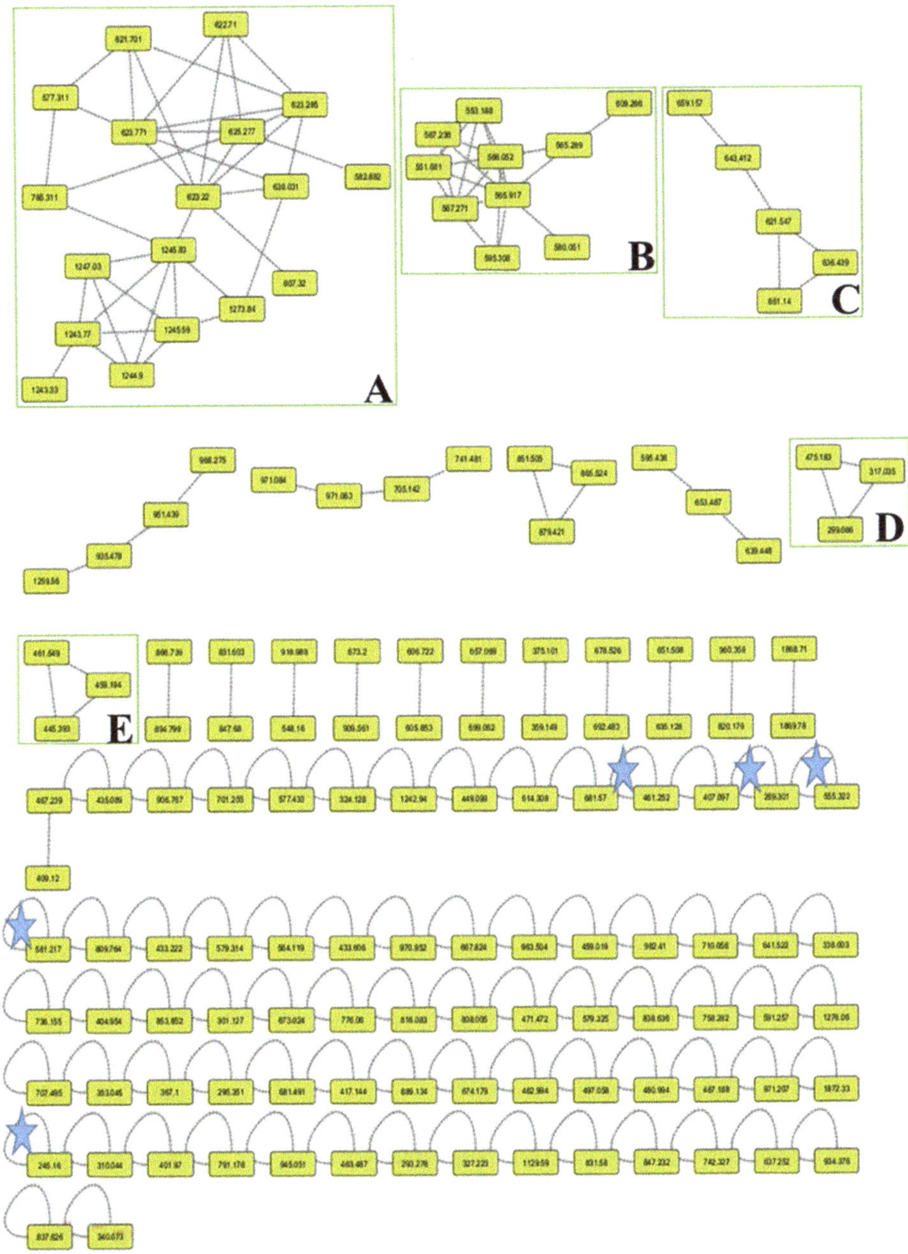

Figure 2. Complete molecular networking (MNW) generated using MS/MS data in (–ve) negative mode from *L. coronopifolia* aerial parts' extract. Nodes are labeled with parent mass.

Table 1. Chemical composition of *L. coronopifolia* Poir. aerial parts' extract using LC-MS/MS.

No.	t_R (min)	[M–H]$^-$	MS^2 (*m/z*)	Tentatively Identified Compound	Ref.
1	1.68	191	111, 173	Quinic acid [a]	[38]
2	1.75	133	89, 71, 115	Malic acid [a]	[39]
3	3.51	211	137, 179	Caffeic acid derivative [b]	
4	4.03	239	149, 179, 221	Eucomic acid [b]	
5	4.34	359	315, 197, 153	Syringic acid 4-*O*-hexoside [b]	[38]
6	4.61	237	115, 121, 137	2-(4-hydroxybenzyl)-malic acid [b]	
7	5.03	375	125, 169, 213	Sinapoyl trihydroxybenzoic acid [b]	
8	6.31	475	197, 359	Rosmarinic acid malate [b]	
9	7.78	461	153, 315	Protocatechuic acid rhamnosyl glucoside [b]	
10	8.95	245	203, 186, 115	*N*-Acetyltryptophan [c]	
11	9.18	289	245	Catechin [d]	[40]
12	9.29	299	115, 133, 183	Methyl trihydroxybenzoic acid malate [b]	
13	13.54	385	223, 179	Sinapic acid 3-*O*-glucoside [e]	
14	13.69	387	433, 225, 179	Dihydrosinapic acid hexoside [e]	
15	14.51	387	433, 225, 179	Dihydrosinapic acid hexoside [e]	
16	15.61	387	433, 225, 179	Dihydrosinapic acid hexoside [e]	
17	17.57	389	227	Resveratrol glucoside [f]	
18	17.72	461	285	Isoscutellarein-8-*O*-glucuronide [#,g]	
19	18.25	359	197, 179, 161, 135	Rosmarinic acid [b]	
20	18.77	389	227	Resveratrol glucoside [f]	
21	19.88	461	285	Luteolin-7-*O*-hexouronide [g]	[6]
22	20.67	593	503, 473, 383, 353	Apigenin di-C-hexoside [g]	[6]
23	25.79	445	269, 175	Apigenin-7-*O*-hexournide I [@,g]	[6]
24	27.05	445	269, 175	Apigenin-7-*O*-hexouronide II [g]	[6]
25	28.62	623	477, 461, 315	Hypolaetin 4′-*O*-methyl ether-*O*-hexoside-*O*-rhamnoside I [g]	
26	33.57	623	477, 461, 315	Hypolaetin-4′-*O*-methyl ether-*O*-hexoside-*O*-rhamnoside II [g]	
27	32.30	623	179, 315, 461	Isorhamnetin *O*-hexoside-*O*-rhamnoside [g]	
28	34.98	607	461, 315, 299	Hypolaetin di-*O*-rhamnoside [g]	
29	35.08	447	285	Luteolin-7-*O*-glucoside [#,g]	[6]
30	37.23	637	491, 461, 315	Hypolaetin 4′-*O*-methyl ether-8-glucuronide-*O*-rhamnoside [g]	
31	38.80	459	283, 268	Acacetin-*O*-hexouronic acid [g]	[41]
32	41.04	577	269	Apigenin-*O*-caffeoyl rhamnoside [g]	
33	41.14	461	299, 283	Methoxy leteolin-7-*O*-hexoside [g]	
34	43.90	651	505, 475, 329	Tricin-*O*-feruloyl rhamnoside [g]	
35	48.62	621	459, 313	Crismaritin-*O*-caffeoyl rhamnoside [g]	
36	51.41	327	171, 229, 327	Unknown	
37	59.10	313	298, 284, 269	Luteolin-7,3′-dimethyl ether [g]	[6]
38	61.18	269	269, 151, 149	Apigenin [g]	[6]
39	62.63	555	403, 429, 327, 299	Binaringenin methyl ether [h]	[37]
40	69.07	551	457, 431, 389	Methoxy amentoflavone [h]	
41	69.42	553	458, 432, 390	Dihydrobilobetin [h]	
42	71.92	283	268, 133	Acacetin [g]	[41]
43	75.90	565	471, 389	Dimethoxy amentoflavone [h]	
44	78.77	609	577, 551, 489, 269	Penta methoxy dihydro biapigenin [h]	
45	81.37	579	533, 485, 389, 268	Kayaflavone [h]	
46	81.81	581	579, 535, 487	Dihydrokayaflavone [h]	

[#] Isolated before from the same species [4]. [@] Confirmed by UV λ_{max}. [a] Organic acid, [b] phenolic acid, [c] amino acid derivative, [d] flavan-3-ol, [e] phenylpropanoic acid, [f] stilbenoid, [g] flavone, and [h] biflavonoid.

3.3. ^{1}H-NMR Analysis of *L. coronopifolia* Extract

The extract was dissolved in deuterated dimethyl sulfoxide (DMSO-*d6*) and introduced into the proton NMR (500 MHz) experiment. The resonated peaks at different chemical shifts explained the kind of protons present in the chemical structures and represented the classes of skeletons in the solution. The 2-phenyl chromen-4-one of the flavone structure bearing substituents at positions 5 and 7 of A-ring was observed as the main skeleton. Through the ^{1}HNMR chart, the downfield protons that appeared as a doublet at δ 7.93 ppm were characterized for H-2′ and 6′ of B-ring, while the protons resonating at δ 6.92 ppm were assigned for H-3′ and 5′ of B-ring, with the *ortho*-coupling constant (J = 9.0 Hz) suggesting that the B-ring bearing the hydroxyl group was at position 4′, while the appearance of a singlet proton at δ 6.83 ppm was characterized for H-3 of the flavone structure. In addition, the downfield shift observed at δ 6.42 and 6.81 ppm for H-6 and H-8 of the A-ring of the flavone structure with *meta*-coupling (2 Hz) or broad singlet was predicted due to the substitution at position 7. The previous interpretation confirmed the presence of the apigenin derivative in the solution with substitution at position 7. In addition, the ABX system of B-ring was observed along with the chart through the coupling constant of resonated protons as a doublet of doublet (*ortho*- and *meta*-coupled), doublet (*meta*-coupled), and doublet (*ortho*-coupled) at δ 7.46, 7.43, and 6.92 ppm, respectively, were assigned, respectively, for H-6′, H-2′, and H-5′ of the B-ring, suggesting that the B-ring was substituted with substituents at positions 3′ and 4′. The rest of H-3, H-6, and H-8 proton signals resonated around the mentioned chemical shifts, which confirmed the presence of a luteolin derivative with 7-*O*-substituents. In addition, the resonated singlet signal at δ of 3.83 ppm was assigned to 4′-O-methylated flavone (i.e., the presence of an acacetin or methyl apigenin derivative in the solution). Moreover, the series of signals between δ 3.14 and 3.51 ppm were attributable to a sugar moiety, and the doublet signal with a coupling constant of 7.2 Hz was distinctive for the anomeric proton of sugar with O-β-D-linkage. Depending on the previously published data and MS assignment, the major attached sugar for the flavone structure is glucuronide moiety.

3.4. Insecticidal Activity

Plant extracts are considered a new ecofriendly and efficient alternative means for controlling mosquitoes. The larvicidal activity of the extract was evaluated against the 3rd instar larvae of *C. pipiens*. The fiducial limits were calculated for LC_{25}, LC_{50}, and LC_{95} at $p < 0.05$ (Table 2). The extract exhibited considerable larvicidal activity against *C. pipiens* larvae where the LC_{50} values after 24, 48, and 72 h of exposure were 52.74, 34.07, and 29.076 µg/mL, respectively. The essential oil from the Egyptian plants exhibited insecticidal activity against the 4th larval instar of *C. pipiens* [42]. Similar activities were reported from essential oils of other *Lavandula* species, among them *L. stoechas* and *L. dentata* [43,44].

Table 2. Larvicidal activity of *L. coronopifolia* extract against the 3rd instar larvae of *Culex pipiens* 24, 48, and 72 h post treatment.

Extract (µg/mL)	24 h Post Treatment	48 h Post Treatment	72 h Post Treatment
LC_{25} (* F.l. at 95%)	20.054 (15.20–24.85)	11.274 (7.55–15.05)	8.668 (5.29–12.21)
LC_{50}(* F.l. at 95%)	52.74 (44.304–62.95)	34.07 (27.43–41.41)	29.076 (22.58–36.062)
LC_{95}(* F.l. at 95%)	557.50 (374.11–975.18)	505.44 (326.011–953.94)	556.28 (341.87–1151.47)
Slope ± SE	1.61 ± 0.147	1.40 ± 0.141	1.28 ± 0.138
$\chi^{2\,a}$	6.4544	1.9897	0.5856
Probability (P)	0.0915	0.5745	0.8997

* Fiducial limits; [a] chi square.

3.5. Biochemical Activity

Insects release several detoxifying enzymes such as esterases, oxidases, and reductases to face and detoxify many invader pesticides [45]. To get an insight into the mechanisms involved, we explored the activities of five different enzymes in the 3rd larval instar of *C. pipiens*. The extract, at a concentration of LC_{50} and exposure time of 72 h, inhibited cytochrome P-450 monooxygenase, acetylcholinesterase, and carboxylesterase by −9.92%, −19.41%, and −25.47%, respectively, compared to the control group (Figure 3), while the treated larvae showed elevation of α-esterases and glutathione S-transferase contents by 15.63% and 37.02% compared to the untreated larvae. This could highlight the significant role of glutathione S-transferase and α-esterases in the detoxification mechanism of the extract. Our results come in agreement with those of the Huang et al. study, which reported an increase in the intracellular glutathione content when the larvae were treated with a polyphenolic-rich extract [46–48].

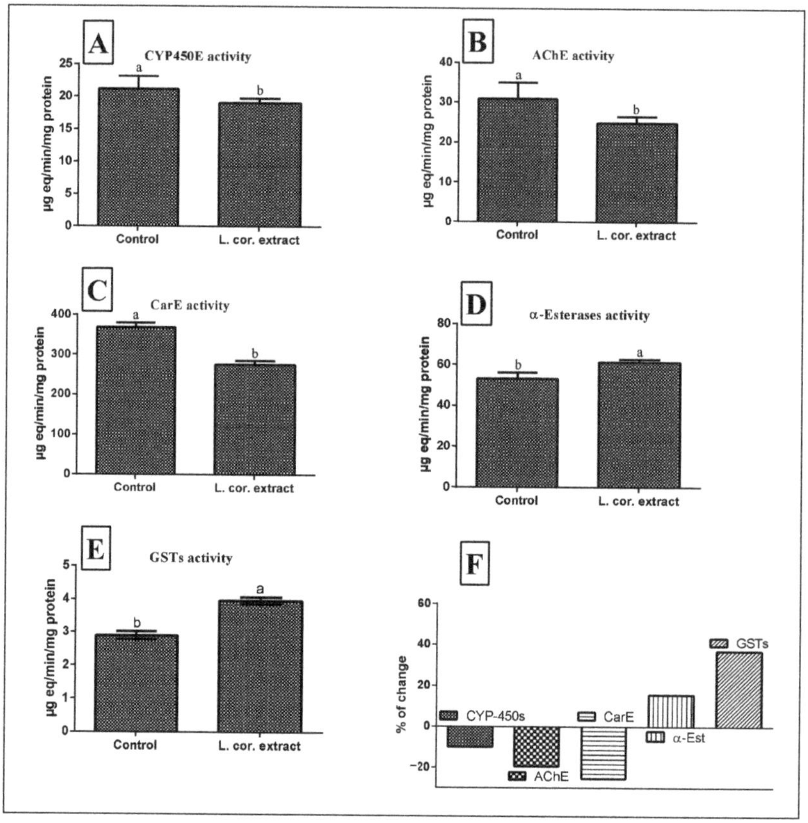

Figure 3. Effects of *L. coronopifolia* extract (L. cor.) on the enzymatic activities of cytochrome P-450 monooxygenase (CYP-450s) (**A**), carboxylesterase (CarE) (**B**), acetylcholinesterase (AchE) (**C**), α-esterases (**D**), and glutathione S-transferase (GSTs) (**E**) in the 3rd larval instar of *C. Pipiens*. (**F**); % of changes. Data were represented as mean ± SE. Lowercase letters above the bars indicate significant differences between different treatment groups (Duncan's multiple range test, $p < 0.01$). Error bars indicate 95% confidence intervals.

As the time of exposure increased, the toxicity of the extract to the 3rd instar larvae increased, followed by a substantial decrease in carboxylesterase (CarE), acetylcholinesterase (AChE), and cytochrome P-450 monooxygenase (CYP450) levels. This suggests the temporary response and neurotoxic effects of the extract. Our results come in agreement with those of the Gershenzon et al. and Salunke et al. studies, which reported the inhibitory effect of natural secondary metabolites for the detoxification enzymes mentioned previously [49–53].

3.6. Microbiological Studies

3.6.1. Antimicrobial Susceptibility, MIC, and MBC

The antimicrobial susceptibility screening was performed by the cup diffusion method. The results showed that all tested *P. aeruginosa* were susceptible to the extract compared to the anti-pseudomonal activity of ciprofloxacin (2 mg/mL; Table 3). The extract displayed moderate activities against all the clinical isolates of *P. aeruginosa* and *P. aeruginosa* ATCC (12924) in the microdilution test (Table 3).

Table 3. Antimicrobial susceptibility, minimum inhibitory concentration (MIC), and minimal bactericidal concentration (MBC) of the *L. coronopifolia* extract and ciprofloxacin against the tested *P. aeruginosa*.

P. aeruginosa Isolates	Zone of Inhibition (mm)		Extract	
	Extract	Ciprofloxacin	MIC	MBC
			mg/mL	
C1	20	1748	0.3125	1.25
C2	20	40	0.3125	1.25
C3	20	50	0.3125	1.25
C4	28	35	0.3125	1.25
ATCC (12924)	26	22	0.1562	1.25

3.6.2. Biofilm Formation and Quantification Assay

Twenty clinical isolates of *Pseudomonas* were examined for their ability to form a biofilm using the microtiter plate method. Out of the tested isolates, four bacteria showed biofilm formation, where the clinical isolate C4 exhibited the strongest biofilm formation, while C1, C2, C3, and *P. aeruginosa* ATCC (12924) showed moderate biofilm formation (Figure 4).

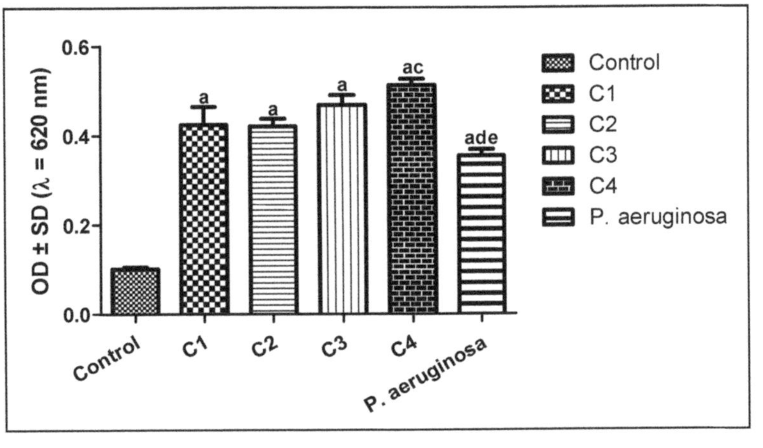

Figure 4. Results of biofilm formation of *P. aeruginosa*. Isolates were classified as negative (OD ≤ ODc), weak (ODc ≤ OD ≤ 2ODc), moderate (2ODc < OD ≤ 4ODc), and strong biofilm production (4ODc < OD). OD = optical density. [a] Significant compared to control at $p < 0.05$. [b] Significant compared to C1 at $p < 0.05$. [c] Significant compared to C2 at $p < 0.05$. [d] Significant compared to C3 at $p < 0.05$. [e] Significant compared to C4 at $p < 0.05$.

3.6.3. Biofilm Inhibition Assay

The biofilm inhibition activity of the extract was evaluated using three concentrations (1/2, 1/4, 1/8 MICs). The extract inhibited the biofilm formation of *P. aeruginosa* in a dose-dependent manner and ranged from 17 to 38% (Figure 5). These considerable antibiofilm properties are similar to that of polyphenol-rich extract from the bark of *Salix tetrasperma* and the leaves of *Annona glabra* and *Gynura procumbens* [54–56]. A recent study by Koely et al. described the antibiofilm activities of *Enydra fluctuans* against *P. aeruginosa*. They also attributed these activities to the presence of several bioactive phenolic compounds, among them kaempferol, quercetin, and luteolin and their glycosides [57]. In addition, different crude extracts from *Arbutus unedo* having high phenolic contents demonstrated comparable antibacterial activity [58].

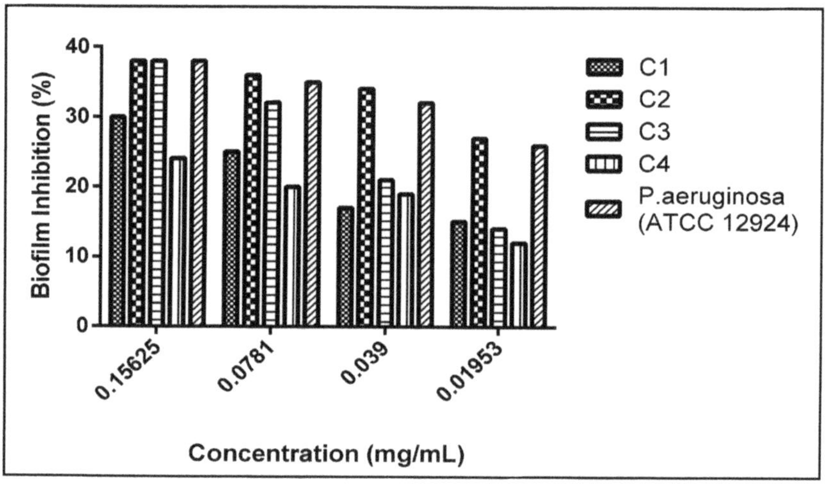

Figure 5. Biofilm formation inhibition activities of *L. coronopifolia* extract against *P. aeruginosa* isolates.

3.6.4. Synergistic Activities

Antimicrobial resistance (AMR) has increased markedly in the recent years and is causing a major threat to patients' treatment. *P. aeruginosa*, for example, has developed antibiotic resistance, and its increasing dissemination is causing severe infections in hospitals. Combinations of plant extracts with antibiotics represent a novel approach to increase their effectiveness and to overcome AMR. In an attempt to explore the synergistic activities of the extract, we combined it with the reference drug ciprofloxacin in a 1:1 ratio. The combination of ciprofloxacin and the extract substantially potentiated the reduction of biofilm from 24%, 20%, and 19% to 53.5%, 48.9%, and 45.26% at 1/2, 1/4, and 1/8 MICs, respectively (Figure 6). Our findings come in agreement with those of Okansi et al. (2013), who reported a synergy when ciprofloxacin was combined with the methanol extract of *Phyllantus muellerianus* leaves (containing flavonoids) against *P. aeruginosa* [59]. Another study described the synergistic activities against *P. aeruginosa* when zingerone extract was combined with the reference drug ciprofloxacin [60].

Coronavirus disease 2019 (COVID-19) has become the utmost and worst public health crisis of our generation. There are several risk factors associated with COVID-19, among them secondary bacterial infections, which in turn lead to serious negative outcomes and fatal clinical complications. To prevent these negative outcomes and secondary bacterial infections, patients with serious illness are treated with antibiotics. As a result, the use of antibiotics has increased, and this will significantly elevate the antibiotic resistance rates [61]. Plant extracts, with diverse secondary metabolites and several molecular targets, alone or as an adjuvant therapy, would not only boost the overall antimicrobial properties but can also work as modifying/modulating agents. This will effectively reduce the use of antibiotics and, therefore, reduce the risk of developing antibiotic resistance [62].

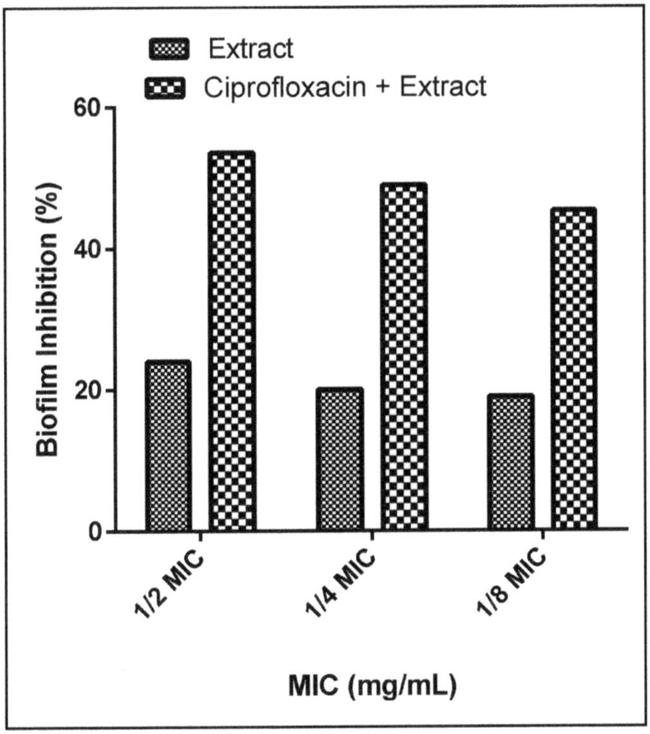

Figure 6. Biofilm formation activity of *L. coronopifolia* extract alone and in combination with ciprofloxacin on C4 isolate.

4. Conclusions

In this study, the LC-MS profiling of *L. coronopifolia* extract revealed 46 secondary metabolites. The extract displayed promising insecticidal activities against 3rd instar larvae of *C. pipiens*. The larvae showed a defensive mechanism by increasing the activities of detoxification enzymes of GSTs and α-esterases, while the toxification of *C. pipiens* was significantly observed through the reduction of CarE, AChE, and CYP450 activities. Moreover, the extract demonstrated promising antibiofilm formation against *P. aeruginosa* alone and when combined with the reference drug ciprofloxacin. To sum up, the wild plant, *L. coronopifolia,* exploits a substantial natural source to control disease carriers and manage resistant bacteria infections. Further studies are needed to evaluate the effects of *L. coronopifolia* extract on the life cycle of *C. pipiens* larvae and to explain the regulatory mechanisms of toxification.

Author Contributions: Conceptualization, M.E., M.M.S., R.R.A.L., M.A.E.R., M.S., L.J.M.A.-H., D.R.A.-H. and S.M.F.; formal analysis, M.E., D.R.A.-H., L.J.M.A.-H., M.S. and M.A.E.R.; methodology, M.E., D.R.A.-H., M.M.S., L.J.M.A.-H., R.R.A.L., S.M.F., M.S. and M.A.E.R.; validation, D.R.A.-H., L.J.M.A.-H., M.E., and M.S.; visualization, M.E., and M.S.; Writing—original draft, M.E., D.R.A.-H., M.M.S., L.J.M.A.-H., R.R.A.L., S.M.F. and M.A.E.R.; Writing—review and editing, M.S. and M.E. All authors have read and agreed to the published version of the manuscript.

Funding: This research did not receive any specific grant from funding agencies in the public, commercial, or not-for-profit sectors.

Institutional Review Board Statement: Not applicable.

Informed Consent Statement: Not applicable.

Data Availability Statement: All data are included at the manuscript.

Conflicts of Interest: There is no conflict of interest.

Sample Availability: Samples of the extract are available from the authors.

References

1. Raju, P.; Arivalagan, P.; Natarajan, S. One-pot fabrication of multifunctional catechin@ ZIF-L nanocomposite: Assessment of antibiofilm, larvicidal and photocatalytic activities. *J. Photochem. Photobiol. B Biol.* **2020**, *203*, 111774. [CrossRef] [PubMed]
2. Bagavan, A.; Rahuman, A.A. Evaluation of larvicidal activity of medicinal plant extracts against three mosquito vectors. *Asian Pac. J. Trop. Med.* **2011**, *4*, 29–34. [CrossRef]
3. Angajala, G.; Ramya, R.; Subashini, R. In-vitro anti-inflammatory and mosquito larvicidal efficacy of nickel nanoparticles phytofabricated from aqueous leaf extracts of Aegle marmelos Correa. *Acta Trop.* **2014**, *135*, 19–26. [CrossRef] [PubMed]
4. El-Garf, I.; Grayer, R.J.; Kite, G.C.; Veitch, N.C. Hypolaetin 8-O-glucuronide and related flavonoids from Lavandula coronopifolia and L. pubescens. *Biochem. Syst. Ecol.* **1999**, *27*, 843–846. [CrossRef]
5. The Plant List (2013). Version 1.1. Available online: http://www.theplantlist.org/ (accessed on 1 January 2021).
6. Contreras, M.D.; Algieri, F.; Rodriguez-Nogales, A.; Gálvez, J.; Segura-Carretero, A. Phytochemical profiling of anti-inflammatory Lavandula extracts via RP–HPLC–DAD–QTOF–MS and–MS/MS: Assessment of their qualitative and quantitative differences. *Electrophoresis* **2018**, *39*, 1284–1293. [CrossRef] [PubMed]
7. Harborne, J.B.; Williams, C.A. *Phytochemistry of the Genus Lavandula*; CRC Press: London, UK, 2002.
8. Lis-Balchin, M. *Lavender: The Genus Lavandula*; CRC Press: Boca Raton, FL, USA, 2002.
9. El-Gendi, O.D.; Kusano, A.; Kusano, G. Two new triterpenic glucosidates from Lavandula coronipifolia in Egypt. *Nat. Med.* **2000**, *54*, 38–41.
10. Elsbaey, M.; Mwakalukwa, R.; Shimizu, K.; Miyamoto, T. Pentacylic triterpenes from Lavandula coronopifolia: Structure related inhibitory activity on α-glucosidase. *Nat. Prod. Res.* **2019**, 1–9. [CrossRef] [PubMed]
11. Messaoud, C.; Chograni, H.; Boussaid, M. Chemical composition and antioxidant activities of essential oils and methanol extracts of three wild Lavandula L. species. *Nat. Prod. Res.* **2012**, *26*, 1976–1984. [CrossRef]
12. Hassan, W.; El Gamal, A.; El-Sheddy, E.; Al-Oquil, M.; Farshori, N. The chemical composition and antimicrobial activity of the essential oil of Lavandula coronopifolia growing in Saudi Arabia. *J. Chem. Pharm. Res.* **2014**, *6*, 604–615.
13. Farshori, N.N.; Al-Sheddi, E.S.; Al-Oqail, M.M.; Hassan, W.H.; Al-Khedhairy, A.A.; Musarrat, J.; Siddiqui, M.A. Hepatoprotective potential of Lavandula coronopifolia extracts against ethanol induced oxidative stress-mediated cytotoxicity in HepG2 cells. *Toxicol. Ind. Health* **2015**, *31*, 727–737. [CrossRef]
14. Ait Said, L.; Zahlane, K.; Ghalbane, I.; El Messoussi, S.; Romane, A.; Cavaleiro, C.; Salgueiro, L. Chemical composition and antibacterial activity of Lavandula coronopifolia essential oil against antibiotic-resistant bacteria. *Nat. Prod. Res.* **2015**, *29*, 582–585. [CrossRef]
15. El-mekkawy, S.; Shahat, A.A.; Alqahtani, A.S.; Alsaid, M.S.; Abdelfattah, M.A.; Ullah, R.; Emam, M.; Yasri, A.; Sobeh, M. A Polyphenols-Rich Extract from Moricandia sinaica Boiss. Exhibits Analgesic, Anti-Inflammatory and Antipyretic Activities In Vivo. *Molecules* **2020**, *25*, 5049. [CrossRef] [PubMed]
16. El-Garawani, I.; Emam, M.; Elkhateeb, W.; El-Seedi, H.; Khalifa, S.; Oshiba, S.; Abou-Ghanima, S.; Daba, G. In Vitro Antigenotoxic, Antihelminthic and Antioxidant Potentials Based on the Extracted Metabolites from Lichen, Candelariella vitellina. *Pharmaceutics* **2020**, *12*, 477. [CrossRef] [PubMed]
17. El Raey, M.A.; El-Hagrassi, A.M.; Osman, A.F.; Darwish, K.M.; Emam, M. Acalypha wilkesiana flowers: Phenolic profiling, cytotoxic activity of their biosynthesized silver nanoparticles and molecular docking study for its constituents as Topoisomerase-I inhibitors. *Biocatal. Agric. Biotechnol.* **2019**, *20*, 101243. [CrossRef]
18. Wang, M.; Carver, J.J.; Phelan, V.V.; Sanchez, L.M.; Garg, N.; Peng, Y.; Nguyen, D.D.; Watrous, J.; Kapono, C.A.; Luzzatto-Knaan, T. Sharing and community curation of mass spectrometry data with Global Natural Products Social Molecular Networking. *Nat. Biotechnol.* **2016**, *34*, 828–837. [CrossRef] [PubMed]
19. Gerberg, E.J.; Barnard, D.R.; Ward, R.A. *Manual for Mosquito Rearing and Experimental Techniques*; American Mosquito Control Association Inc.: Mount Laurel, NJ, USA, 1994.
20. World Health Organization. *Guidelines for Laboratory and Field Testing of Mosquito Larvicides*; World Health Organization: Geneva, Switzerland, 2005.
21. Amin, T. Biochemical and Physiological Studies of Some Insect Growth Regulators on the Cotton Leafworm *Spodoptera littoralis (Boisd.)*. Ph.D. Thesis, Faculty of Science, Cairo University, Giza, Egypt, 1998.
22. Simpson, D.; Bull, D.; Lindquist, D. A semimicrotechnique for the estimation of cholinesterase activity in boll weevils. *Ann. Entomol. Soc. Am.* **1964**, *57*, 367–371. [CrossRef]
23. Van Asperen, K. A study of housefly esterases by means of a sensitive colorimetric method. *J. Insect Physiol.* **1962**, *8*, 401–414. [CrossRef]

24. Habig, W.H.; Pabst, M.J.; Jakoby, W.B. Glutathione S-transferases the first enzymatic step in mercapturic acid formation. *J. Biol. Chem.* **1974**, *249*, 7130–7139. [CrossRef]
25. Hansen, L.; Hodgson, E. Biochemical characteristics of insect microsomes: N-and O-demethylation. *Biochem. Pharmacol.* **1971**, *20*, 1569–1578. [CrossRef]
26. El-Sayed, N.R.; Samir, R.; Abdel-Hafez, J.M.; Ramadan, M.A. Olive Leaf Extract Modulates Quorum Sensing Genes and Biofilm Formation in Multi-Drug Resistant Pseudomonas aeruginosa. *Antibiotics* **2020**, *9*, 526. [CrossRef]
27. Cooper, K. Theory of antibiotic inhibition zones in agar media. *Nature* **1955**, *176*, 510–511. [CrossRef]
28. Moideen, M.M.J.; Raffick, M.M. Antibacterial, antifungal activity and phytochemical analysis of Sanseviveria roxburghiana root. *IJ Phytopharm.* **2012**, *3*, 21–26.
29. Briceland, L.L.; Pasko, M.; Mylotte, J. Serum bactericidal rate as measure of antibiotic interactions. *Antimicrob. Agents Chemother.* **1987**, *31*, 679–685. [CrossRef] [PubMed]
30. Wikler, M.A. Methods for dilution antimicrobial susceptibility tests for bacteria that grow aerobically: Approved standard. *CLSI (NCCLS)* **2006**, *26*, M7-A7.
31. O'Toole, G.A.; Kolter, R. Initiation of biofilm formation in Pseudomonas fluorescens WCS365 proceeds via multiple, convergent signalling pathways: A genetic analysis. *Mol. Microbiol.* **1998**, *28*, 449–461. [CrossRef]
32. O'Toole, G.A. Microtiter dish biofilm formation assay. *J. Vis. Exp.* **2011**, *47*, e2437. [CrossRef] [PubMed]
33. Stepanović, S.; Vuković, D.; Hola, V.; Bonaventura, G.D.; Djukić, S.; Ćirković, I.; Ruzicka, F. Quantification of biofilm in microtiter plates: Overview of testing conditions and practical recommendations for assessment of biofilm production by staphylococci. *Apmis* **2007**, *115*, 891–899. [CrossRef]
34. Stepanović, S.; Vuković, D.; Dakić, I.; Savić, B.; Švabić-Vlahović, M. A modified microtiter-plate test for quantification of staphylococcal biofilm formation. *J. Microbiol. Methods* **2000**, *40*, 175–179. [CrossRef]
35. Duncan, D.B. Multiple range and multiple F tests. *Biometrics* **1955**, *11*, 1–42. [CrossRef]
36. Vukics, V.; Guttman, A. Structural characterization of flavonoid glycosides by multi-stage mass spectrometry. *Mass Spectrom. Rev.* **2010**, *29*, 1–16. [CrossRef] [PubMed]
37. Yao, H.; Chen, B.; Zhang, Y.; Ou, H.; Li, Y.; Li, S.; Shi, P.; Lin, X. Analysis of the total biflavonoids extract from Selaginella doederleinii by HPLC-QTOF-MS and its in vitro and in vivo anticancer effects. *Molecules* **2017**, *22*, 325. [CrossRef] [PubMed]
38. Sun, L.; Tao, S.; Zhang, S. Characterization and quantification of polyphenols and triterpenoids in thinned young fruits of ten pear varieties by UPLC-Q TRAP-MS/MS. *Molecules* **2019**, *24*, 159. [CrossRef]
39. Spínola, V.; Pinto, J.; Castilho, P.C. Identification and quantification of phenolic compounds of selected fruits from Madeira Island by HPLC-DAD–ESI-MSn and screening for their antioxidant activity. *Food Chem.* **2015**, *173*, 14–30. [CrossRef]
40. Yuzuak, S.; Ballington, J.; Xie, D.-Y. HPLC-qTOF-MS/MS-based profiling of flavan-3-ols and dimeric proanthocyanidins in berries of two muscadine grape hybrids FLH 13-11 and FLH 17-66. *Metabolites* **2018**, *8*, 57. [CrossRef]
41. Li, Q.; Wang, L.; Dai, P.; Zeng, X.; Qi, X.; Zhu, L.; Yan, T.; Wang, Y.; Lu, L.; Hu, M. A combined strategy of mass fragmentation, post-column cobalt complexation and shift in ultraviolet absorption spectra to determine the uridine 5′-diphospho-glucuronosyltransferase metabolism profiling of flavones after oral administration of a flavone mixture in rats. *J. Chromatogr. A* **2015**, *1395*, 116–128. [PubMed]
42. Helmy, N.; Bakr, R.F.; Nawwar, G.A.; Helmy, O.M. Biochemical effects of some agricultural waste extracts against Culex pipiens (Diptera: Culicidae). *Egypt. Acad. J. Biol. Sci. C Physiol. Mol. Biol.* **2010**, *2*, 75–81. [CrossRef]
43. Dris, D.; Tine-Djebbar, F.; Soltani, N. Lavandula dentata essential oils: Chemical composition and larvicidal activity against Culiseta longiareolata and Culex pipiens (Diptera: Culicidae). *Afr. Entomol.* **2017**, *25*, 387–394. [CrossRef]
44. Traboulsi, A.F.; Taoubi, K.; El-Haj, S.; Bessiere, J.M.; Rammal, S. Insecticidal properties of essential plant oils against the mosquito Culex pipiens molestus (Diptera: Culicidae). *Pest Manag. Sci.* **2002**, *58*, 491–495. [CrossRef]
45. Waliwitiya, R.; Nicholson, R.A.; Kennedy, C.J.; Lowenberger, C.A. The synergistic effects of insecticidal essential oils and piperonyl butoxide on biotransformational enzyme activities in Aedes aegypti (Diptera: Culicidae). *J. Med. Entomol.* **2012**, *49*, 614–623. [CrossRef]
46. Huang, C.-S.; Lii, C.-K.; Lin, A.-H.; Yeh, Y.-W.; Yao, H.-T.; Li, C.-C.; Wang, T.-S.; Chen, H.-W. Protection by chrysin, apigenin, and luteolin against oxidative stress is mediated by the Nrf2-dependent up-regulation of heme oxygenase 1 and glutamate cysteine ligase in rat primary hepatocytes. *Arch. Toxicol.* **2013**, *87*, 167–178. [CrossRef]
47. Oppenoorth, F.; Rupes, V.; ElBashir, S.; Houx, N.; Voerman, S. Glutathione-dependent degradation of parathion and its significance for resistance in the housefly. *Pestic. Biochem. Physiol.* **1972**, *2*, 262–269. [CrossRef]
48. Motoyama, N.; Dauterman, W. Interstrain comparison of glutathione-dependent reactions in susceptible and resistant houseflies. *Pestic. Biochem. Physiol.* **1975**, *5*, 489–495. [CrossRef]
49. Gershenzon, J.; Fontana, A.; Burow, M.; Wittstock, U.; Degenhardt, J. Mixtures of plant secondary metabolites: Metabolic origins and ecological benefits. In *the Ecology of Plant Secondary Metabolites: From Genes to Global Processes*; Cambridge University Press: Cambridge, UK, 2012; Volume 56.
50. Salunke, B.; Kotkar, H.; Mendki, P.; Upasani, S.; Maheshwari, V. Efficacy of flavonoids in controlling Callosobruchus chinensis (L.) (Coleoptera: Bruchidae), a post-harvest pest of grain legumes. *Crop Prot.* **2005**, *24*, 888–893. [CrossRef]
51. Djeridane, A.; Yousfi, M.; Nadjemi, B.; Maamri, S.; Djireb, F.; Stocker, P. Phenolic extracts from various Algerian plants as strong inhibitors of porcine liver carboxylesterase. *J. Enzym. Inhib. Med. Chem.* **2006**, *21*, 719–726. [CrossRef] [PubMed]

52. Barbosa Filho, J.M.; Medeiros, K.C.P.; Diniz, M.d.F.F.; Batista, L.M.; Athayde-Filho, P.F.; Silva, M.S.; da Cunha, E.V.; Almeida, J.R.; Quintans-Júnior, L.J. Natural products inhibitors of the enzyme acetylcholinesterase. *Rev. Bras. Farmacogn.* **2006**, *16*, 258–285. [CrossRef]
53. Praveena, A.; Sanjayan, K. Inhibition of acetylcholinesterase in three insects of economic importance by linalool, a monoterpene phytochemical. *Insect Pest Manag. A Curr. Scenar.* **2011**, 340–345.
54. Galvão, S.S.L.; Monteiro, A.; Siqueira, E.P.; Bomfim, M.R.Q.; Dias-Souza, M.V.; Ferreira, G.F.; Denadai, A.M.L.; Santos, Á.R.; Lúcia dos Santos, V.; Souza-Fagundes, E.M. Annona glabra flavonoids act as antimicrobials by binding to Pseudomonas aeruginosa cell walls. *Front. Microbiol.* **2016**, *7*, 2053. [CrossRef] [PubMed]
55. Mostafa, I.; Abbas, H.A.; Ashour, M.L.; Yasri, A.; El-Shazly, A.M.; Wink, M.; Sobeh, M. Polyphenols from Salix tetrasperma impair virulence and inhibit Quorum sensing of Pseudomonas aeruginosa. *Molecules* **2020**, *25*, 1341. [CrossRef]
56. Nain, Z.; Mansur, F.J.; Syed, S.B.; Islam, M.A.; Azakami, H.; Islam, M.R.; Karim, M.M. Inhibition of biofilm formation, quorum sensing and other virulence factors in Pseudomonas aeruginosa by polyphenols of Gynura procumbens leaves. *J. Biomol. Struct. Dyn.* **2020**, 1–15. [CrossRef]
57. Koley, T.K.; Khan, Z.; Oulkar, D.; Banerjee, T.; Singh, A.; Karkute, S.G.; Banerjee, K. Coupling the high-resolution LC–MS characterisation of the phenolic compounds with the antimicrobial and antibiofilm properties of helencha (*Enydra fluctuans* Lour.). *J. Food Sci. Technol.* **2021**, 1–11. [CrossRef]
58. Mrabti, H.N.; Bouyahya, A.; Ed-Dra, A.; Kachmar, M.R.; Mrabti, N.N.; Benali, T.; Shariati, M.A.; Ouahbi, A.; Doudach, L.; Faouzi, M.E.A. Polyphenolic profile and biological properties of Arbutus unedo root extracts. *Eur. J. Integr. Med.* **2021**, *42*, 101266. [CrossRef]
59. Ofokansi, K.C.; Attama, A.A.; Uzor, P.F.; Ovri, M.O. Evaluation of the combined antimicrobial activity of the leaf extract of Phyllantus muellerianus with ciprofloxacin. *J. Pharm Technol. Drug Res.* **2013**, *2*, 16. [CrossRef]
60. Kumar, L.; Chhibber, S.; Harjai, K. Zingerone inhibit biofilm formation and improve antibiofilm efficacy of ciprofloxacin against Pseudomonas aeruginosa PAO1. *Fitoterapia* **2013**, *90*, 73–78. [CrossRef] [PubMed]
61. Grau, S.; Echeverria-Esnal, D.; Gómez-Zorrilla, S.; Navarrete-Rouco, M.E.; Masclans, J.R.; Espona, M.; Gracia-Arnillas, M.P.; Duran, X.; Comas, M.; Horcajada, J.P. Evolution of Antimicrobial Consumption During the First Wave of COVID-19 Pandemic. *Antibiotics* **2021**, *10*, 132. [CrossRef]
62. Tiwari Pandey, A.; Pandey, I.; Zamboni, P.; Gemmati, D.; Kanase, A.; Singh, A.V.; Singh, M.P. Traditional Herbal Remedies with a Multifunctional Therapeutic Approach as an Implication in COVID-19 Associated Co-Infections. *Coatings* **2020**, *10*, 761. [CrossRef]

MDPI
St. Alban-Anlage 66
4052 Basel
Switzerland
Tel. +41 61 683 77 34
Fax +41 61 302 89 18
www.mdpi.com

Molecules Editorial Office
E-mail: molecules@mdpi.com
www.mdpi.com/journal/molecules

www.ingramcontent.com/pod-product-compliance
Lightning Source LLC
Chambersburg PA
CBHW061450210926
51850CB00008B/1674
9783036569932